Digital Development of the European Union

David Ramiro Troitiño • Tanel Kerikmäe •
Ondrej Hamuľák
Editors

Digital Development of the European Union

An Interdisciplinary Perspective

Springer

Editors
David Ramiro Troitiño
Jean Monnet Chair on Digital Europe and Future Integration
Tallinn University of Technology
Tallinn, Estonia

Tanel Kerikmäe
Department of law
Tallinn University of Technology
Tallinn, Estonia

Ondrej Hamuľák
Faculty of Law
Palacký University Olomouc
Olomouc, Czech Republic

Department of Law, School of Business and Governance
Tallinn University of Technology
Tallinn, Estonia

ISBN 978-3-031-27314-8 ISBN 978-3-031-27312-4 (eBook)
https://doi.org/10.1007/978-3-031-27312-4

This Springer imprint is published by the registered company Springer Nature Switzerland AG
The registered company address is: Gewerbestrasse 11, 6330 Cham, Switzerland

This volume was prepared with the cooperation of Tallinn University of Technology, School of Business and Governance, Department of Law, Estonia, and Palacký University Olomouc, Faculty of Law, Czech Republic, on behalf of project no. 20-27227S "The Advent, Pitfalls and Limits of Digital Sovereignty of the European Union" funded by the Czech Science Foundation (GAČR).

Foreword

In the 2021 survey published by the Swiss business school the Institute of Management Development and the Singapore University of Technology and Design, Singapore topped the Smart City Index, which measures how economic and technological aspects of smart cities are balanced against the "humane dimensions" of smart cities such as the quality of life, the environment, and inclusiveness, for the third year running. But even as Singapore is set on transforming itself to become a Smart Nation, we can do well to take a page from other countries, both developed and developing, and learn from their experiences and journeys in harnessing technology to transform how their peoples and businesses live, work, and play.

This monograph, *The Digital Future of the European Union*, is one such invaluable source of information. An examination of this excellent monograph reveals various case studies that allow us to learn about the controversies, challenges, and risks that the EU has faced and is facing in the context of digital transformation. By shedding light on significant political, social, and economic impact caused by the modern digital transformations on European societies, while coherently and systematically examining both present and future digital solutions and their practical implementations in the EU, the monograph both records and analyses the journey undertaken by the EU. Many of the lessons have immediate and direct application to Singapore. For instance, we could compare Singapore's relatively new cybersecurity laws with the EU's cybersecurity regulations (*Kasper, Osula*, 'Spill over' and 'fail forward' in the EU's cybersecurity regulations). The extraterritorial impact of EU's AI Act will also influence international relations both politically and commercially (*Outeda, Cacheda*, Artificial intelligence: a reading from European politics). And the rising new digital economy will certainly impact the trade relations between countries (*Kerikmäe, Troitiño*, The Digital world market and the European Union), especially since Singapore is EU's second largest trading partner in the Association of Southeast Asian Nations (ASEAN) in 2020.

To paraphrase John Donne, "no man [or country] is a [digital] island entire of itself." In a world which has become even more interconnected with the advent of the digital revolution, this monograph has a lot of insights to offer its readers. We need to look no further than the COVID-19 pandemic, which has greatly accelerated the adoption and use of public health and tracking technologies in societies, but we still understand very little about the impact of these technologies on our personal health and the management of our personal information. It serves as yet another sobering reminder that we should not forget our principles and our humanity even as we—as individuals, as societies, and as a collective species—embark on our digital transformation journey.

The Centre for Technology, Robotics, Artificial Intelligence and the Law, TRAIL, National University of Singapore, Singapore, Singapore

Daniel Seng

Contents

Part II Law

Part III Politics

Part I
General Aspects

The Digital Future of the European Union

David Ramiro Troitiño, Tanel Kerikmäe, and Ondrej Hamuľák

1 Introduction

This academic monograph focuses on the digital developments of the European Union. It consistently develops the theme of digital evolution in the EU at a crucial moment for the future organization from a multidisciplinary perspective. The book presents the main fields of digital development in the near future within the Union and has the participation of prestigious international authors in the chosen field. At a time of transformation in the model of society, digitization has seen its implementation accelerated by the recent COVID-19 pandemic. Therefore, understanding and promoting the correct digital development is a priority for scientific dissemination and for the interest of readers.

The main themes are digital developments in fields of action of the European Union, such as politics, economy, law, international relations, and social aspects. The objective is understanding and foreseeing the fast development on the digital agenda that can change completely the perception of the European Union, its influence over the citizen lives, and the external scope of the organization. The

D. Ramiro Troitiño (✉)
Jean Monnet Chair on Digital Europe and Future Integration, Tallinn University of Technology, Tallinn, Estonia
e-mail: david.troitino@taltech.ee

T. Kerikmäe
Department of law, Tallinn University of Technology, Tallinn, Estonia
e-mail: tanel.kerikmae@taltech.ee

O. Hamuľák
Faculty of Law, Palacký University Olomouc, Olomouc, Czech Republic

Department of Law, School of Business and Governance, Tallinn University of Technology, Tallinn, Estonia
e-mail: ondrej.hamulak@upol.cz

D. Ramiro Troitiño et al. (eds.), *Digital Development of the European Union*,
https://doi.org/10.1007/978-3-031-27312-4_1

methodology depends on each chapter as it is a multidisciplinary book. Nevertheless, each chapter will follow a similar structure with an introduction to the field, state of art, most problematic aspects, and future possibilities of implementing digital reforms in the EU, including conclusions on their impact on the EU development.

It is difficult currently to find a book providing a general view, coherent and compact enough to understand the process of digitalization in the European Union. The novelty of this book is the analysis of the impact of the digital revolution on the EU development, understanding the European Union as a process that started in the 1950s and is still under development without its final shape. Therefore, digitalization can change the essence of the Union as digital tools allow new reforms, advances, and reforms not possible before.

2 Content

The Digital revolution is already a reality. It is affecting all kinds of aspects of our everyday life, our States and the European Union. We need to understand it from a coherent approach; what the digital revolution can offer the European Union as a peace project based on political, economic, and social aspects? The Digital influence on the development of the European Union will be interconnected in several fields that as a whole will result in a much more integrated Union with more tools for improving the living conditions of the EU citizens and getting closer to a deeper Union. On external approach, the digital revolution of the EU will defend the Europeans from new threats. The globalization of the digital world means the incapacity of each individual States protecting their citizens effectively from the negative aspects of digitalization. Therefore, it is the responsibility of the European Union to address it from a European perspective (Troitino, 2013). This book analyzes the main areas where the EU can, and should act, for creating an efficient and protective digital space in Europe, where the citizens are protected and free to enjoy the endless possibilities of a process that will change the world. The comprehensive approach allows the reader to understand the whole process without being a specialist on any specific area included in the publication. The future is here and it is the responsibility of the European Union to address it, looking for prosperity and defending the European conception of society. European values must be incorporated into the digital revolution.

The book is divided into three different areas covering the most relevant issues in connection with the digitalization of the European Union, general aspects, Law and Politics. The first area of study includes an eclectic selection of researches to provide a broad vision on the digital impact on the integration process in the European Union. First, global challenges and priorities are defined, and then specific fields are studied. The selection of each one of them is based on their potential to increase the development of the European Union on digital basis. Artificial Intelligence, cybersecurity (Kasper & Vernygora, 2021), digital health (Sogomonjan, 2021), the Digital Single Market, digitalization-focused or human-centered business contracts, Digital taxes (Künnapas, 2021), EU competition law goals and digital economy and Ethics and new Technologies, conform the spine of the digitalization process within the European Union (Kesa & Kerikmäe, 2020). Each of the aspects has been analyzed under an EU perspective linked with the convenience to Europeanize the policy or actions.

The field of Law is relevant in the monograph because of the urgent issue of regulating the digitalization process within the European Union. There is a necessity to establish common rules in the European Union to control and rationalize the implementation in new technologies (Hamulak et al., 2021). It affects the national law of the Member States when they already legislated on any specific field affected by the digitalization of the European Union, but also is a process linked with gray areas where there is no legislation and the big tech companies move in the shadows of the society. Therefore, it means that the European Union must make an effort to cover these areas with sensitive legislation promoting innovation, allowing economic development and protecting the social values of the EU (Steible, 2021). In addition, the Law area of the book focuses partially on the issue of legislation and sovereignty, analyzing the role of the EU as Sovereign in the Digital World and the Role of EU Law in the Regulation of Modern Technologies.

The last part of the book offers a vision of the political implications of the digital evolution of the European Union, including several aspects as e-governance at the EU level, the implications of the i-voting in the European Parliament elections and the importance of Digital European citizenship (Ivic, 2018), the implication of the three main EU institutions in the digitalization process, and the relevance of Artificial Intelligence from a political perspective (Cancela & González, 2021). It presents EU politics from a digital perspective, offering alternatives to the existing political models based on efficiency and equality.

All the authors of the book aim for the same target, a common understanding on the most relevant technical revolution in the last decades and its impact on the European integration process, the creation of a common Europe. Both represent novelty, innovation, and their combination will definitely affect the organization of society substantially.

References

Cancela Outeda, C., & González Cacheda, B. (2021). Mecanismos digitales para la participación en partidos políticos de la CPLP. *População e Sociedade*.

Hamulak, O., Kiss, L. N., & Gábriš, T. (2021). 'This content is not available in your Country'a general summary on geo-blocking in and outside the European Union. *International and Comparative Law Review, 21*(1), 153–183.

Ivic, S. (2018). European philosophical identity narratives. *Cultura, 15*(1), 125–145.

Kasper, A., & Vernygora, V. (2021). The EU's cybersecurity: A strategic narrative of a cyber power or a confusing policy for a local common market? *Deusto Journal of European Studies*.

Kesa, A., & Kerikmäe, T. (2020). Artificial intelligence and the GDPR: Inevitable nemeses? *TalTech Journal of European Studies, 10*(3), 32.

Künnapas, K. (2021). Legal engineering of the anti-abuse rule in ATAD: Architecture of the regression tree model. *TalTech Journal of European Studies, 11*(2), 65–82.
Sogomonjan, M. (2021). Challenges and opportunities for e-mental health policy: An Estonian case study. *Contemporary Social Science, 16*(2), 185–198.
Steible, B. (2021). Hacer respetar los valores de la UE. In *Anuario del Boletín de la Academia de Yuste: Reflexiones sobre Europa e Iberoamérica* (pp. 105–108). Fundación Academia Europea e Iberoamericana de Yuste.
Troitino, D. R. (2013). *European integration: Building Europe*. Nova Science Publishers.

Priorities and Challenges: The Digital Transition in the European Integration

David Ramiro Troitiño

Abstract The European Union is a distinctive organization in the global arena. It integrates different areas of the Member States, forming a common sovereignty. The European building process focuses on the priorities of the actors involved, as citizens, institutions, and Member States. Parallel to new solutions, new problems are generated by the process and required new solutions. Presently digitalization is a priority that needs to be addressed. Consequently, the European Union focuses on the implementation of common actions ruling the digital aspects relevant to the organization. The following chapter focus on the digital priorities of the EU and the future solutions to the current problems offered by the process of digitalization.

1 Introduction

The European Commission's priorities on digitalization include several technologies and digital tools whose world market is huge in terms of economic activity. Consequently, the European Union companies need to access the world market with the right competitiveness and legal protection from the EU institutions. Parallelly, the digital transformation of society is influencing the interaction between European citizens and the common rules to live in society. Therefore, the European Union (EU) needs to respond to the new social requirements of the citizens and provide the right environment for European companies to compete in the digital world with chances to succeed. The EU faces a double challenge to adapt its structures to social and economic requirements dictated by the current digital reality.

This contribution integrates the activities of the Jean Monnet Chair "Digital Europe and Future Integration" (DEFI).

D. Ramiro Troitiño (✉)
Jean Monnet Chair on Digital Europe and Future Integration, Tallinn University of Technology, Tallinn, Estonia
e-mail: david.troitino@taltech.ee

D. Ramiro Troitiño et al. (eds.), *Digital Development of the European Union*,
https://doi.org/10.1007/978-3-031-27312-4_2

This chapter analyzes what are the real necessities of the European Union's consequence of the socioeconomic digitalization o the world, and the digital answers required to solve any dysfunctionality. The identification of the challenge leads to a multidisciplinary approach consequence of the variety of fields affected by the process of digitalization.

Regarding the theoretical frame, the three main theories of integration, Neo-functionalism, Federalism and Intergovernmentalism provide the skeleton of the research, the evolutionary path to respond to the necessities of citizens and companies within the Union digital development (Ramiro Troitiño, 2022b). Consequently, methodology includes traditional theories of integration applied to digital developments, challenges and answers in the European Union from a multidisciplinary implementation.

The European Union's digital transition produces relevant expectations among the economic agents who hope to benefit directly from it, with a significant growth Gross Domestic Product of the whole European Union in the coming years. The process does not require a long transitional period as the benefits are expected in 2–3 years' time. Complementary, the transitional economy toward digitalization has sped up the process, thanks to the influence of the measures taken in the context of the last pandemic provoked by COVID-19. The need to face the mobility restrictions intensified the digital transition in the European Union. In addition, China and the United States of America have invested considerable resources in their own digitalization, forcing the European Union to develop its own digital model in order to reassure its independence and expand its influence in a field dominated by these two geopolitical giants.

On public affairs, the digital transition offers the EU countless opportunities to improve its functional capacity in good governance. Information and communication technologies (ICTs) permit the European public administration to work in a translucent and efficient way, contributing to the elimination of corruption. It helps to overview the common governance with open access to all the citizens, avoiding misuse of common funds or nepotism. In addition, the digital tools provide European Union citizens with more chances to be active in the EU political life, bringing closer the institutions to the people and easing the involvement of the citizens, including minorities and other crowds at the risk of marginalization, in public affairs.

The idea of increasing the involvement of the citizens via digitalization in the EU politics has a deep impact fostering deeper integration in the organization, closing the gap between the institutions and the Union without the intervention of the Member States. It will generate a common political identity between the European citizens, allowing the transfer of further national sovereignty to the European level. The European Communities started as a peace project where the nation had to be separated gradually from politics, a steady transfer of sovereignty from the national to the European level. The fathers of Europe thought of political nationalism as the main threat to the socioeconomic stability of Europe; therefore, they promoted the creation of a new level of sovereignty trespassing national limits (Ramiro Troitiño, 2022c). The way chosen was based on low politics in order to prevent the opposition of the majority of the European citizens, still currently embodied in a strong nationalism. Nevertheless, the integration has reached great levels after a long process initiated in the 1950s and high politics have become a target. Therefore, digital transition can help deepen the integration at a political level. It can create a common political digital space, reducing distances and facilitating the participation of political subjects in common affairs regardless the nationality (Rivas-de-Roca & García-Gordillo, 2022). The political future of Europe involves removing obstacles to create a common political environment.

Digital transition fosters the optimization of resources and dropping costs at all levels. Hence, the EU pursues speeding up the evolution, already unavoidable, in order to maximize the benefits of the process increasing economic performance and rationalizing public expenditure. The world, and the European Union, are facing a new form of Industrial Revolution based on digital developments. New technologies such as data analyses, Artificial Intelligence (AI), automated systems, cloud computing, the Internet of things (IoT), robotics, 3D printing, or 5G, present such a great variety of opportunities with the potential to change European society at all levels of social and economic interactions (Outeda, 2020).

2 Digital Agenda for Europe

The Lisbon Strategy, and the further amendment of 2005, were out-of-date in 2010; consequently, the European Commission produced a new tactic, the Digital Agenda for Europe (DAE). The plan is fostering a digital revolution in the context of the EU by significant developments in crucial fields. In the economic area, the most successful in terms of integration, the Commission proposed to advance in the consecution of the Digital Single Market, implementing a European digital economic area with no national restrictions, promoting the digital free movement of goods, people, services, and capital. The European common digital area will also be increasingly effective in protecting European consumers. It is especially relevant in terms of data protection (Andraško et al., 2021).

Presently, data management depends on global companies, beyond the effective control of most countries. A European common management could empower the organization with the required means to challenge the domination of the big tech companies on such a sensible issue. Hence, integrating data management at the European level will be more effective, in terms of the protection of consumers, than the individual actions of each Member State of the organization.

The creation of a digital space allowing private and economic actors the accessibility to products and services, regardless of the location within the European Union territory, is a normalization of the existing dysfunctionality of the market. European economic integration has been a story of success since the first European Community, the European Coal and Steel Community (1951) created a partial common market. The evolution of economic integration with further European Treaties enlarged the market, eliminating national barriers or measures of equivalent effect. The next border to be overtaken is the digital economy, evolving economic integration to the next level of integration. It will abolish new obstacles to European trade and prevent the creation of new barriers as the digital economy is quickly growing.

The DAE includes a more effective use of Information Technology Tools (ICT). It will help to endorse employment, economic growth, internal and external competition, and further investment (Ramiro Troitiño, 2022a). European Commission would like to eliminate existing obstacles and prevent barriers that could stop digital development, in a dual strategy. The European institution estimates that 90% of European companies are small and medium-sized enterprises (SMEs), and just 7% of them sell their products outside of their national markets. The creation of a Digital Common Market will allow a radical change in the existing trade paradigm and expand the European economy to levels inconceivable before. Consequently, a common digital market will breed a successful common economy avoiding the previous analytic obstacles to integration.

European Commission predicts that Digital Common Market will generate additional 415 billion euros to the European Union economy and an annual increase of 1% in the GDP. Additional reports promoted by the EC reveal a very optimistic view on the consequences of the Digital Single Market, as 1.3 million new jobs by 2025. This date is significantly important because of the devastating economic impact of the pandemic and the war in Ukraine for European economic performance. Therefore, the integration of the digital market generates new commercial prospects, promotes innovation, and inspires the creation of new businesses (Borowiecki et al., 2021).

The roadmap presented by the European Commission to accomplish the integration in the digital economy includes three main pillars:

1. Digital Access: Including internet trade, distribution and logistics of goods all over the European market, elimination of geo-blocking, copyright legislation at the European level and with international impact, and a standard type of Value Added Tax (VAT) for economic actors benefiting from the European Digital Market (Bunn, 2021).
2. Stable frame supporting technological developments: The EC pretends to implement common European legislation affecting telecommunications tech companies and digital information media.
3. Digital socioeconomic aspects: Data market for sale, or for implementation of new technologies based on algorithms, is a priority for the European Commission. These types of technologies required a large amount of data to feed the algorism's effectiveness.

These pillars present differentiated areas of interaction and affect a large amount of actors involved in the European integration, as citizens, civil society, companies, and even Member States governments, complicating the digital transition process, but not preventing it.

3 Obstacles to Digital Transition

The Digital Single Market grants abundant chances. Yet, it still must deal with relevant challenges that could handicap its progress and brake the European digital transition. A delay in such a delicate transition will have an important impact on the international relations situation, as several contenders are trying to impose their models by huge investments in digital solutions for economy and social interaction (Steible, 2018). The European Union needs to abolish the national existing barriers as a first step toward a common digital economy. Other competitors, as China or the USA do not have to fight against internal differences as they have integrated markets, a clear disadvantage for the European Union in terms of internationalization of its digital model. The European market is a reality in other fields, then, the EU needs to adapt the legislation from analytical to digital aspects, and a reform in the European Treaties is needed to provide legal security to the digital transition (Kajander et al., 2020). Previous treaties faced coetaneous challenges with common initiatives. Therefore, the EU has successfully solved previous problems with new treaties; just following the existing pattern will allow the elimination of national differences.

The European Union faces a relevant challenge in its digital transition because of the power of big tech companies. The EU was a pioneer in the attempt to subject the activity of the powerful technology companies of the USA to some type of regulation, because they have become an enormous parallel power that defies the rules of the market and government controls because of their size and global scope. Somehow, they could become the new Fruit Company controlling banana republics unless they are effectively supervised by a powerful and effective action. Currently, the US authorities themselves have taken the lead in these aspects, although until now the USA has been complacent with these companies. American federal government is changing its previous policy and pretends to be the technological empire led by Google, Amazon, Facebook, and Apple, the big four. The enormous business concentration of the large American technology companies that currently completely dominate the digital economy is worrying. The European Union faces a practical monopoly with the capacity to corner EU competition law (Parvin & Attar, 2022). In addition, these digital dominations pose a threat to the health of democracy by conveying large portions of the information that circulates around the world. Nobody regulates the digital world except the technology companies themselves, which use the trillions of data from citizens around the world for their own benefit, and whose tentacles cover communication, advertising, commerce, finance, and health, among many others (Kerikmäe & Särav, 2017). Apart from the commercial manipulation of this data, the political use of the networks and false information, which would be at the height of many of the current populisms, is also of great concern.

The Member States of the European Union lack the capacity to control these kinds of companies that operate in larger market units beyond national borders. Therefore, it is a priority for the European Union to actively seek a minimum mechanism of surveillance and a common approach to reassure obedience to the common European rules by all the economic agents, including the digital operators. Following the example set by the US government that the end of the era of self-regulation has begun for large technology companies, for the sake of greater protection of citizens and freedom from the power of monopolies. Nevertheless, the political battle to gain control over them will be difficult in the European Union because they have in their hands enormous economic and, above all, technological power, which transcends borders, allowing them to foster differences between Member States, and consequently stopping the integration required for their control (Varshney et al., 2021).

An additional obstacle that requires in-depth coordination is the fast development of the digital world. The Member States are readjusting constantly to the new developments in the digital economy, generating national legislation according to their own understanding of the situation. It is not a common strategy, but a patch to prevent greater evils. Nevertheless, it is generating new obstacles for common integration in the European level, as much national legislation exists, more difficult will be the elaboration of common rules. The European Commission in order to prevent this situation promotes integration before a significant proliferation of national legislation in digital aspects will affect integration in the future. The European Commission estimates that digital integration will add €415 billion to the EU economy, increase labor participation, and have many other positive effects. Additionally, a strong digital Europe could generate international influence and protect Europeans from the domination of technological corporations under the influence of totalitarian governments or just be guided by the maximization of profits without social contributions to the development of the society (Hamulak, 2015).

4 Working System

Digitalization offers great benefits for public service that will influence positively the vision of the citizens toward the EU institutions. The organization has an opportunity to improve the life quality of its citizens, simultaneously increasing popular support for the European integration process. As the symbiosis between nation and State was strong and centered the political and public life of Europe for at least two centuries, the European Union faced this problem since its foundation with the ECSC. The EU is a supranational entity looking to exercise the prerogatives of a common sovereign project, but constantly nationalism stands against this process, hopelessly slowing it down. The supporters of the nation as the last receiver of sovereignty advocate for a minimum functional transfer of sovereignty. Therefore, their idea about the European Union is based on cooperation rather than integration, a free association of Member States willing to collaborate with each other without impositions. Nevertheless, nationalists tend to appropriate for the nation the sovereignty that in fact belongs to the citizens of their States (not to nationals, nor the State itself). The citizens, as full political subjects, are the real holders of their freedom, and their decision to share it creates the common sovereignty that needs to be managed by the State. The common rules proposed by the politicians (managers of sovereignty) required the acceptance of the citizens in an exercise of institutional loyalty from a politics, or patriotism from nationalism. The combination of both loyalties, institutional loyalty and patriotism has proved to be the more effective way to guarantee the loyalty of the citizens toward political institutions (Ivic, 2012).

Nevertheless, the State is not necessarily the last recipient of the sovereignty; neither is its owner, but its manager. The issue of sovereignty at the European Union level could be solved with the implementation of digital tools to address the problems of the citizens. If their problems are effectively solved, their loyalty to the European Union will be stronger. Consequently, digital solutions can increase support for the EU among Europeans who even currently are confused about the role of the EU in their lives. The EU cannot compete with emotions emanating from the national concept, but using rationality can be more successful in many occasions. Therefore, the European Union needs effectiveness in its actions to attract the required loyalty from the EU citizens to move forward in the integration. As an example on how the EU can be more effective than the Member States individually, current initiatives, as the free movement of people and goods within the EU, created road systems that transcend State borders and require common management for their correct operation. The digitization of supervision would allow great advances and numerous benefits for citizens such as models to predict problems with the quality of the road, traffic accidents, speed problems, profitability, etc.

The impact of the Digital economy will depend on the capacity to generate data for operative modeling. Therefore, the dependence on reliable data is fundamental; putting European companies ahead of other international actors because of the data capacity of the European industry exceeds those of any other part of the world. Nevertheless, this advantage is temporal, since the generation of data is an international priority and the investments are relevant globally. The EU emphasizes its priority on the investment generating data due to its potential and the current necessities. The idea regarding data management from a European perspective is linked with scales. The generation of data is more profitable and effective on a European rather than State level. Subsequently, it is appropriate that the European Union will manage the issues related to data generation and management in the close future. The Member States are just able to generate insignificant databases, lacking efficiency. As an illustrative sample, the Common Agricultural Policy, under the European power, requires a large amount of data to increase its effectiveness and support properly the European rural socio and ecosystem. It has a need for analysis to determine the status and quality of land used for farming and ranching that is highly efficiently analyzed by satellite (McCampbell et al., 2022). Consequently, it would be wiser to use a European satellite for all the Member States than one national satellite for each partner.

The generation of data is also essential for an effective treatment of the health system, including the analysis of the medical needs of any patient and finding the location where the patients can be best served according to the available resources. The application of a digital arrangement at the European health level will generate visible benefits for the citizens, like the improvement of health, the optimization of resources, or a more rational planning. Even if the challenges are abundant because of the full identification of national States with their health systems as a supreme part of their national identity, the benefits at the European level are considerable. European Union faces great obstacles in terms of digital integration but offers possibilities to move forward in the integration process because of the benefits it offers to its citizens, the real holders of sovereignty (Elena-Bucea et al., 2021).

5 Challenges

The digital economy is a new marvel as reflects the outstanding growth of the online platforms used by workers not having full-time job but flexible hours. The collaborative economy platforms specializing on digital tasks do not involve physical attendance, neither closeness between workers nor their clients. Therefore, they are clearly online Labor Markets separated from traditional labor. Information technology (IT) specialists to taxi drivers or food deliverers, including all age groups, are affected by the new online economy that shows a growing and cross-border tendency. The supranational essence of the digital economy generates countless issues regarding State control by the national authorities. It is a dysfunctionality affecting the system easily solved with a common management in a wider area of action. Therefore, integration at the European level of the digital economy does not involve any undermining of the national sovereignty of the Member States. Individually, they lack the necessary authority to implement effectively such sovereignty in the context of digital transformation (Celeste, 2021). The European leadership, controlling and regulating, the digital economy is not just preferable but a requirement for the correct function of the political and social system of all its members.

The European Commission, providing data that reinforces the need for common action, has analyzed the employment conditions in the digital economy. The report focuses on the restrictions pertinent to ride-sharing or meal-delivery applications, taking into consideration the numerous impact factors or stakeholders. The regulation within the European Union does not count with a common position, some Member States support firm labor regulations, while others outline "a third category of workers" and struggle to attract innovators. Member States, such as Austria, Denmark, Finland, France, Germany, Sweden, and Switzerland, consider employees in the digital economy as personnel of the company with all legal implications. It entails the payment of social taxes, and the application of usual labor national legislation without the presence of a new third category of workers (Künnapas, 2016). Equally, in these States, there is robust pressure from the national trade unions to normalize all labor relations alike. Another group of Member States, including Benelux, Czech Republic, Hungary, Italy, Iceland, Norway, Poland, Portugal, and Spain, include certain restrictions on the digital economy in relation to labor regulation. It is possible to foresee a tendency toward the creation of common standards. While these countries are involved in legal development, as expected future judicial sentences on labor regulations linked to the digital economy are expected in the coming years (Erstad et al., 2021). The number of cases presented before the national courts is relevant and will have an impact on the national systems. Parallelly, there is a new emerging category of employees having restricted labor rights, for example, maximum number of hours or health insurance, but not the full range of social benefits for workers. Obviously, it has generated a heated debate about its social and economic consequences. Finally, Member States such as Bulgaria, Cyprus, Croatia, Estonia, Greece, Latvia, Lithuania, Romania, and Slovakia have not sufficiently defined their labor legislation on the status of workers of the digital economy. Basically, these States have equated workers to self-employed entrepreneurs or workers. Consequently, the required social taxes are not a concern of the digital platforms but the workers themselves (De Gregorio, 2021).

These divergences in an emerging digital market create a distortion of the Single European Market of the European Union. Although each Member State has prerogatives to implement the labor policy it deems appropriate, these online platforms function at a European Union level. Accordingly, companies generate a distortion in the European economic system that could considerably affect the competitiveness of the economies of the Member States (Grisold et al., 2021). A common harmonization ruling the labor conditions of the online platforms is extremely unlikely because of the connections between labor policy and national social system. Nevertheless, the obvious necessities can be addressed by introducing minimum common standards all over Europe; a basic agreement of the Member States to be implemented all over the European Union in the face of the challenge posed by the digital economy, helping to the foundation of a Common European Digital Market (Cabral et al., 2021). Shared position of minimum agreements will reduce differences between the European economies and promote integration at the European level. At the same time, permitting national differences will present a more flexible position and more likely to be supported by the Member States.

6 European e-Democracy

The European Parliament is the central institution of the European Union in terms of democracy. Its progression has been continuous from its humble origins as a simple consultative assembly, to a fully democratic parliament representing the citizens of the European Union. Consequently, the EP participates actively in the legislative process of the European Union standing equally to the Council for the representation of the Member States. The progressive path of the European Parliament presents constantly a number of obstacles that could be straightforwardly solved from a digital action. The European Parliament is located in several cities (Brussels, Luxembourg, and Strasbourg), but logically should be located in only one place, Belgium, as it is the center of decision-making of the Union. The digitization of parliamentary bureaucracy and services could condense these locations to two cities, or even just one. It will help to reduce the traveling of MEPs, their teams, and the transport of tons of papers in reports.

The elections to the European Parliament are important for the democratic health of the European Union. Any improvement in this field will affect positively the whole Union, as it will improve the connection between citizens and politicians. The current European elections are organized according to the national electoral rules with the consequent national differences affecting the European level. However, if the European Parliament is chosen by Europeans, and represents the EU citizens, there should be a common method in the European elections. The application of a single electoral system based on a digital voting system could remedy this problem, promoting integration by supporting equality among all Europeans regardless of their location. It will create a compressive electoral system all over the European Union territory, harmonizing the electoral legislation into one European common approach (Małkowska et al., 2021). The digital tools could also help to generate real European political parties, European referendums, reduce mismanagement, corruption and nepotism, and increasing the effectiveness of European democracy. Therefore, digitalization already offers solutions to some of the democratic deficits of the European Union, but political will is needed for its implementation.

The European Parliament itself is a relevant actor in the process of digitalization because it has a significant influence on the legislative process of the EU. Due to the approval of the Lisbon Treaty, relevant reforms were implemented; co-decision system was renamed as the ordinary legislative procedure and became the main decision-making procedure for the approval of European Union legislation. This system is used to pass new EU legislation in more than 80 policies under the responsibility of the Union. If a legislative draft is disallowed at any stage of the procedure, or if the European Parliament and the Council cannot reach a common agreement, the legislative initiative will be rejected. Europe's digital revolution requires an agreement between European necessities and Member States' national interests through this system.

In addition, the European Parliament counts with the Panel for the Future of Science and Technology (STOA). Many of the issues coming before the MEPs have a scientific or technological dimension that increases the difficulty of decision-making. This specific panel of the EP helps parliamentarians to understand the impact of new technologies and the process of digitalization. The growing need for legislators and policy-makers to rely on independent, impartial, and accessible information about digital transition is answered by this panel. It shows the importance the EP is giving the process that will change the European Union itself, preparing the legislators to understand opportunities provided by digital economies, but also the risks they entail and the ethical implications of the process.

The STOA Panel forms an integral part of the structure of the European Parliament. It is composed of 27 Members of the European Parliament who are nominated by 11 permanent Committees of the Parliament, implying the horizontal influence of the digital transition on all the relevant areas of parliamentarian responsibility. Its priority areas for the period from 2019 to 2024 are artificial intelligence, new technologies and the Green Deal, digitalization and quality of life. These three pillars are analyzed from a coherent perspective based on science, ethical challenges, economic impact, and legal consequences. STOA provides the Parliament's Committees and other parliamentary bodies the independent information required for the assessment of the impact of possibly introducing or promoting new technologies and identifying, from the technological point of view, the options for the best courses of action to take. This specific unit emphasizes the importance given by the EP to digital transition.

7 Conclusions

The EU is facing unprecedented challenges in terms of evolution due to the incredible technological development, dealing with powerful obstacles that threaten integration maximizing common actions that could provide the EU with international relevance, economic development, and internal cohesion.

The European Union owns a powerful scientific and industrial fabric on which to build. It includes leading research laboratories and excellent universities, plus numerous innovative businesses. EU has developed a comprehensive legal frame protecting consumers while promoting innovation. In addition, the European Union is making progress in establishing a Digital Single Market. Therefore, the main elements are already there for Europe to become a leader in the current international technological revolution, in its own way, based on its values and traditions. The digital transition described in this chapter shows the path that Europe wants to pursue, including the necessity to integrate forces at the European Union level to guarantee a homogeneous digital transition. It is imperative to highlight the European Union's commitment to digital innovation and its convergence with the Union's values (Ciarli et al., 2021).

Dangerously, there is a certain inability of national States to share their precious national sovereignty at the European level, but most likely this resistance will be overcome by the complexity of the digital world, which will separate the convenience of common management from the populist debate of nationalism. Nevertheless, logic is not the whole lot in social relations and the European Union should keep the track of its current strategy.

The future is digital, and Europe counts on the right tools to implement it, but requires a decisive action from the EU to provide a common frame for a common development with the support of the Member States. The digital transition will change the European Union accelerating integration and maximizing the benefits of the union between Europeans regardless of their nationality.

References

Andraško, J., Mesarčík, M., & Hamuľák, O. (2021). The regulatory intersections between artificial intelligence, data protection and cyber security: Challenges and opportunities for the EU legal framework. *AI & Society, 36*(2), 623–636.

Borowiecki, R., Siuta-Tokarska, B., Maroń, J., Suder, M., Thier, A., & Żmija, K. (2021). Developing digital economy and society in the light of the issue of digital convergence of the markets in the European Union countries. *Energies, 14*(9), 2717.

Bunn, D. (2021). The European Commission and the taxation of the digital economy. *Dostupno na*. https://taxfoundation.org/european-commission-eu-digitallevy

Cabral, L., Haucap, J., Parker, G., Petropoulos, G., Valletti, T. M., & Van Alstyne, M. W. (2021). The EU digital markets act: A report from a panel of economic experts. In L. Cabral, J. Haucap, G. Parker, G. Petropoulos, T. Valletti, & M. Van Alstyne (Eds.), *The EU digital markets act*. Publications Office of the European Union.

Celeste, E. (2021). Digital sovereignty in the EU: Challenges and future perspectives. *Data Protection Beyond Borders: Transatlantic Perspectives on Extraterritoriality and Sovereignty*, 211–228.

Ciarli, T., Kenney, M., Massini, S., & Piscitello, L. (2021). Digital technologies, innovation, and skills: Emerging trajectories and challenges. *Research Policy, 50*(7), 104289.

De Gregorio, G. (2021). The rise of digital constitutionalism in the European Union. *International Journal of Constitutional Law, 19*(1), 41–70.

Elena-Bucea, A., Cruz-Jesus, F., Oliveira, T., & Coelho, P. S. (2021). Assessing the role of age, education, gender and income on the digital divide: Evidence for the European Union. *Information Systems Frontiers, 23*(4), 1007–1021.

Erstad, O., Kjällander, S., & Järvelä, S. (2021). Facing the challenges of 'digital competence' a Nordic agenda for curriculum development for the 21st century. *Nordic Journal of Digital Literacy, 16*(2), 77–87.

Grisold, T., vom Brocke, J., Gross, S., Mendling, J., Röglinger, M., & Stelzl, K. (2021). Digital innovation and business process management: Opportunities and challenges as perceived by practitioners. *Communications of the Association for Information Systems, 49*(1), 27.

Hamulak, O. (2015). Is charter of the fundamental rights of the EU taking social rights seriously? *European Studies: The Review of European Law, Economics and Politics, 2*, 14–28.

Ivic, S. (2012). EU citizenship as a mental construct: Reconstruction of postnational model of citizenship. *European Review, 20*(3), 419–437.

Kajander, A., Kasper, A., & Tsybulenko, E. (2020). Making the cyber mercenary–autonomous weapons systems and common article 1 of the Geneva conventions. In *2020 12th international conference on cyber conflict (CyCon)* (Vol. 1300, pp. 79–95). IEEE.

Kerikmäe, T., & Särav, S. (2017). Paradigms for automatization of logic and legal reasoning. In *Law and logic: Contemporary Issues* (Vol. 205, p. 222). Duncker & Humblot.

Künnapas, K. (2016). Maksukohustuse täitmise preventiivne tagamine enne maksukohustuse tuvastamist: ettevaatuspõhimõte maksumenetluses. *Tartu Ülikool kirjastus*.

Małkowska, A., Urbaniec, M., & Kosała, M. (2021). The impact of digital transformation on European countries: Insights from a comparative analysis. *Equilibrium: Quarterly Journal of Economics and Economic Policy, 16*(2), 325–355.

McCampbell, M., Schumann, C., & Klerkx, L. (2022). Good intentions in complex realities: Challenges for designing responsibly in digital agriculture in low-income countries. *Sociologia Ruralis, 62*(2), 279–304.

Outeda, C. C. (2020). La inteligencia artificial: una lectura desde la política. In *Inteligencia artificial: de la discrepancia regional a las reglas universales: integración de percepciones políticas, económicas y legales* (pp. 319–344). Aranzadi Thomson Reuters.

Parvin, F., & Attar, S. (2022). The law of the European union and the challenge of property rights in data in the era of digital economy.

Ramiro Troitiño, D. (2022a). George of Poděbrady: The role of external stimulus to the European integration. In *The European Union and its political leaders* (pp. 5–16). Cham: Springer.
Ramiro Troitiño, D. (2022b). Jean Monnet: Neofunctionalism at work in the European integration. In *The European Union and its political leaders* (pp. 121–140). Cham: Springer.
Ramiro Troitiño, D. (2022c). François Mitterrand: French leadership in the European Union. In *The European Union and its political leaders* (pp. 233–242). Cham: Springer.
Rivas-de-Roca, R., & García-Gordillo, M. (2022). Understanding the European public sphere: A review of pending challenges in research. *European Politics and Society, 23*(3), 380–394.
Steible, B. (2018). EU support to domestic prosecution of violations of international humanitarian law. *UNIO–EU Law Journal, 4*(1), 51–66.
Varshney, A., Garg, N., Nagla, K. S., Nair, T. S., Jaiswal, S. K., Yadav, S., & Aswal, D. K. (2021). Challenges in sensors technology for industry 4.0 for futuristic metrological applications. *Mapan, 36*(2), 215–226.

'Spill Over' and 'Fail Forward' in the EU's Cybersecurity Regulations

Agnes Kasper and Anna-Maria Osula

Abstract The European Union's cybersecurity policy spans across domains from electronic communications, payment services and fight against cybercrime to cyberdiplomacy and building cyber defence capabilities. While cybersecurity is a horizontal policy area, synergies among its parts are missing or incomplete and the related challenges are addressed in silos at the supranational EU level. Cybersecurity is linked to various other policy areas (often those at the heart of the single market), which are textbook cases of how integration in one functional area 'spills over' into another one, hence such dynamics of integration are well-explainable in terms of neofunctionalism. However, cybersecurity is also linked to issues within the areas of freedom, security and justice, as well as Common Foreign and Security Policy (CFSP), where 'pure' neofunctionalist explanations face significant constraints and the 'failing forward' framework was proposed by scholars to yield deeper and valuable insights. This chapter investigates the applicability of the 'failing forward' framework to the case of the EU's cybersecurity policy by focusing on key regulatory instruments. For the discussion in this study the history of cybersecurity policy is divided into cyber eras of 'cyber sleeping', 'cyber awakening', 'cyber reacting' and 'cyber powering'. The authors aim to chart the course of integration dynamics in cybersecurity policy throughout these eras, resulting in insights on efforts to advance

The contribution by A. Kasper is part of the cooperation within Jean Monnet Network 'European Union and the Challenges of Modern Society' (611293-EPP-1-2019-1-CZ-EPPJMO-NETWORK). The European Commission support for the production of this publication does not constitute an endorsement of the contents which reflects only the views of the authors, and the Commission cannot be held responsible for any use which may be made of the information contained therein.

A. Kasper (✉)
Department of Law, Tallinn University of Technology, Tallinn, Estonia

Present Address: Law Branch, NATO Cooperative Cyber Defence Centre of Excellence, Tallinn, Estonia
e-mail: agnes.kasper@taltech.ee

A.-M. Osula
Centre for Digital Forensics and Cyber Security, Tallinn University of Technology, Tallinn, Estonia

D. Ramiro Troitiño et al. (eds.), *Digital Development of the European Union*,
https://doi.org/10.1007/978-3-031-27312-4_3

synergies, and deepening and widening of integration in EU cybersecurity, and ultimately on shaping European integration.

1 Introduction

The growing dependence of societies on information and communication technologies (ICTs) has brought along the increasing complexity of cybersecurity policies in the European Union (EU) and its Member States (MSs). Security concerns in this context first emerged and were addressed in the EU legislation on electronic communications[1] and personal data protection[2] starting in the mid-1990s. Today, the EU's cybersecurity policy and related legislation cover a vast number of areas and have also significantly expanded in depth. Cybersecurity is no longer perceived as a technical challenge but as a strategic one. This, matched with the *sui generis* nature of the EU, its principles of conferral, subsidiarity, proportionality and its ability to legislate within its exclusive and shared competence areas can leave scholars and policymakers puzzled in their efforts of trying to further responses to identified cybersecurity challenges.

Integration in the area of cybersecurity has its roots in specific policy areas (telecom, privacy), where security aspects were rather incidental and contextual with a narrow scope—corresponding to the competences of the EU at the time. However, while at the end of 1990s the EU had limited competences relating to criminal matters, it was closely following and observing the development of the Council of Europe (CoE) Convention on Cybercrime.[3] From thereon, great leaps in the EU cybersecurity policy have taken place, and as argued in this article, are roughly following the timeline of impactful events, such as serious cyber attacks and revelations about the misuses of technology. We can observe how new threats, the complex geopolitical setting and increasing insecurity in the wider EU region have pushed MSs to agree on a shared EU vision and further common action. We can also witness the progressive deepening of cooperation, the widening of tasks and instruments in the domains beyond what was agreed upon following each previous crisis situation.

Today, the EU cybersecurity policy has a strong intergovernmental character, which may lead to incomplete or sub-optimal reforms, but this domain is also deeply rooted in economic integration and the single market where neofunctionalism seems to have taken its course and spillover effects have been the most visible. In order to explain integration dynamics in the field of cybersecurity as a whole, the failing forward framework proposed by Jones et Al.[4] is applied, which is expected to yield insights on how the comprehensive cybersecurity policy is moving towards a truly integrated one.

---

[1]Commission (1995).

[2]Directive 95/46/EC of the European Parliament and of the Council of 24 October 1995 on the protection of individuals with regard to the processing of personal data and on the free movement of such data.

[3]Council of Europe, Convention on Cybercrime, 2001, ETS No. 185.

[4]Jones et al. (2016).

Considering the above starting points (telco, privacy, cybercrime), spillover effects are discernible in policy areas close to the 'heart' of the EU single market and sometimes beyond. Examining responses to major crises, especially when related to the Common Security and Foreign Policy (CSFP) and Common Security and Defence Policy (CSDP), provides a fertile ground for addressing the failing forward approach and to draw conclusions on how integration takes place within the complex cybersecurity policy domain.

We proceed as follows. First, we briefly describe the failing forward framework and its added value to the established theoretical approaches regarding integration. Then we examine the integration dynamics in cybersecurity policy areas throughout four distinct eras starting from the inception of the policy at the EU level. We characterize these eras as 'cyber sleeping', 'cyber awakening', 'cyber reacting' and 'cyber powering' and we build primarily on EU regulatory instruments, since these are clear manifestations and evidence of integration.

2 Failing Forward Framework and Cyber Eras

European integration is often explained in the terms of two main schools, in accordance with theories of intergovernmentalism and neofunctionalism. The former focuses on the games among states, whereas the latter considers the interplay of societal actors, in particular supranational institutions.[5] Both schools explain integration as a result of cooperation and competition among their respective actors.[6] On the one hand, intergovernmentalism posits that interdependence among states shapes their bargaining, where a typical outcome is the lowest common denominator, i.e., delegation of just enough authority to common institutions in order to respond to a problem.[7] On the other hand, a key concept in the neofunctionalist approach is the 'spillover' among policies, where integration in one area triggers further integration in others. Policy spillovers are engineered in supranational activism by altering the preferences and tactics of national elites and non-state actors.[8] These perspectives have been used as alternatives, and they describe integration dynamics with varying success depending on the policy sector.

Jones et al. bridged the above-mentioned traditions and argued that integration, in some circumstances, proceeds through a pattern of failing forward in cycles of four consecutive phases.[9] Accordingly, the lowest common denominator deals lead to half-baked institutions, and this incompleteness and resulting failures unleash spill-

[5]Hooghe and Marks (2019) pp. 1114–1115.

[6]Ibid.

[7]Ibid. p. 1116.

[8]Ibid. p. 1115.

[9]JJones et al. (2021) pp. 1519–1520.

over effects that in turn spark a crisis.[10] It has been suggested that European policymakers are more likely to fail forward when they face an immediately life-threatening challenge,[11] and others have criticized the partial applicability of the framework, pointing out that in the area of foreign policy, crises are largely sparked by exogenous factors.[12]

Jones et al. propose that three key factors are to be assessed: the intensity of the policy crisis, the (existing) competences of the EU in the policy area, and the costs of not granting more authority to the supranational institutions.[13] In the context of cybersecurity, we have identified three main junctures when the EU has faced major problems sparked by exogenous factors and which were followed by significant shifts in the Union's cybersecurity policy. We use these junctures to construct cyber eras for the analysis of the above factors and integration dynamics. Moreover, having pointed to overall external causes of crises in advance, it is expected that fail forwards can be detected easier.

The four cyber eras can be divided as follows. From the mid-1990s until 2007, in the era of cyber sleeping, network and information security was considered a predominantly technical and economic issue due to costly disruptions and rising trends in profit-oriented cybercrime.[14] The first major turning point was the 2007 cyberattacks against Estonia,[15] which captured the attention of EU MSs and clearly indicated a new dimension of coordinated attacks against private and government IT systems as a potential new economic, political and military weapon.[16] From around 2008 to 2013 many advanced economies realized the gravity of cybersecurity challenges, a period which Estonia's first cyber ambassador called as cyber awakening.[17] Cybersecurity proved to be a strategic issue including all EU competence areas, although the focus remained predominantly on threats emanating from the economic sphere. The second turning point took place around 2013, when the true dimension behind nation-states potential was revealed.[18] From 2014 to 2019, in the era of cyber reacting, cybersecurity challenges intensified, and although financial motivation remained the main factor behind cyberattacks,[19] the EU's focus shifted towards addressing the employment of the cyber domain for geopolitical aims by adversaries. The third major turning point and the era of cyber powering was accelerated by the COVID-19 pandemic, which pushed for abrupt technological adoption and transformation, thereby expanding the attack surface in

[10] Ibid. p. 1522.

[11] Ibid. p. 1532.

[12] Rabinovych (2021).

[13] Jones et al. (2021) p. 1525.

[14] Commission (2001, 2006).

[15] Kaska et al. (2010).

[16] Council (2008).

[17] Tiirmaa-Klaar (2018) p. 18.

[18] ENISA (2013) p. 36.

[19] ENISA (2020)

cybersecurity,[20] as well as the increasingly hostile security environment which has in general required the EU to commit to making 'a quantum leap forward' and increase its capacity and willingness to act, strengthen resilience, and invest more and better in defence capabilities.[21]

In the next chapters we note the key events characteristic to each era, analyze the major strategic and regulatory instruments and point out interlinks between the fail-forward factors.

3 Cyber Sleeping (1994–2007)

The cyber sleeping era is characterized by the realization that the diffusion of ICT is causing a disruption raising to the level of a revolution, thus it raised the question of whether this process is going to last and intensify, and to what extent policy (and regulation) could or should go along (Fig. 1).

ICT-related security concerns first arose in the legal instruments regulating personal data protection and services and goods on the single market. Articles 16 and 17 of the 1995 Personal Data Protection Directive (Directive 95/46/EC) laid down general obligations to ensure confidentiality of processing and to implement technical and organizational measures to enhance the security of processing. In the field of provision of services, two main areas were addressed. First, being concerned with the key role of intermediary service providers in the development of the single market, the e-commerce directive (Directive 2000/31/EC) enacted a regime for exempting providers of information society services from liability for illegal activity within their system/services under certain circumstances, and prohibited to impose general obligation to monitor or actively seek for indicators of illegal activities. Second, the Framework Directive (Directive 2002/58/EC) of the telecommunication package, which provided for the liberalization of the sector's market, imposed general obligations on electronic communications service providers to implement security measures and to maintain confidentiality of the communications transmitted in their systems. A palpable deliverable in the new Network and Information Security (NIS) policy of the EU[22] was the establishment of ENISA, the European Network and Information Security Agency in 2004, which had an advisory role and was tasked to assist the Commission, MSs, cooperate with business community in order to help them to meet the requirements of NIS, thereby ensuring the smooth functioning of the internal market. This focus of the NIS policy on the internal market remained constant and it significantly expanded in scope both horizontally (to other sectors) and vertically throughout the years, which will be addressed in the next sections.

---

[20] Ibid.

[21] Council (2022) p. 6.

[22] Commission (2001).

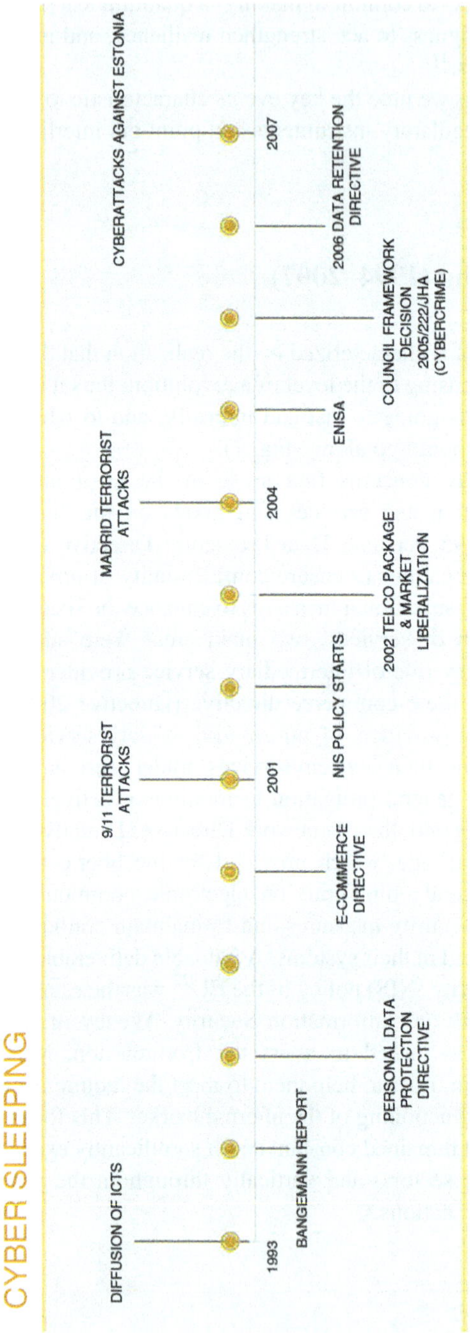

Fig. 1 Cyber sleeping

The first expansion of the NIS policy into new areas took place at the crossroads of e-privacy and criminal justice. The e-Privacy Directive (Directive 2002/58/EC) of the telco package had a very turbulent history as it clearly prescribes that traffic data related to communications must be erased when it is no longer necessary for the transmission; however it allows exceptions such as for the purposes of investigating and prosecuting serious crimes. In the deep shock and aftermath of the 2004 terrorist bombings at the Atocha railway station in Madrid the EU was very quick to adopt the data retention directive (Directive 2006/24/EC), making the exception in the e-Privacy Directive a community-wide rule. The event triggered an intense crisis (Hamulák, 2018), since the European Council in its Declaration on Combating Terrorism of 25 March 2004 announced nearly an entire strategy with a long list of planned legislative measures relating to cooperation in the area of criminal justice.[23] The then-perceived costs of not moving forward with the integration have been presented as considerable since new quasi-anonymous communication technologies make it much harder to fight crime and terrorism.[24] The new solution, originally proposed by France, Ireland, UK and Sweden, was sharply criticized by the European Data Protection Supervisor[25] due to its disproportionate rules, but was nevertheless adopted in 2006. This points to the conclusion that the Data Retention Directive expresses an incomplete policy—that later leads to its cancellation in 2014 by the ECJ (discussed in Sect. 4).

The Madrid terrorist attacks led to a wave of reforms in the broader criminal justice area, but also in the fight against cybercrime, and the Council of the EU soon made its first move by adopting a harmonizing measure, the Council Framework Decision 2005/222/JHA emphasizing that 'attacks against information systems [. . .] constitute a threat to the achievement of a safer information society'.[26] This measure was clearly based on the recent CoE Convention on Cybercrime, but the Framework Decision took over only the provisions relating to the substantive criminal laws (that also in a simplified form) and did not touch on the procedural aspects. These were new competencies for the EU at that time, and which were clearly limited to what is necessary to respond to the threats of terrorism and rising cybercrime.

NIS and data security issues were steadily infiltrating EU policies from the mid-1990s. The Union had its first attempt in 2001 to create a specific policy addressing this area, and subsequent terrorist attacks both on the global scene (9/11) as well as in Europe (2004 Madrid, 2005 London) led to a crisis where decisions in the intergovernmental pillar of the EU had to be taken and resulted in newly acquired competences in the field of criminal justice, including in cybercrime and criminal investigations. However, these solutions have probably been incomplete, at least when it comes to cyber issues, as argued earlier. The EU's policy documents also point out that 'information systems and communication networks

[23] Council (2004).

[24] Commission (2005).

[25] European Data Protection Supervisor (2005).

[26] Preamble point (2).

have become a critical factor for other infrastructures (e.g. water and electricity supply) and other markets (e.g. the global finance market)',[27] but this dimension remained underdeveloped until the 2007 large-scale cyberattacks against Estonia rang the emergency bell.

4 Cyber Awakening (2008–2013)

The cyber awakening era can be characterized by denoting of the transcendence of ICT-security issues beyond the technical dimension and connecting them with EU-level policies as well as regulatory frameworks. This was not an easy puzzle, but rather a million-piece one that had to be digested and made sense of (Fig. 2).

During this period, the EU's understanding of NIS and data security has step-by-step developed into a comprehensive understanding of cybersecurity. In the core single market aspects the changes just trickled down and besides the 2009 overhaul of the telecom framework, where the e-Privacy Directive's security provisions were amended and deepened (Directive 2009/136/EC), the EU started to wander into new fields. The Council adopted a new measure on critical infrastructure protection (Council Directive 2008/114/EC (CIP Directive)), but—more importantly for the cybersecurity policy—it was taken to the next level by the Commission communication on critical information infrastructure protection in 2009 (CIIP Communication), which addressed the problem of large-scale cyberattacks.[28]

The CIP Directive was a response to the arising functional need that had previously been underdeveloped (see Sect. 3). However, the CIP Directive did not sufficiently take into account the threats arising from the use of ICTs in critical infrastructures, therefore it may be qualified as common denominator and incomplete policy. Supranational activity in the form of emphasizing the specifics related to dependencies on ICTs is pointed out in the CIIP Communication, which however remained soft-law, but also led to the NIS Directive proposal[29] later in 2013. The NIS proposal ambitiously aimed for giant leaps regarding MSs' cybersecurity policies and cooperation within the EU, as well as security requirements and obligations imposed on market players, which included way more than critical infrastructures. This road from CIP to NIS engineered by the Commission was not paved with policy failures, but rather fits the description of spillover according to the neofunctionalist approach. However, the story of the NIS continues in the next era and Sect. 4.

In the field of cybercrime, the rising trends in financially motivated cybercrime, for example the surfacing of a series of banking malware, such as Zeus,[30] as well as

[27] Commission (2001).

[28] Commission (2009).

[29] Commission (2013a).

[30] See https://en.wikipedia.org/wiki/Zeus_(malware).

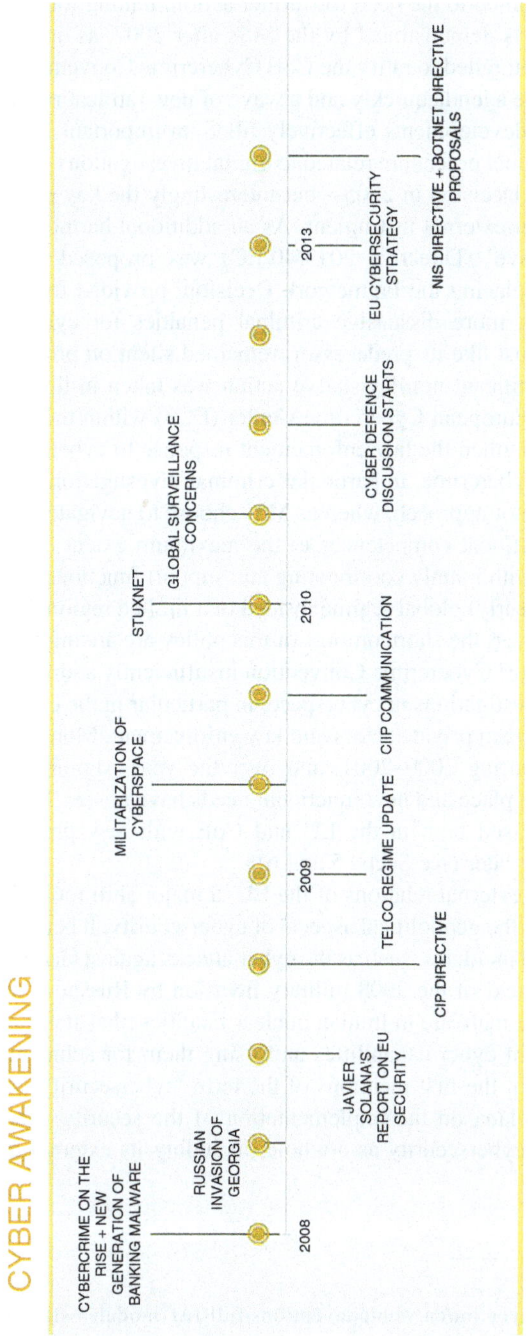

Fig. 2 Cyber awakening

the previously mentioned 2007 attacks against Estonia and difficulties in following investigations[31] pointed to the need for further action. Failing forward dynamic and significant activity is demonstrated by the MSs after 2007, as until that time many countries signed, but failed to ratify the CoE Cybercrime Convention. However, this issue returned to the agenda quickly and a wave of new ratifications followed in this cyber era.[32] Such developments effectively filled an important gap related to harmonization of criminal procedure related to digital investigations—as left untouched by the Framework Decision in 2005—but interestingly the key part of the solution was provided by an external instrument. As an additional harmonizing instrument, the 'Botnet directive' (Directive 2013/40/EC) was proposed 2013, which was building on and replacing the Framework Decision, provided for further harmonization in requiring more dissuasive criminal penalties for cybercrime. Yet, the Botnet Directive, just like its predecessor, remained silent on procedural questions, however, very significant non-legislative action was taken in this respect with the foundation of the European Cybercrime Center (EC3) within the Europol in 2013, which aims to strengthen the law enforcement response to cybercrime in the EU.[33] This approach to cybercrime, in particular criminal investigations, reflects a lowest common denominator approach, whereas MSs choose to navigate so as to keep their sovereignty and national competences to the maximum extent and leave supranational institutions with mainly coordinating and support functions. This is not to say that opting for a (nearly) global regime instead of a limited regional one is incorrect. In hindsight, however, the shortcomings of this policy are arising from the systemic problem that the CoE Cybercrime Convention insufficiently addresses some aspects of cross-border investigations in cyberspace, in particular in the case of direct cross-border contact between private actors and law enforcement. Moreover, the Convention was drafted during 2000–2001, and over the years significant technological changes have taken place and new functional needs have arisen. These have, in later stages, been addressed both in the EU and CoE with new proposals relating to electronic evidence later (see Sects. 5 and 6).

With respect to external relations of the EU, a major shift took place in terms of attention turning to the geopolitical aspects of cybersecurity. It became clear in a line of cyber events and incidents, such as the cyber attacks against Georgian information systems in the context of the 2008 military invasion by Russia or the discovery in 2010 of the Stuxnet malware in Iranian nuclear facilities, that states are increasingly acquiring significant cyber capabilities and using them for achieving their foreign policy goals. One of the first mentions of the term 'cybersecurity' was in the 2008 report of Javier Solana on the implementation of the security strategy, where he made it clear that cybersecurity as a whole, including its external dimensions that

[31] Tikk et al. (2010).

[32] See https://www.coe.int/en/web/conventions/full-list?module=signatures-by-treaty&treatynum=185.

[33] See https://www.europol.europa.eu/about-europol/european-cybercrime-centre-ec3.

transcend the single market, was becoming a strategic and national security issue.[34] The 2009 Commission Communication on CIIP started to link the internal and external, public and private policy aspects of the cyber domain.[35] Cautious discussions on the EU's approach to cyber defence started as early as 2010 and engaged a limited scope of topics and remained hesitant in terms of agreeing on concrete commitments.[36] Finally, in 2012 the Council welcomed the pooling and sharing projects supported by the European Defence Agency, including areas of Cyber Defence, and raised the cyber issue again in the context of defence industry and market.[37] Having in place the competences for the key building blocks of an umbrella policy, by 2013 the Commission put together a package including the proposal for the NIS Directive, the Botnet Directive and the first Cybersecurity Strategy of the EU[38] embracing the whole-of-the-society approach, and stretching the EU action further into increasingly sensitive policy domains.[39] The strategy had three pillars—NIS, law enforcement and defence—which operate at two levels, within the EU and national frameworks.[40] Reflecting the concern for misuses of cyberspace for censorship and mass surveillance, the key new elements in the strategy were, compared to previous narrower policies, the clear inclusion of the defence domain, but also the incorporation of the protection of fundamental rights, freedom of expression, personal data and privacy as an underlying principle of the entire domain.

While we acknowledge that for the purposes of our analysis the complex cyber policy, as such, and EU Cybersecurity Strategy can hardly ever be considered as complete, and thus, will almost always fulfill the criteria of incompleteness in terms of the fail forward approach, the next chapters can nevertheless detect some relevant sub-domains for further analysis.

5 Cyber Reacting (2014–2019)

The era of cyber reacting can be characterized by the deepening and widening of cybersecurity policy, as well as the increasing integration among its pillars. Key legal frameworks were adopted in this era, yet to be perfected later. On the policy front, the 2013 cybersecurity strategy pillars of NIS, law enforcement and defence were redefined into the overlapping resilience building, deterrence and external policy pillars by the updated 2017 strategy (Fig. 3).[41]

[34] Council (2008).

[35] Commission (2009).

[36] Robinson (2014).

[37] Council (2012).

[38] Commission (2013b).

[39] Kasper and Vernygora (2021) pp. 51, 67.

[40] Commission (2013b).

[41] Commission (2017a).

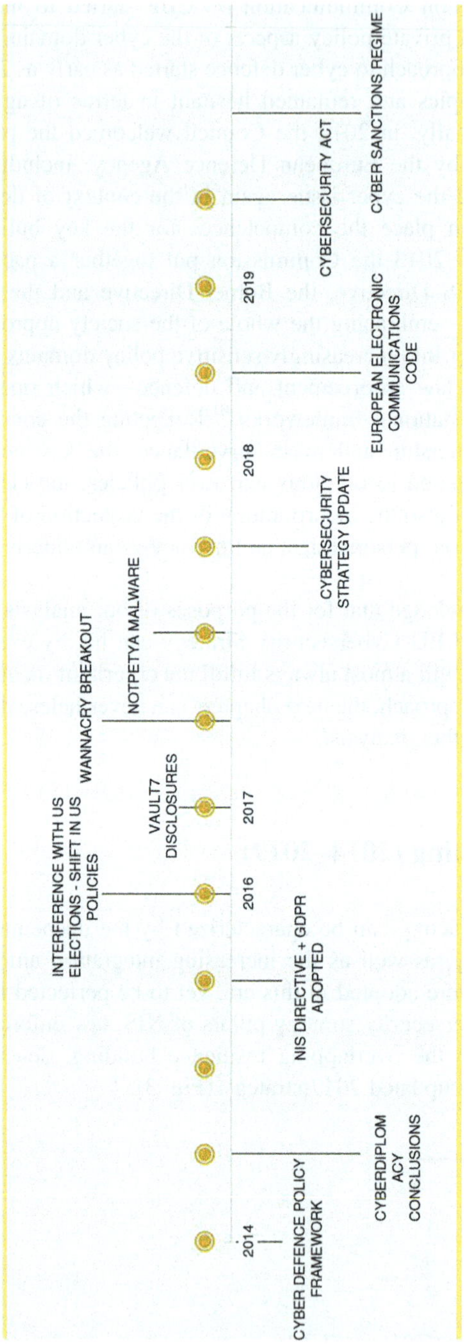

Fig. 3 Cyber reacting

Resilience building in the single market included targeting specific issues and/or sectors. The text of the NIS Directive adopted in 2016 reflects a lowest common denominator approach from two perspectives. First, the original proposal's ambitious provisions for close cooperation among the MSs were toned down to accepting obligations to compose national NIS strategies, furthering technical cooperation among national CSIRTs and creating a forum for voluntary information sharing (NIS Group) for MSs. The plans for establishing a secure information sharing system, early warning system, coordinated response plan, union NIS cooperation plan[42] were stricken from the text during NIS Directive (Directive (EU) 2016/1148) negotiations. Secondly, the proposed broad reach of the directive, where it was to impose security requirements and obligations on public administrations and 'market operators' was narrowed down significantly (Ramiro Troitiño et al., 2020). The adopted text features digital service providers only—e-commerce platforms, search engines and cloud services—in addition to the operators of essential services, which is a very close category to critical infrastructure operators (see also in Sects. 3 and 4). These changes tell about the incompleteness of the solution whereas significant criticism by supranational institutions was related to existing obstacles to information sharing and too narrow coverage.[43] However, the NIS directive does not operate in isolation and it is complemented by a series of other instruments, both legally binding and non-binding. As an example for a non-binding instrument and supranational activism, the Commission put forth a recommendation on coordinated response to large-scale cybersecurity incidents and crises,[44] an element that was excluded from the final text of the adopted NIS Directive. While the 2017 Wannacry incident[45] tested the EU's response capabilities to large-scale attacks, it also included the 'lucky' early discovery of a kill-switch by the technical community, thus any conclusions that the coordination worked well needs to be taken with a pinch of salt. Nevertheless, the attack served as another wake-up call, and in the sub-domain of cooperation may exhibit the features of the fail forward dynamics.

It should also be noted that the NIS/resilience pillar includes new legal instruments such as the Cybersecurity Act of 2019 (Regulation (EU) 2019/881), turning ENISA into the permanent Cybersecurity Agency of the EU, also tasked with establishing a cybersecurity standard and certification framework for products and services and the integration of detailed cybersecurity-related provisions into sectoral regulations. EU also took concrete steps regarding the security requirements and assurance levels in trust services and electronic identification systems in the eIDAS Regulation (Regulation (EU) No 910/2014), security standards of financial services in the second Payment Service Directive (Directive (EU) 2015/2366). Equally, the telecom regulatory regime underwent a reform and consolidation, and the European

[42] Commission (2013a, 2013b).

[43] Commission (2017a).

[44] Commission (2017b).

[45] See https://www.enisa.europa.eu/news/enisa-news/wannacry-ransomware-first-ever-case-of-cyber-cooperation-at-eu-level.

Electronic Communications Code (EECC, Directive (EU) 2018/1972) now details extensive security-related requirements. This list is merely representative of the efforts, giving a good indication of the breadth and depth of the policy developments, but does not cover all relevant instruments. The continuous and relatively smooth expansion of issues belonging under the umbrella of resilience certainly points to the neofunctionalist dynamics and spillover effect. Since no major hesitation can be detected, the integration in this policy area may be attributed to the understanding of the gradually increasing functional needs which are conditioned by the diffusion of ICT technologies, rather than the failure of stopgap reforms. However, we do not exclude the possibility that if each sub-domain is examined separately, the conclusion may be different.

As a parallel stream under the cyber awakening era, the data protection and privacy aspects took on their own independent life, yet still closely related to cybersecurity policies. The General Data Protection Regulation (GDPR, Regulation (EU) 2016/679) is often seen as a key cybersecurity tool, even if its scope is limited to personal data only. The GDPR and its 'accompanying' Police Directive (Directive (EU) 2016/680) undoubtedly signify a milestone in the EU's data protection policy. So is the activity of the European Court of Justice (ECJ) in striking down the Data Retention Directive in the *Digital Rights Ireland* case in 2014.[46] This judgment started a crisis in the area of criminal justice cooperation and led to a saga of data retention cases. Also, in terms of the fail-forward approach, the invalidation of the Data Retention Directive was a clear instance of an institutional failure and negative spillover as several national laws based on the directive were rolled back, while others kept their data retention regimes intact.[47] The resulting intense discussion and the lack of an acceptable EU-wide solution point at an incomplete cycle of fail forward. However, this presumes that the proposed e-Privacy Regulation and/or another instrument will in the future address the question in 'forward' terms and the EU will find a common denominator. The quest for that is still ongoing and proceeds with baby steps as the ECJ is addressing elements of the data retention regime, such as in the recent *Prokuratuur* case,[48] but not the entire regime as such.

The data retention saga has the closest links to the fight against cybercrime which is part of another pillar in the strategy now better referred to as deterrence. The 2017 Cybersecurity strategy uses the term deterrence in dual meaning,[49] for a framework that is credible and dissuasive for would-be cybercriminals and attackers, both for state and non-state perpetrators. In respect of criminal investigations, the EU has also moved forward, addressing shortcomings in current frameworks relating to criminal procedure and electronic evidence. It introduced the e-Evidence proposal in 2019

[46] Court of Justice of the European Union (2014).

[47] EDRI (2019).

[48] Court of Justice of the European Union (2021).

[49] Kasper and Mölder (2020).

and the EU was also given a mandate to negotiate the 2nd Protocol to Cybercrime Convention,[50] which is now finalized and open for signatures.

The cyber reacting era was the time for consolidation of cybersecurity policy as a whole and was marked by initiating several efforts in cyber diplomacy and cyber defence. Following the US elections in 2016, and also learning from the intensifying efforts of adversaries to try to influence election results,[51] the EU's overall attention turned to strategic autonomy. Increasingly louder voices also emphasized the dependence of the EU on external providers, in terms of software in the USA and in terms of hardware in China. This triggered the conceptualization of technological sovereignty.[52] The EU MSs were effectively stripped of the possibility to remain passive or internally focused, or that so at the price of Europe being sidelined in global politics.

While the EU and its MSs were faced with complex questions in the intersection of internal/external, public/private, civilian/military policies as excellently demonstrated by NonPetya,[53] and the attacks against OPCW,[54] the non-military nature of the EU steered the focus toward non-military responses. Based on the cyberdiplomacy conclusions,[55] the EU ministers did not hesitate much to adopt (Decision (CFSP) 2019/797) and apply a regime of cyber sanctions.[56] Yet, the integration dynamics reach beyond and do not miss the cyber defence aspects, which is already far removed from the initial idea of ensuring the proper functioning of the single market, as started in the mid-1990s. This approach becomes way harder to reconcile with the idea of the EU as merely economic cooperation. Yet, it also points back to the reasons why this economic cooperation was devised in the first place—and that is clearly stated in the Paris Treaty of 1951—to prevent future bloody conflicts.

Cyber defence has always been a delicate domain for wider EU coordination since the EU security and defence policy is predominantly a competence of the MSs and there has been a lot of discussion on the need for a more common approach.[57] While the EU has worked hard on defining effective policies and legal frameworks for ensuring cyber security in general, the common vision towards cyber defence has not been well defined. In this regard, a big step forward was the EU Cyber Defence Policy Framework (CDPF), adopted in 2014,[58] seeking to improve MSs' capabilities, reinforcing the protection of critically relevant Communications and

[50] Council (2019).

[51] US Department of Justice (2019).

[52] Luukas and Osula (2020).

[53] NonPetya malware was based on a vulnerability stockpiled by the USA and disclosed in the 2017 Vault7 revelations by Wikileaks.

[54] BBC (2018).

[55] Council (2015).

[56] Council (2020).

[57] Lazarou and Dobreva (2019); Commission (2020a).

[58] Council (2014).

Information Systems, training and education and furthering cooperation among relevant stakeholders both in the EU and outside. Importantly, the CDPF recognized cyberspace as critical to the EU Common Security and Defence Policy (CSDP), now recognized as part of the framework of the EU Global Strategy, and made a push towards mainstreaming cybersecurity into EU crisis management. This step reflected the challenges of integrating cyber defence issues with existing EU policies and mechanisms as well as underlined the need for greater cooperation between military and civilian approaches.

Soon enough, there was a visible development in finding common ground on the urgency of cybersecurity among the MSs, as the Global Strategy for the EU's Common Foreign and Security Policy (CFSP), adopted in 2016, defined the EU's principles, priorities and instruments for addressing various security challenges such as hybrid and cyber threats and building the capacities of partner countries. The 2016 strategy committed to increasing the EU's focus on cybersecurity and acted as a comprehensive framework over various cybersecurity-related regulatory and policy instruments which is essential for ensuring a coherent approach among different pieces of the cybersecurity jigsaw puzzle. This was complemented by the update of the EU CDPF in 2018 which clearly recognized cyberspace as the fifth domain of operations and once again underlined the need for robust and resilient cyber operational capabilities.[59]

The cyber reacting era is defined by many relatively low-intensity crises. While unexpected (?) revelations in cyber-surveillance, intensifying problem of cybercrime (that came along with the diffusion of ICTs) and known, yet somewhat ignored processes of militarization of cyberspace brought along a comprehensive response by the EU, it also remains a fact that MSs are responsible for cybersecurity, and the EU has its expanding, yet limited role therein. However, fail-forward episodes can be pinpointed at the crossroads of the cyber resilience and deterrence sub-domains. The cybersecurity policy in this cyber reacting era has still been criticized for its siloed approach and that it lacks linkages between its pillars.[60] Indeed, some side effects of the GDPR have caught law enforcement by surprise—for example, when some internet registrars outside the EU decided to avoid the EU rather than bother to implement the GDPR.[61] This resulted in the sudden unavailability of IP lookup which is a turn-to service in criminal investigations. Another area, where the siloed approach was criticized, is the neglected information exchange between private and public sectors. This is related to the realization that the 2017 Wannacry and NotPetya attacks were built on the NSA-stockpiled vulnerability information,[62] which was stolen. Similarly, in a series of disclosures code-named as Vault7 by Wikileaks the CIA's hacking arsenal was leaked,[63] also including vulnerability information that

[59] Council (2018).

[60] Bendiek et al. (2017).

[61] European Data Protection Board (2018).

[62] ENISA (2017).

[63] See https://wikileaks.org/ciav7p1/.

could have been fixed by the software producers. These incidents prompted responses toward addressing the entanglement of various sub-domains of cybersecurity, but also regarding cyber threat intelligence and information sharing.[64] The process of integration within the cybersecurity policy sub-domains started, which is qualitatively different from the deepening and widening of the policy, therefore the real forwardness of this cyber era lies therein.

6 Cyber Powering (2020–. . .)

We define the cyber powering era by the start of the COVID-19 pandemic and its consequences. While this is certainly not an internally-induced affair, the so-far discussed cybersecurity policy elements did experience significant changes. The cyber powering era is characterized by the decompilation of cybersecurity-related law/policies and deep-level integration into sub-domain regulations (Fig. 4).

While the 2020 Cybersecurity Strategy of the EU[65] did not rearrange the structure of how we see cybersecurity policy, it clearly furthered the emphasis on geopolitics. Areas or sub-domains, where failing forward may explain integration, are overwhelming in volume and extent. The 2020 strategy emphasized the role of the private sector and industry and initiated processes that cut deep into the heart of the single market, such as the adoption of the long-awaited implementing acts of the Radio Equipment Directive (RED, Directive 2014/53/EU),[66] the announcement of the e-IDAS revision, NIS2 proposal,[67] CER proposal,[68] AI proposal,[69] Cyber Resilience Act[70] and many others. The 2020 cybersecurity strategy is a clearly geopolitically engaging external policy instrument, yet still remains rooted in the single market, which should not come as a surprise for an EU policy.

In the NIS/resilience dimension, the NIS2 proposal turns back to its 'missed' elements and incompleteness. As recently approved by the Council, the new breath of the NIS2 Directive is way beyond its predecessor, now including the public sector and more broadly even some small operators.[71] Yet, this dynamic calls upon the missed opportunities back in 2013. Although it took significant time to sort out and put into place NIS obligations, it did not need a full-fledged crisis to make this

[64] ENISA (2017).

[65] Commission (2020b).

[66] RED applies to equipment that makes use of radio waves, and many of our smart devices indeed use wifi, bluetooth and other technologies, that are based on data exchange facilitated by radio waves.

[67] Commission (2020c).

[68] Commission (2020d).

[69] Commission (2021).

[70] Commission (2022).

[71] Commission (2020c).

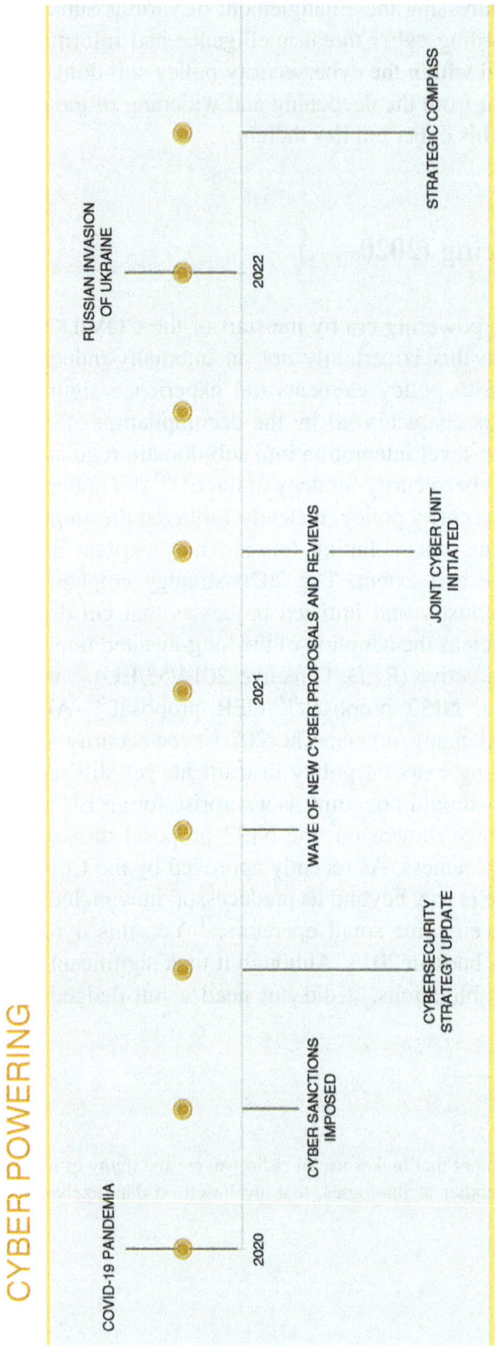

Fig. 4 Cyber powering

happen. However, in the field of electronic communications, in parallel to the uptake of cybersecurity concerns in the EECC, the union is also introducing the concern for external threats at the strategic level into the NIS/resilience pillar. For example, the 5G toolbox now provides a set of measures which should be prioritized in risk mitigation plans at the national and union levels.[72] Another example of this integration dynamic without expansion or deepening is the creation of the Joint Cyber Unit, composed of a key agency from each pillar of the cybersecurity policy, ENISA, EC3 and EDA. The unit will be tasked with strengthening the IT capabilities of defense communities in the field of cybersecurity and law enforcement agencies in cooperation with civilian and diplomatic communities.[73]

As far as data protection and cybercrime goes, developments are way more diverse and the lowest common denominators lead back to the data retention question where the EU has still not come up with a new proposal. The main ongoing discussion is about the costs and benefits of mass surveillance (data retention of traffic data of electronic communication), where states find themselves in-between complying with seemingly impossible rules or risking to uphold a more security-oriented regime that, from an internal viewpoint, balances between needs and fundamental freedoms (Kerikmäe et al., 2019). While it has been said long ago that data retention should be proportional, the current divide on acceptable and unacceptable practices leads us to conclude that this policy field is a good candidate for examining the fail-forward approach. However, so far we are yet to see the 'forward' part.

The era of cyber powering also brought along the application of the cyber sanctions regime and a new quest for EU's global actorness, aiming to substantively engage in the ongoing discussions on the application of international law in cyberspace, also by proposing a potential common position on how international law applies to cyberoperations.[74] However, the recent Russian aggression has also led to the reconsideration of the role of the EU as a security actor.

While the EU had done considerable progress in identifying the main security threats and threat actors targeting the EU with cyber as well as hybrid tools, including disruptive technologies, disinformation, and other non-military sources of influence, the Union has still struggled for a long time to determine the strategic common direction to guide its actions in the geopolitically contested environment. With the adoption of the new EU Security Union Strategy in 2020 we can see the EU stance becoming more concrete and focused on a coordinated response, but this still lacks a genuine common voice in cyber defence. The strategy reiterates that in facing contemporary complex security threats, EU citizens cannot be protected only through MSs acting on their own and that the Union will substantially enhance its

[72] NIS Cooperation Group (2020).

[73] Bendiek and Kettemann (2021).

[74] Commission (2020c).

resilience and ability to counter hybrid threats, cyberattacks and foreign information manipulation and interference.[75]

In March 2022, emphasizing the need for an accelerated response for geopolitical challenges, the EU adopted the Strategic Compass which presents the MSs' common vision about the EU as a security provider. Among other security domains, the strategy calls for further developing the EU Cyber Defence Policy to allow for better preparedness for and response to cyberattacks and includes a number of actions towards boosting cyber defence research and education, cooperation, capability development and generally enhancing the EU's cyber posture.[76] The document also recognizes the need to further invest in and develop new technologies, notably quantum computing, Artificial Intelligence and Big Data, which would help the EU to achieve comparative advantages, including in terms of cyber-responsive operations and information superiority.[77] The Strategic Compass came about a month after the beginning of the Russian invasion of Ukraine and represents a great leap in terms of common voice and readiness for common action, perhaps a fail-forward pattern, compared to previous policies of the EU.

It is no question that the reaction to the COVID-19 pandemic was disruptive. Despite its significant negative effects, many of the adopted steps may be characterized as failing forward due to open questions left in previous policies, and/or leaving no reasonable choice to roll back the achievements.

7 Conclusions

Looking back, the EU has over the time increased its role in providing a multidisciplinary and integrated response to security threats, as well as leading the development and harmonization of MSs' capabilities. This process started decades ago and by today a new discipline is created that at least partly from the practical work undertaken in the private sector.

In order to explain the dynamics of this process, we used the toolboxes of neofunctionalist and intergovernmental theories of integration, also adopting the prism of the recently proposed fail-forward approach that bridges the two main schools of thought. In order to better structure our analysis, we divided the history of EU cybersecurity policy into cyber eras (cyber sleeping, cyber awakening, cyber reacting, cyber powering) and examined the main pillars of the policy for spillovers and fail forwards.

We found several episodes of fail forward related to the policy pillars of cybersecurity. We also identified continuous and steady spill-overs especially in the NIS/resilience pillar. The first fail-forward episode may be the suddenly intensifying

[75] Commission (2020a) pp. 1, 34.

[76] Council (2022) pp. 12, 35.

[77] Ibid. p. 45.

cooperation and acquiring of new competences by the EU in the field of criminal justice, including cybercrime and digital investigations following the 2004 Madrid terrorist attacks. The then-created institutions were incomplete and did not respond well to the need to protect critical infrastructures against large-scale cyberattacks; therefore, the EU failed forward to address this sub-domain following the lessons from the 2007 cyberattacks against Estonia. This was also accompanied by a new wave of ratifications of the CoE Cybercrime Convention which can be seen as a response to a crisis; however, this proceeded with the caution of the intergovernmental approach relying on an external instrument. The 2013 NIS directive is a clear example of supranational engineering efforts based on the 2009 CIIP Communication, thereby illustrating the fail forward cycle within the first two cyber eras. Proceeding to cyber reacting era, we can see how the NIS directive was incomplete, with the example of the sub-domain of information sharing, which was toned down during negotiations. However, the 2017 Wannacry and NotPetya outbreaks made it clear that further steps are needed, and now, in the cyber powering era, the NIS2 is approved with a broader scope, and additional policy measures are underway (such as in the sub-domain of coordinated vulnerability disclosure policies). The changing nature of cyberthreats was also recognized in the era of cyber reacting and it led to the process of horizontal integration within cybersecurity policy pillars, integrating cybersecurity aspect into foreign and defence policies, but also taking note of the relevance of external policy within internally-focused pillars.

We conclude that integration in the field of cybersecurity proceeds on at least two tracks, and the fail-forward patterns may be detected in some areas, whereas continuous and gradual spillover can be observed in others. We also realized that none of the great schools of European integration theorize the dynamic of increasing entanglement among policy areas, but which does not necessarily mean either deepening competences or spillover into neighboring policy areas. Synchronization of existing policies as a response to entanglement dynamics is qualitatively different from other dimensions of integration. Enhanced coordination instruments, such as the 5G toolbox or the Joint Cyber Unit, potentially represent another 'forward' direction of integration, where the question is not about new competences of the EU, but about creating new synergies by linking policy responses within existing competences.

References

BBC. (2018). How the Dutch foiled Russian 'cyber-attack' on OPCW, 4 October 2018. https://www.bbc.com/news/world-europe-45747472.

Bendiek, A., Bossing, R., & Schulze, M. (2017). The EU's revised cybersecurity strategy—half-hearted Progress on far-reaching challenges. German Institute for International and Security Affairs, SWP comments 47, November 2017.

Bendiek, A., & Kettemann, M. C. (2021). *Revisiting the EU cybersecurity strategy: A call for EU cyber diplomacy. SWP comment, 16/2021.* Stiftung Wissenschaft und Politik—SWP—Deutsches Institut für Internationale Politik und Sicherheit. https://doi.org/10.18449/2021C16

Commission. (1995). *European Commission, Europe and the global information society: Recommendations to the European council: Conference G7—Raport BANGEMANN*. Publications Office.
Commission. (2001). Network and information security: Proposal for a European policy approach. COM (2001), 298 final, Brussels.
Commission. (2006). Commission communication of 31.5.2006, a strategy for a Secure Information Society—Dialogue, partnership and empowerment. COM(2006) 251 final.
Commission. (2005). Commission MEMO/05/328 Brussels, 21 September 2005.
Commission. (2009). Communication on critical information infrastructure protection—protecting Europe from large scale cyber-attacks and disruptions: enhancing preparedness, security and resilience. COM(2009) 149 final.
Commission. (2013a). Proposal for a Directive of the European Parliament and of the Council concerning measures to ensure a high common level of network and information security across the Union. COM (2013) 48.
Commission. (2013b). Communication of 7.2.2013 on cybersecurity strategy of the European Union: An Open, Safe and Secure Cyberspace, JOIN(2013) 1 final.
Commission. (2017a). Commisson Joint Communication of 13.9.2017 on Resilience, Deterrence and Defence: Building strong cybersecurity for the EU, JOIN(2017) 450 final.
Commission. (2017b). Commission Recommendation (EU) 2017/1584 of 13 September 2017 on coordinated response to large-scale cybersecurity incidents and crises.
Commission. (2020a). Communication from the Commission to the European Parliament, the Council, the European Economic and Social Committee and the Committee of the Regions on the EU Security Union Strategy. COM/2020/605 Final.
Commission. (2020b). Joint Communication to the European Parliament and the Council, The EU's cybersecurity strategy for the digital decade. JOIN/2020/18 final.
Commission. (2020c). Proposal for a Directive of the European Parliament and of the Council on measures for a high common level of cybersecurity across the Union, repealing Directive (EU) 2016/1148. COM/2020/823 final.
Commission. (2020d). Proposal for a Directive of the European Parliament and of the Council on the resilience of critical entities. COM/2020/829 final.
Commission. (2021). Proposal for a regulation of the European Parliament and of the Council Laying Down Harmonised Rules on Artificial Intelligence (Artificial Intelligence Act) and Amending Certain Union Legislative Acts. COM/2021/206 final.
Commission. (2022). Cyber resilience act—new cybersecurity rules for digital products and ancillary services. https://ec.europa.eu/info/law/better-regulation/have-your-say/initiatives/13410-Cyber-resilience-act-new-cybersecurity-rules-for-digital-products-and-ancillary-services_en.
Council. (2004). Council Declaration of 25 March 2004 on Combating Terrorism, Brussels. https://www.consilium.europa.eu/uedocs/cms_data/docs/pressdata/en/ec/79637.pdf.
Council. (2008). Report on the implementation of the European security strategy—providing security in a changing world—on the implementation of the European security strategy 11.12.2008. S407/08.
Council. (2012). "3199th Council Meeting Foreign Affairs" (16062/12 (Presse 467) Brussels) https://data.consilium.europa.eu/doc/document/ST-16062-2012-INIT/en/pdf.
Council. (2014). EU Cyber Defence Policy Framework, 18 November 2014, 15585/14 https://www.europarl.europa.eu/meetdocs/2014_2019/documents/sede/dv/sede160315eucyberdefencepolicyframework_/sede160315eucyberdefencepolicyframework_en.pdf.
Council. (2015). Council Conclusions on Cyber Diplomacy, 11 February 2015, 6122/15.
Council. (2018). Council of the European Union, EU Cyber Defence Policy Framework (2018 update).
Council. (2019). Council gives mandate to Commission to Negotiate International Agreements on e-Evidence in Criminal Matters. Press release 6 June 2019. https://www.consilium.europa.eu/

en/press/press-releases/2019/06/06/council-gives-mandate-to-commission-to-negotiate-international-agreements-on-e-evidence-in-criminal-matters/.
Council. (2020). EU imposes the first ever sanctions against cyber-attacks, Press release 30 July 2020. https://www.consilium.europa.eu/en/press/press-releases/2020/07/30/eu-imposes-the-first-ever-sanctions-against-cyber-attacks/.
Council. (2022). A strategic compass for security and defence—for a European Union that protects its citizens, values and interests and contributes to international peace and security, 21 March 2022, 7371/22.
Court of Justice of the European Union. (2014). Judgment of the Court (Grand Chamber), 8 April 2014 Digital Rights Ireland Ltd v Minister for Communications, Marine and Natural Resources and Others and Kärntner Landesregierung and Others. Joined Cases C-293/12 and C-594/12.
Court of Justice of the European Union. (2021). Judgment of the Court (Grand Chamber) of 2 March 2021. Case C-746/18—Prokuratuur.
EDRI. (2019). EU member states willing to retain illegal data. January 16, 2019. https://edri.org/our-work/eu-member-states-willing-to-retain-illegal-data-retention/.
European Data Protection Board. (2018). Letter to ICANN of 5th July 2018, EDPB-85-2018. https://edpb.europa.eu/sites/default/files/files/news/icann_letter_en.pdf.
European Data Protection Supervisor. (2005). Opinion of the European Data Protection Supervisor on the proposal for a Directive of the European Parliament and of the Council on the retention of data processed in connection with the provision of public electronic communication services and amending Directive 2002/58/EC (COM(2005) 438 final) (2005/C 298/01).
ENISA (2013). ENISA threat landscape 2013, 11 December 2013. https://www.enisa.europa.eu/publications/enisa-threat-landscape-2013-overview-of-current-and-emerging-cyber-threats.
ENISA. (2017). ENISA threat landscape report 2017, 15 January 2018. https://www.enisa.europa.eu/publications/enisa-threat-landscape-report-2017.
ENISA. (2020). ENISA threat landscape 2020, The year in review—from January 2019 to April 2020. https://www.enisa.europa.eu/topics/threat-risk-management/threats-and-trends/enisa-threat-landscape/enisa-threat-landscape-2020.
Hamulák, O. (2018). La carta de los derechos fundamentales de la union europea y los derechos sociales. *Estudios constitucionales, 16*(1), 167–186.
Hooghe, L., & Marks, G. (2019). Grand theories of European integration in the twenty-first century. *Journal of European Public Policy, 26*(8). https://doi.org/10.1080/13501763.2019.1569711
Jones, E., Kelemen, R. D., & Meunier, S. (2016). Failing forward? The Euro crisis and the incomplete nature of European integration. *Comparative Political Studies, 49*(7). https://doi.org/10.1177/0010414015617966
Jones, E., Kelemen, R. D., & Meunier, S. (2021). Failing forward? Crises and patterns of European integration. *Journal of European Public Policy, 28*(10). https://doi.org/10.1080/13501763.2021.1954068
Kaska, K., Talihärm, Osula, A. M., & Tikk, E. (2010). Developments in the legislative, policy and organisational landscapes in Estonia since 2007. *International Cyber Security Legal and Policy Proceedings*, 40–60. NATO CCDCoE.
Kasper, A., & Vernygora, V. (2021). The EU's cybersecurity: A strategic narrative of a cyber-power or a confusing policy for a local common market? *Cuadernos Europeos de Deusto, 65*, 29–71. https://doi.org/10.18543/ced-65-2021
Kasper, A., & Mölder, H. (2020). The EU's common security and defence policy in facing new security challenges and its impact on cyber defence. In D. Ramiro Troitiño, T. Kerikmäe, R. de la Guardia, & G. Pérez Sánchez (Eds.), *The EU in the 21st century*. Springer. https://doi.org/10.1007/978-3-030-38399-2_15
Kerikmäe, T., Troitiño, D. R., & Shumilo, O. (2019). An idol or an ideal? A case study of Estonian e-governance: Public perceptions, myths and misbeliefs. *Acta Baltica Historiae et Philosophiae scientiarum, 7*(1), 71–80.
Lazarou, E., & Dobreva, A. (2019). Security and Defence. European Parliamentary Research Service.

Luukas, I., & Osula, A. M. (2020). The technological sovereignty dilemma—and how new technology can offer a way out. *European Cybersecurity Journal, 6*(1), 24–35.
NIS Cooperation Group. (2020). Cybersecurity of 5G networks—EU Toolbox of risk mitigating measures, CG publication 01/2020.
Ramiro Troitiño, D., Kerikmäe, T., Guardia, R. M. D. L., & Pérez Sánchez, G. Á. (2020). EU in twenty-first century, does crisis mean opportunity?. In *The EU in the 21st century* (pp. 3–9).-Cham: Springer.
Rabinovych, M. (2021). Failing forward and EU foreign policy: The dynamics of 'integration without membership' in the eastern neighbourhood. *Journal of European Public Policy, 28*(10). https://doi.org/10.1080/13501763.2021.1954066
U.S. Department of Justice. (2019). Report on the Investigation into Russian Interference in the 2016 Presidential Election, Volume I of II, Special Counsel Robert S. Mueller, III. https://www.justice.gov/archives/sco/file/1373816/download.
Robinson, N. (2014). EU cyber-defence: A work in progress, EUISS. https://www.files.ethz.ch/isn/182329/Brief_10_Cyber_defence.pdf.
Tikk, E., Kaska, K., & Vihul, L. (2010). International cyber incidents—legal considerations. *NATO CCDCoE, 2010*, 14–34. https://ccdcoe.org/uploads/2018/10/legalconsiderations_0.pdf
Tiirmaa-Klaar, H. (2018). Two generations of EU cybersecurity strategies. In J. Rehrl (Ed.), *Handbook of cybersecurity* (pp. 8–26). Federal Ministry of Defence of the Republic of Austria.

Online Sale of Pharmaceuticals: Liberalization of EU Law in the Context of Transnational Criminal Law

Melita Sogomonjan and Teodoro Forcht Dagi

Abstract This research offers a critical review of the relevance and influence of the liberalization of EU law concerning the online sale of pharmaceuticals in the context of transnational criminal law.

The Internet has enabled and accelerated the distribution of a class of drugs called substandard, spurious, falsely labeled, falsified, and counterfeit medicines (SSFFC) through online sales. This situation exposes many unsuspecting individuals to enormous public health risks. Both individuals and widespread criminal networks are involved. Both the buyers and the purveyors of illicit medications and pharmaceutical ingredients contribute to this problem. The liberalization of pharmaceutical sales in the digital marketplace has contributed to the growth of pharmaceutical crime and provided expedients through which criminal groups have circumvented liability.

Several steps would help mitigate these risks at national, regional, and international levels. These include improving control and verification mechanisms for the manufacture of pharmaceuticals and raw pharmaceutical materials supplied in the EU, toughening the penalties for criminal distribution of SSFFCs, and educating the population at large about their dangers.

1 Introduction

The purpose of this research is to evaluate the influence of liberalization of EU law to transnational criminal law concerning online sale of substandard, spurious, falsely labeled, falsified, and counterfeit medicines (SSFFC). In view of the fact a liberal

M. Sogomonjan (✉)
Tallinn University of Technology, Tallinn, Estonia
e-mail: melita.sogomonjan@taltech.ee

T. Forcht Dagi
Neurosurgery at Mayo School of Medicine and Science, Rochester, MN, USA

Queen's University Belfast, Belfast, UK
e-mail: dagi.teo@mayo.edu

D. Ramiro Troitiño et al. (eds.), *Digital Development of the European Union*,
https://doi.org/10.1007/978-3-031-27312-4_4

market may pose the risk of opportunities for criminal networks to smuggle medicines, this research will: (1) critically review the EU regulatory system for medicines, (2) analyze how fragmentation of criminal legislation at the EU level regarding online sale of SSFFC medicines lessens its effects in countering against such egregious offences and (3) explain how ratification of MEDICRIME Convention can facilitate harmonization of substantive laws within Community. The research concludes with a review of the powers of the EU to legitimately reduce the risks of online sales of pharmaceuticals and brings out new dimensions for further research.

Globally, an estimated number of SSFFC medications exceeds 10% of pharmaceutical market (Berkot, 2012) resulting in millions of deaths annually (Orton et al., 2019; WHO, 2018). Furthermore, illicit trade of SSFFC medicines results in increased costs associated with healthcare, regulation, and enforcement, as well as investments in anti-counterfeiting technologies such as the adoption of new track-and-trace systems (TRACIT, 2022) and lawsuits for pharmaceutical companies (Magdun, 2021).

Commerce in SSFFC medications is a serious crime posing danger to patients' safety worldwide. These products are sold without a license. They do not have meaningful therapeutic effects. SSFC antibiotics may lead to antimicrobial resistance[1] and to the emergence of drug-resistant strains of diseases due to an insufficiency of bioavailable active ingredients (Harris et al., 2009; WHO, 2018).

Moreover, SSFFC medicines may be adulterated with toxic chemicals that lead to serious adverse reactions and deaths (Harris et al., 2009).

1.1 Dimensions of the Problem

According to a 2008 report carried out by the European Alliance for Access to Safe Medicines, 62% of ostensibly legitimate medications purchased online were either substandard or counterfeit. Furthermore, at least 90% of online pharmacies were unauthorized and were found not to require prescription for prescription-only medicines (EAASM, 2008). In 2011, counterfeit medicines accounted for nearly a quarter of all counterfeit goods seized in the EU (Lechleiter, 2012). A study in 2016 showed that 97% of online pharmacies were detected in violation of national applicable legislation and failing to comply with safety requirements (LegitScript, 2016).

Over 50% of all SSFFC medicines are lifesaving-related medicines (Mackey et al., 2015). The most common types of counterfeit pharmaceuticals seized by customs between 2014 and 2016 were antibiotics (>35%), followed by treatments for sexual dysfunction (>15%) and painkillers (>10%) (OECD/EUIPO, 2020).

[1]To note, antimicrobial resistance only can lead to 10 million deaths per year by 2050 (Breman, 2019).

Other important classes of medicines that have been affected by the SSFFC epidemic include oncological drugs, and treatments for Hepatitis C (EC, 2018), respiratory, cardiovascular and digestive conditions, vaccines, (Kubic, 2008; EUROPOL, 2020; Balfour, 2021) insulin, cholesterol drugs, antihistamines (Liang, 2007; Cockburn et al., 2005), vitamins, weight loss and hair loss medicines, human growth hormone, anti-malarials, anabolic steroids, anti-inflammatories, analgesics, opioids, sedatives and other psychiatric drugs, and fertility medications (INTERPOL, 2014; Hall et al., 2017).

1.2 *Effects of the Global Health Crisis During COVID-19*

The growth of online pharmacies was facilitated by the global health crisis during COVID-19. This crisis resulted in supply chain issues, limited mobility and lockdowns, and created new opportunities for illicit trade (Magdun, 2021; Wechsler, 2012).

The majority of customers in the EU are not aware that they might have been purchasing SSFFC medicines sold illegally (Lavorgna, 2015), that many of them may have been expired, and that most websites selling medicines online are neither regulated nor lawful (OECD/EUIPO, 2020). Online sales of SSFFC medicines are lucrative mainly due to low prices.

1.3 *Criminality of SSFC Sales*

Deception along these lines is criminal. Intentionally denying or preventing access to original medicines on a systematic and widespread basis is considered as a transnational organized crime against humanity (Attaran et al., 2011).

E-commerce of SSFFC medicines in the EU qualifies as one of the 'Euro crimes' because it is against the interests of the EU; it crosses borders; and it is sufficiently notorious by itself (Miettinen, 2013). Such 'Euro crimes' are subject to mutual recognition in the EU and depend upon mutual trust in criminal justice systems across the Member States (Miettinen, 2013).

Manufacturing and distribution of SSFFC medicines may fall under crimes defined in Art. 2 (2) of the Framework Decision (2002/584/JHA) on the European arrest warrant (EAW). These include:

- Participation in a criminal organization
- Computer-related crime (or "cyber-enabled" crime as defined by Interpol)
- Counterfeiting and piracy of products
- Illicit trafficking in narcotic drugs, psychotropic and hormonal substances, and other growth promoters
- Corruption and fraud in the healthcare sector (Hall et al., 2017)

The EAW system has advanced the effectiveness of cross-border cooperation in investigation and combat of serious and organized crimes, but EU criminal law itself is not legislated nor enforced at the national level and its effectiveness for criminal sanctions and procedures depends on the legal and judicial systems of Member States (Miettinen, 2013).

The conditions allowing implementation of the EAW for offences not included in the 32 categories for which the verification of double criminality is excluded require that the crime in question constitute a punishable offence under the national law of the country executing the EAW as well as the EAW itself. Furthermore, under the national law of the member state, a penalty of at least 4 months imprisonment is required (Turudić, 2018).

These relatively lax provisions allow criminal organizations involved in falsification and counterfeiting medicines, but successfully keeping a low profile to avoid law enforcement authorities internationally (INTERPOL, 2014).

A MEDICRIME Convention was released to combat against the falsification and counterfeiting of medical products and similar crimes posing serious threats to public health. The MEDICRIME Convention is the first binding international treaty to criminalize offences considered dangerous to public health. It aims to protect the rights of victims and promote cooperation at both national and international levels (Alarcón-Jiménez, 2015).

While the MEDICRIME convention has been in force since 2016, as of 27 May 2022 only 21 countries have ratified it. The reason may be that around 20% of WHO Member States have developed adequate domestic drug regulatory systems and legislation (Alarcón-Jiménez, 2015; Negri, 2016), and so the convention is seen to be redundant or superfluous.

1.4 Law Enforcement Considerations

Law enforcement has not been particularly effective. Only one European country has indicated that the number of illegal online pharmacies registered in their country has decreased due to enforcement action (INTERPOL, 2014). For the most part, EU states have not instituted robust monitoring and supervision. Enforcement is hampered by the limited capacity of law enforcement force to investigate and prosecute organized crime (Seelke et al., 2011).

1.5 Growth of Illicit SSFFC Sales

As a result, the sales of SSFFC medicines are twice as quickly as those of legitimate pharmaceuticals, and are expected to maintain a 20% annual growth rate annually in near future (Acri, 2018). Unsurprisingly, the profits earned from illegal online pharmacies worldwide are substantial (Lechleiter, 2012; Seelke et al., 2011) and contribute to the growth of corruption (INTERPOL, 2014). Illicit pharmaceuticals are even more profitable than the illicit drug trade.

1.6 Recommendation

Inasmuch as e-commerce of SSFFC medicines is rapidly turning into a global crime issue, harmonization and integration of several international legal instruments at pan-European and international level is required (Negri, 2016). Moreover, unified procedures are needed to combat illegal practices, and to prevent, control, and counter the threats evolving from SSFFC medicines entered the legal supply chain. This is necessary even in countries having regulations supporting traceability (Di Giorgio, 2015). A robust and enhanced pharmacovigilance system is required at pan-European and an international level to protect vulnerable patients from the unintended and irreversible adverse effects of SSFFC medicines. Finally, it is critical to strengthen international cooperation and mutual collaboration in criminal matters, as demonstrated in "Operation Volcano" or, as it is better known "the Herceptin case" (Di Georgio, 2015).

2 The Unity and the Divergence of EU Pharmacological Legislation to Prevent Entering SSFFC Medicines in European Digital Single Market

2.1 The Historical Development of EU Pharmaceutical Legislation to Safeguard Public Health

The European regulatory system for medicines aims to ensure availability of and access to high-quality, effective, and safe medicines in the EU, both for human and veterinary use. In 1994, the EU ratified the European Pharmacopoeia Convention and a Protocol to the Convention on the Elaboration of a European Pharmacopoeia that became the first legally binding document establishing the standard requirements for assessing and controlling the quality of medicines and the substances used to apply for marketing purposes. Since then, legal framework governing the use and marketing of pharmaceuticals has gradually enlarged.

The number of medicines for human use sold at a distance increases yearly. The legal framework in the field of pharmaceuticals has also grown and includes the following legislation (in chronological order):

- Directive 92/28/EEC on the advertising of medicinal products for human use
- Directive 98/34/EC to establish a procedure for the provision of information in the field of technical standards and regulations including inter alia

   - technical specifications of a medicinal product for human use (e.g., level of quality, performance, and safety), and
   - the requirements applicable to the medicinal product (e.g., the name under which the product is sold, terminology, symbols, testing and test methods, packaging, marking or labeling and conformity assessment procedures)

- Directive 2000/31/EC regarding certain legal aspects of information society services, in particular electronic commerce, in the Internal Market or Directive on electronic commerce or e-Commerce Directive aiming to create a legal framework to ensure the free movement of information society services between Member States;
- Directives 2001/83/EC and 2004/27/EC, as amended, on the Community code relating to medicinal products for human use to be placed on the market in Member States to safeguard public health and without prejudice to hinder the development of the pharmaceutical industry or trade in medicinal products in the Community, and Directive 2001/82/EC on the Community code relating to veterinary medicinal products
- Regulation (EC) No 726/2004 to lay down Community procedures for the authorization, supervision, and pharmacovigilance of medicinal products for human and veterinary use, and to establish a European Medicines Agency
- Regulation (EC) No 1394/2007 on advanced therapy medicinal products
- Directive 2010/84/EU on pharmacovigilance, amending Directive 2001/83/EC, aimed to protect public health without prejudice to impede the free movement of safe medicinal products within the EU
- Regulation (EU) No 1235/2010, amending Regulation (EC) No 726/2004 and Regulation (EC) No 1394/2007, as regards pharmacovigilance of medicinal products for human use
- Directive on Falsified Medicines 2011/62/EU or FMD, amending Directive 2001/83/EC providing measures to prevent the entry into the legal supply chain of falsified medicinal products
- Directive 2012/26/EU, amending Directive 2001/83/EC as regards pharmacovigilance
- Regulation (EU) No 1027/2012, amending Regulation (EC) No 726/2004, as regards pharmacovigilance
- Delegated Regulation (EU) 2016/161, supplementing Directive 2001/83/EC providing specific details of the safety features that are applied and required on the packaging of medicines when entering the European medicines market

In cooperation with the European Union, Member States coordinate health policy and programs independently with respect to:

- The areas related to public health improvement
- The prevention of physical and mental illness and diseases
- Countering sources of danger to physical and mental health (Art. 168 (2) of TFEU)

The Union may and should support Member States in reducing drug-related health damage and combating major health scourges including serious cross-border threats to health (Art. 168 (1) of TFEU), inter alia through mechanisms such as establishing a centralized authorization procedure and verification system around falsified medicines entering the EU market. Online sale of SSFFC medicines is one of the major cross-border crimes causing serious threats to health.

2.2 *The Limits of EU Legitimate Power to Establish Quality Standards and Safety Measures for Manufacturing and Distribution of Pharmaceuticals Including at a Distance*

The fundamental objective of EU regulatory system for medicines is to harmonize the highest level of public health protection. Nevertheless, the EU does not have a harmonized system establishing quality and safety standards for pharmaceuticals (Abma, 2016). These standards are developed in response to major public health needs (WHO, 2022) in accordance with recommendations by

- WHO governing bodies
- The International Conference of Drug Regulatory Authorities
- The WHO Expert Committee on Specifications for Pharmaceutical Preparations (ECSPP)
- International organizations
- United Nations agencies and other WHO programs and activities, or in response to major public health needs

Member States are responsible for establishing their own standards of quality and safety for the manufacturing and sale of pharmaceuticals, but these standards are required to meet common safety measures.

Measures designed to protect and improve human and public health, and to combat the major cross-border health scourges and serious threats to health in particular, may be adopted by the European Parliament and the Council without harmonization with the laws and regulations of the Member States (Art. 168 (5) of TFEU).

These measures, however, are insufficient to countervail the supply of SSFFC medicines in the context of criminal law. Furthermore, regulation of online sale of pharmaceuticals to the public by means of information society services remain under the jurisdiction of the Member States (Directive 2011/62/EU). Member States should not pose restrictions impeding the functioning of the EU internal market without necessity, even though such restrictions might be justified by the protection of public health (Directive 2011/62/EU).

In *Apothekerkammer des Saarlandes and Others* v *Saarland*, 2009, the Court of Justice of the European Union (ECJ) recognized that EU law "does not detract from the power of the Member States to organise their social security systems and to adopt <...> provisions intended to govern the organisation of health services such as pharmacies" (Case C-171/07, p. 18). However, "Member States must comply with Community law, in particular, the provisions of the Treaty on the freedoms of movement, including freedom of establishment. Those provisions prohibit the Member States from introducing or maintaining unjustified restrictions on the exercise of those freedoms in the healthcare sector" (Case C-171/07, p. 18). To assess the compliance of national laws with EU law, ECJ emphasized that "the health and life of humans rank foremost among the assets and interests protected by the Treaty and that it is for the Member States to determine the level of protection which they wish to afford to public health and the way in which that level is to be achieved. Since the level may vary from one Member State to another, Member States must be allowed discretion" (Case C-171/07, p. 19).

2.3 Measures Preventing Entering SSFFC Medicines in European Single Market

A wide range of medicines subject to prescription are substandard, spurious, falsely labeled, falsified, and counterfeit. The term "falsified" was adopted by the EU to refer to medical products that usually contain substandard or falsified ingredients, ingredients that include wrong doses of active substances and ingredients with no therapeutic effects or no active substances. The EU distinguished falsified medicinal products from other illegal medicines and those violating intellectual property rights (Directive 2011/62/EU).

Thus, counterfeit medicines are considered medicines intended to violate intellectual property rights and to protect an unauthorized, counterfeited registered trademark (Buckley et al., 2013). Although no international definition of counterfeit medicine has yet emerged, the meaning of counterfeit medicines is much broader and includes any medicine manufactured and distributed with the intent to deceive (Buckley et al., 2013; Negri, 2016).

According to the FMD definition, a "falsified medicinal product" includes any medicinal product falsely representing its identity, source or history. It excludes unintentional quality defects resulting from manufacturing or distribution errors (Directive 2011/62/EU).

By identity, falsification of a product can be found through its packaging and labelling, name or composition and any of the ingredients including excipients and the strength of those ingredients (Directive 2011/62/EU).

By source, falsification applies to a manufacturer of the product, country of manufacturing and country of origin or product's marketing authorization holder (Directive 2011/62/EU).

By history, falsification of the records and documents relating to the distribution channels used is considered (Directive 2011/62/EU).

To prevent serious risks of threat to public health due to these products entering the legitimate supply chain and to ensure that medicines are safe and that the trade in medicines is rigorously controlled, all medicines must be authorized before they can be placed in the EU internal market. Authorization of medicines can be accomplished either through centralized procedure by the European Commission or according to the principle of mutual recognition, decentralized and national procedures by the national authorities of EU Member States.

As a rule, evaluation of generic medicines and those available without a prescription is done through national authorization procedure.

Scientific evaluation of medicines that contain a new active substance and are intended to treat acquired immune deficiency syndrome, cancer, neurodegenerative disorder, diabetes, auto-immune diseases, and other immune dysfunctions including viral diseases, and medicines derived from biotechnology and genetic engineering, and advanced-therapy medicines, such as gene-therapy, somatic cell-therapy or tissue-engineered medicines and orphan medicines must be done through centralized authorization procedure (Regulation (EC) No 726/2004). The centralized authorization procedure is based on the objective scientific criteria of quality, safety and efficacy of the medicines concerned excluding economic and other considerations (Regulation (EC) No 726/2004). This requirement applies also to all medicines of major therapeutic, scientific, or technical innovation including the interests and legitimate expectations of patients at the EU level (Regulation (EC) No 726/2004). The unity and harmonization of EU pharmaceutical legislation is focused on developing pharmacovigilance systems capable of achieving high standards of public health protection for all medicinal products and strengthening the verification requirements applicable to the manufacturer of the medicinal product.

2.3.1 Historical Notes

The European Medicines Agency

The European Medicines Agency (EMA) was established in 1995 with responsibility for the scientific assessment of the application, supervision, and pharmacovigilance of medicines prior to marketing medicines in the internal market of the EU. The EMA is responsible also for monitoring any undesirable medication effects to ensure the rapid withdrawal from the EU market of any medicinal product presenting a negative risk-benefit balance under normal conditions of use (Regulation (EC) No 726/2004).

It is up to the Member States to enforce the provisions relating to pharmacovigilance and to ensure that effective, proportionate, and dissuasive penalties are applied to marketing authorization holders for non-compliance with pharmacovigilance obligations (Regulation (EC) No 726/2004).

The European Medicines Verification System

The European Medicines Verification System (EMVR) was developed in 2019 to prevent falsified medicines from entering the legal supply chain and to detect potential falsifications on a basis of end-to-end verification of prescription medicines (EMVO, 2023). Having a strong quality system capable to detect falsified medicines results in greater trust in parallel imports of pharmaceuticals even from regions with higher risk (Abma, 2016). But verification of imported pharmaceuticals alone is insufficient to protect public health. Therefore, verification of substances and raw materials is required.

In the context of an alarming increase of falsified medicinal products now being detected in the EU, FMD provides measures to counter against medicine falsifications. These measures include obligatory safety features allowing verification of the authenticity of medicinal products and identification of individual supplied packs as well as providing evidence of tampering regardless of a way of supply inter alia through online sale (Directive 2011/62/EU).

As a rule, medicinal products subject to prescription should bear the safety measures. However, certain medicinal products subject to prescription are excluded from the requirement to bear the safety features through a delegated act following a risk assessment (Directive 2011/62/EU). The risk assessment includes considerations related to the price of the medicinal product, previous cases of falsified medicinal products being reported in the EU and in third countries, the implications of a falsification for public health, including the specific characteristics of the products concerned, and the severity of the conditions intended to be treated (Directive 2011/62/EU).

Where an objective assessment shows a risk of falsification which leads to serious consequences, medicinal products not subject to prescription might arguably also be subject to safety measures (Directive 2011/62/EU).

Other measures to countervail medicine falsifications include a common, EU-wide logo to identify legal online pharmacies, tougher rules on import of active pharmaceutical ingredients and strengthened record-keeping requirements for wholesale distributors (Directive 2011/62/EU). Nevertheless, all these measures are not intended for commercialization nor do they include anti-counterfeiting technology. This gap is part of what allows the circulation of SSFFC medicines in the EU market and beyond to be possible (Abma, 2016).

A dual track-and-trace system together with anti-counterfeiting technologies such as spectroscopy and handheld NIRS scanners (Trenfield et al., 2019; Wilson et al., 2017; Dégardin et al., 2016) as well as RFID chips and forensic markers such as taggants and nano-encryption (Dahl et al., 2015) would, with a high degree of probability, strengthen verification of medicines identity.

2.4 Requirements to Prevent the Entry of Falsified Pharmaceuticals into the Legal Supply Chain in the EU Market Through Online Sale

EU Directive on falsified medicines stipulates conditions under which Member States shall ensure that online sale of pharmaceuticals is offered to the public by means of information society services (Directive 2011/62/EU). These conditions include:

- Authorization requirement of a person supplying medicines online in accordance with national legislation of the Member State in which that person is established; and
- Authorization of medicines to comply with the national legislation of the Member State of destination (Directive 2011/62/EU).

There is also a requirement to provide information regarding the classification of the pharmaceutical aimed for an online sale as well as the name and permanent address of the place of activity from where pharmaceuticals are supplied and information on the starting date of the activity of offering online sale (Directive 2011/62/EU). Finally, the address of the website used for the online sale of pharmaceuticals and all relevant information necessary to identify that website (e.g., EU-wide logo) must be easily accessible (Directive 2011/62/EU).

Additional notification and documentation measures are required for all participants involved in the supply chain (distributors, regardless of the way of handling the medicinal products, brokers involved in the sale or purchase of medicinal products without selling or purchasing those products themselves, and without owning and physically handling the medicinal products) (Directive 2011/62/EU). However, to identify online pharmacies that operate legally, first and foremost the public, and especially vulnerable groups of people (e.g., people suffering from huge pain or mental diseases) should be provided with comprehensive information regarding the dangers of the risks related to online sale of illegal medicines. Second, individuals should be sufficient enough informed on the functioning of the common EU-wide logo throughout the Union that must be present in the websites of the Member States and the European Medicine Agency.

These websites offering online sale of pharmaceuticals should contain information regarding the contact details of the competent authority or the authority of the Member State where the person offering online sale of pharmaceuticals is established (Directive 2011/62/EU). Furthermore, a hyperlink to the website providing information on the national legislation of the Member State or establishment applicable to the offering of online sale of pharmaceuticals, including information on the fact that there may be differences between Member States regarding the classification of medicinal products and the conditions for their supply must be provided (Directive 2011/62/EU).

The purpose of the common logo, the list of persons offering online sale of pharmaceuticals as well as their website addresses, and the risks related to medicinal products supplied illegally must be disclosed (Directive 2011/62/EU). It is important that websites offering online sale of pharmaceuticals be linked with a website of a European Medicine Agency, providing information on the Union legislation applicable to falsified medicinal products as well as hyperlinks to the Member States' websites and to persons authorized for online sale of pharmaceuticals to the public (Directive 2011/62/EU). Lastly, the common EU-wide logo should be available on every page of these websites and be recognizable throughout the Union (Directive 2011/62/EU). They should enable identification of the Member State wherein the person offering online sale of pharmaceuticals to the public is established (Directive 2011/62/EU).

EU pharmaceutical legislation provides measures to prevent entry into the legal supply chain of falsified medicinal products by requiring authorization of medicines prior to marketing in the EU. Introducing safety features to allow the identification and authentication of pharmaceuticals including all other actors involved in the supply chain is part of the substantial requirements. However, regardless of the type of authorization requirement and provisions stipulated by the EU regulations, each nation has to establish its own quality and safety standards for pharmaceuticals to import and export medicines (Abma, 2016). Measures (e.g., safety features) that Member States need to follow in order to prevent entering of SSFFC medicines into the legal supply chain are not specifically designed to assess the potential falsified character of the pharmaceuticals (Directive 2011/62/EU). Consequently, FMD accelerates the verification of pharmaceutical authenticity including information sharing between Member States. It does not guarantee circulation of SSFFC medicines per se in EU pharmaceutical market or beyond.

It has been argued that lack of standardization demonstrates the weakness of EU law and enables criminal organizations to enter EU market and supply SSFFC medicines (Abma, 2016). As a practical matter, illicit SSFFC medicines reach EU market primarily through online, "hidden", and illicit trading routes (Lavorgna, 2015; Bate, 2012) regardless of the existence of effective track-and-trace systems to control security and safety of legal pharmaceutical supply chain. On the other hand, the divergence of EU pharmaceutical legislation supports the prerogative of Member States to prohibit the use of medicines in their territory if those medicines infringe objectively defined concepts of public policy and public morality. As a rule, the use of prohibited medicines is subject to criminal investigation.

3 Fragmentation of EU Criminal Law in Combating Online Sale of SSFFC Medicines

3.1 Background and Sources of the Pharmaceutical Crimes

The supply of SSFFC medicines, including further distribution at a distance, has been recognized as a global problem since states demonstrated an emergent interest in participation in "Operation Pangea" in 2008. The SSFFC medicines market is considered as "transnational organised crime" by international law enforcement organizations (INTERPOL, 2014). Nevertheless, research on criminal law on combating online sale of SSFFC medicines is limited. There are many different approaches to the topic, resulting in a complex but fragmented view.

In a study conducted by Hall et al. (2017) procedures supplying illicit medicines vary by size, reach, organization, and legality whereas participants involved in often belong to poorly structured network. In other studies, online sale of SSFFC medicines qualifies as pharmaceutical crime, organized crime, illicit trafficking of psychotropic and hormonal substances (Negri, 2016), murder or terrorism (Breman, 2019; Cannon, 2015; Blackstone et al., 2014; Vander Beken & Balcaen, 2007) and a crime against humanity (Attaran et al., 2011). The latter category includes crimes against human health and against the economy (Breman, 2019).

All these crimes are, by nature, serious. However, according to the legal definition of a "serious crime" established in Article 2 of the United Nations Convention against Transnational Organized Crime (or widely known as Palermo Convention of 12 December 2000), a serious crime refers to an offence punishable by a imprisonment of 4 years or more (United Nations, 2002). Obviously, online sale of SSFFC medicines in countries with no specific legislation in this field or with lenient criminal codes and penalties is not considered to be a serious crime, irrespective of its consequences.

People involved in and responsible for online sale of SSFFC medicines can be considered as "organised criminal group" should they form a structured group of three or more persons that exists for a period of time and acts in concert to commit one or more serious crimes or offences <...> in order to obtain, directly or indirectly, a financial or other material benefits. Subsequently, online sale of SSFFC medicines does not legally fall under the concept of a "serious crime" hence allowing culprits to extend their criminal network and continue making a huge profit.

3.2 *Inadequacy of Drug Regulations and Incapacity of Enforcement Agencies to Monitor Circulation of SSFFC Medicines in the EU and Beyond*

Until recently, the import of raw materials to EU countries lacked a proper regulation and was not a subject for criminal investigation, except in cases where huge number of illicit narcotic substances were imported. In many cases, these materials, however, were found to be falsified or counterfeited. As a result, a medicine manufactured on a basis of illicit raw material can easily enter the legal supply chain in the EU as both transit zone and end-user market (Hall et al., 2017). About 15% of the global export of pharmaceuticals between 2014 and 2016 was German—one of the main EU exporters of pharmaceuticals (OECD/EUIPO, 2020). At that time, Germany and

Switzerland were also the main sources of SSFFC medicines in the EU. These medicines were traded worldwide—nearly 10% of global customs seizures for counterfeit pharmaceuticals (OECD/EUIPO, 2020). However, the highest number of arrests of individuals engaged in manufacturing counterfeit medicines in the EU occurred in Spain (OECD/EUIPO, 2020).

Furthermore, barriers to the circulation of illicit SSFFC medicines in the EU market can be circumvented, despite legal and regulatory safeguards, when a licensed medicine is imported from countries where SSFFC medications are prevalent (e.g., India, China, Singapore, Pakistan, and Russia) and reaches an EU country where the same medicine is neither licensed nor controlled under domestic legislation (Hall et al., 2017; OECD/EUIPO, 2020).

In short, while the legal purveyance of SSFFC medicines is still a crime on the basis of both its commercial and pharmaceutical implications, but the boundary between legal and illegal pharmaceutical market is, in fact, blurred (Hall et al., 2017).

3.2.1 The Internet

The Internet[2] has significantly facilitated the criminal market through easy access to and trade in SSFFC medicines (Treadwell, 2012). The Internet has also affected the operations of criminal networks, including ways in which they build new business association with other criminal organizations and (potential) clients. The Internet has also served as a marketing resource in the sense that it allows supplier to observe clients' purchasing preferences. These factors facilitate the promotion of certain counterfeit pharmaceuticals (Lavorgna, 2015).

Consumers often turn to online sales to avoid the stigmas associated with certain medical conditions and to avoid discussing sensitive health topics with healthcare professionals and family members. Online pharmaceutical sales also often offer lower costs, speed and convenience of transaction, and, often, the ability to purchase a medication without the need for a prescription (OECD/EUIPO, 2020).

There are two ways to purchase SSFFC medicines online: the "dark web" and the "freely accessible surface web" (OECD/EUIPO, 2020). But even though crime and commercial activity are interrelated, both, neither the consumer and not the supplier are necessarily engaged in criminal activities when they trade on the Internet (van Duyne et al., 2005). Thus, legal online pharmacies may turn out unwittingly to sell SSFFC pharmaceuticals because deficiencies or malfunctions of the track-and-trace systems fail to verify pharmaceuticals prior to their entering the market.

Consumers engage in criminal behavior when they themselves start to act as knowing commercial resellers of illicit pharmaceuticals with deceptive intent and (Lavorgna, 2015) with intent to harm others (Hock et al., 2019).

[2]Within the first e-commerce dated in 1970s, access to both legitimate and illegal medications increased drastically in the USA (Power, 2013) and later in the UK (George, 2006).

Sixty-six percent of countries worldwide have no clear legislation adequately regulating or prohibiting online sale of pharmaceuticals (WHO, 2011). In countries that do regulate online sales, the tendency is to prohibit sales rather than to allow them, except in European region countries, the USA, and Western Pacific (WHO, 2011, p. 40). Interestingly, countries where SSFFC medicines can be easily manufactured generally prohibit online sale of pharmaceuticals. Countries that implemented strict control policies such as:

- Official accreditation and track-and-trace systems
- Online registries to direct consumers to legitimate websites
- Online pharmacies

are more likely to permit online purchase of medications originating from other countries. In countries, where online sale of prescription-only pharmaceuticals is prohibited, regulatory authorities do not allow anyone to sell them through e-commerce (Jing, 2016; Hock et al., 2019). The legal consequences for breach of the law primarily include seizure of goods, consumer fine or prosecution (WHO, 2011).

The lack of adequate sanctions for criminal activity associated with the manufacture and distribution of SSFFC medicines together with the fragmentation of criminal law in this domain creates a major global enforcement issue (Breman, 2019). The sale of SSFFC medicines provides criminal organizations easy access to a very large and profitable market (Cannon, 2015; Abma, 2016).

3.3 Risks Associated with Prioritizing the Principle of Free Movement of Goods Across the EU

The principle of free movement of goods and a number of related provisions of secondary law has been brought up in the context of online pharmaceutical sales. In *Deutscher Apothekerverband eV* v *0800 DocMorris NV and Jacques Waterval* 2003, (widely known as Doc-Morris case), the court ruled that online sale of pharmaceuticals in Member States could be prohibited only in the event the prohibition covered prescription medicines. It was argued that online sale of pharmaceuticals which are not subject to prescription in the Member State does not create any additional health risks, and its prohibition would, therefore violate community law (Case C-322/01, pp. 112 & 115). At the same time, sale of medically unjustified pharmaceuticals could lead to a waste of public financial resources (Case 171/07, p. 60). The court concluded that quantitative restrictions of medicines imported into a Member State in which they are authorized, where they had been previously obtained by a pharmacy in another Member State from a wholesaler in the importing Member State, must not prevent the lawful sale of pharmaceuticals over the Internet (Case C-322/01, pp. 76, 146 & 149). It was also noted that there is no evidence for possible abuse for persons who wish to acquire non-prescription medicines unlawfully, purchase in a traditional pharmacy is more difficult compared to online pharmacy (Case C-322/01, p. 114).

Many countries have regulated pharmaceutical crime under the category of intellectual property or by criminalizing trafficking of illicit medicines, narcotics, and psychotropic substances, or on the grounds of fraud (INTERPOL, 2014). Still, gaps in criminal law directly addressing pharmaceutical crime (e.g., manufacturing, distribution, supply, and sale including online sale of SSFFC medicines), weak enforcement systems and inadequate penalties applied under general criminal codes contribute to the growth of pharmaceutical crime and proliferation of criminal networks (INTERPOL, 2014; Abma, 2016) as well as evasion from prosecution (Hock et al., 2019). The lack of legislation specifically defining and addressing crimes related to manufacturing and distribution of SSFFC medicines result in inefficient and weak judicial actions (Breman, 2019).

Lack of harmonized legislation and mutual recognition in that field between Member States as well as inconsistent consequences for breach of law worldwide reflected the need for greater cooperation and development of political (Ivic & Troitiño, 2022) and legal framework to counter online sale of illicit pharmaceuticals (WHO, 2011). To this end, a comprehensive Model Law on Medicine Crime by Attaran (2015) provides a legal framework for strengthening national laws and guidance on criminalization against manufacturing and distribution (including trafficking and sale) of SSFFC medicines both in-person and online as well as establishing principles for punishing perpetrators and creating tools to encourage whistle-blowers to cooperate with law enforcement agencies. In addition, the Model Law on Medicine crime provides incentives for governments to strengthen drug regulatory capacity (Attaran, 2015).

3.4 Facilitating States' Immunity and Enhancing the Capacity of National Criminal Laws Through the Adoption of MEDICRIME Convention

The MEDICRIME Convention encourages states to fill the legislative gaps in addressing SSFFC medicines with more specific and effective provisions to impose criminal indictment around pharmaceutical crimes. The Convention creates a comprehensive international legal framework that makes the manufacture and sale of counterfeit medicines, including online sale and similar crimes involving threats to public health, illegal in and of themselves.

The Convention harmonizes definitions and promotes the legislative measures necessary to combat illicit manufacturing of SSFFC medicines. It focuses on prevention, on the protection of victims, and on combating all forms of marketing of SSFFC medicines as well as promotion of a cooperation at national and international levels (Ramiro Troitiño et al., 2022). It does not, however, concern itself with intellectual property rights or with offering to or attempting to supply illegal product;

falsification of a broad range of documents used in these offenses; possession of products, materials, accessories or falsified documents with intent to commit criminal acts; and with intentionally placing products on the market without authorization.

In accordance with MEDICRIME Convention, individuals engaged in online sale of illicit medicines are to be prosecuted regardless of the country where the act was committed. However, prosecution can be challenged if individuals involved in the illegal online sale of pharmaceuticals are based outside of a regulatory authorities' jurisdiction (Terry, 2017; Art. 4 (2) of the UN Convention against transnational organized crime, 2004). In this regard, online sale of illicit pharmaceuticals first and foremost shall be clearly conceptualized if such offence and following penalty are subject to approximation and harmonization of the Member States' legal systems that in fact varies on mutual recognition based on the presumption of mutual trust. Or online sale of SSFFC medicines shall be considered as an act that can be qualified as offence universally—*mala in se*—or a crime that is wrong just because it is prohibited by the law—*mala prohibita*. While EU Member States are responsible for implementing effective, proportionate and dissuasive penalties (Hamulák, 2018), countries decide by themselves what type of a sanction or penalty should be adopted for culprits involved in trafficking of illegal medicines.

As seen, a comprehensive legal framework addressing criminalization of online sale of SSFFC medicines is proceeding slowly. It depends largely on the capacity of states to adopt and integrate the provisions of international treaties within national laws. While it is clear that regulations need to be strengthened to prosecute criminals involved in online sale of SSFFC medicines, it is yet unclear whether customers intentionally buying online pharmaceuticals from illegal websites should be considered victims or criminals.

Children and the elderly are primarily deemed victims of this crime because they cannot control, in general, the medicine they are given, nor are they necessarily informed about falsification of medicines, or control where their medicine comes from (Magdun, 2021).

However, ordinarily, intentionally buying prescription-based pharmaceuticals from illicit pharmacies or illegal websites that do not require prescription is in fact contribution to the growth and expansion of a criminal network, especially since customers are free to decide *what* and *where from* to purchase medicines.

Children, the elderly, and women taking birth control medications are more vulnerable to be affected by SSFFC medicines since they purchase medicines they can afford and what they trust (Magdun, 2021). In this sense, it is not only correct but even unethical to apply the principle of caveat emptor. That principle does not apply. Furthermore, many consumers are deceived by the "high-quality packaging and appearance" of the SSFFC medicine that in fact is almost impossible to differentiate from fake medicines (Magdun, 2021).

The root of the problem lies in the needs of customers. Governments would be wise to think about and consider the increased benefits of medicines derived from innovative technologies with reasonable prices that consumers will actually be able to afford (Magdun, 2021). It follows that no society can ignore the influence of

SSFFC medicines nor can it totally shut down importation, despite having the safest and strongest anti-counterfeiting innovative technologies.

The normative paradigm in European criminal law model is today mutuality, rather than reciprocity, the aim of which is to "foster value sharing beyond the borders of the nation State, as long as the basic fundamental principles revolve around the person as an end, rather than a means" (Fichera, 2013). Thus, EU criminal law is based on the adoption of mutual recognition measures stipulated in Article 2 (2) of the Decision Framework on European Arrest Warrant, according to which national judicial authority "will have to trust that the fundamental values of a foreign legal system are essentially the same as those that are relied upon in its own legal system" (Fichera, 2013).

These measures include:

- Recognition of a product of a foreign criminal system
- The reduction of the number of grounds for refusal in international cooperation
- Elimination of the requirement of double criminality according to which a request must refer to an act which has been qualified as an offence or a crime both in the requested country where a suspect is being held and a country requesting for the suspect to be handed over or transferred to stand trial for a list of 32 categories of criminal offences (extradition is allowed only for offenses alleged as crimes in both jurisdictions) and for which cooperation should be smoother

Most countries criminalize online sale of SSFFC medicines on the basis of counterfeiting and deception with intent to harm (Hock et al., 2019). However, the existing legal frameworks are fundamentally limited by territorial boundaries and this can be partially explained since Decision Framework on European Arrest Warrant is aimed to establish secure free market through the mechanism of integration rather than cooperation (Fichera, 2013). In addition, not all EU countries have signed and ratified the MEDICRIME Convention that would strengthen the relevant criminal law both in the EU and beyond.

Irrespective of several arguments supporting EU criminal law, it has been argued that the EU is inappropriate to legitimately exercise a direct competence in criminal law and further inroad into domestic criminal field is unjustifiable for at least two reasons (Fichera, 2013). First, the EU "is not (yet) a political community where a proper debate on right and wrong can be conducted in a public forum" and second, the EU is "unable to forge a meaningful relationship with its citizens because its values as a community are not communicated to them" (Fichera, 2013). Nevertheless, the EU direct competence in pharmaceutical crime, in particular, in the context of online sale of illegal pharmaceuticals—SSFFC medicines—can be highly relevant.

4 Conclusion

Availability and accessibility to safe medicines of high quality is essential to enhanced longevity and improvement in quality of life and care. With the first e-commerce, access to medicines significantly improved, but only in parallel with increased risks and threats to the health and life of millions of people. The rapid growth of online non-licensed pharmacies and illegal websites providing SSFFC medicines gave birth to a new organized criminal group/network conducting serious crimes on a basis of a systematic and widespread level.

To ameliorate this problem, it will be important to start by educating the general public to the problems of SSFFC. Second, the criminal legal frameworks that are limited to the territorial boundaries of each state and the impact of EU Decision Framework on European Arrest Warrant must be reviewed and modulated to overcome their weakness in the face of crimes that have different interpretations in different states or ones not directly included in the framework. At the same time, more countries need to ratify the MEDICRIME Convention—the only (yet) international binding treaty that criminalizes, inter alia, the distribution of SSFFC medicines over the Internet.

Strengthening cooperation between states and making national jurisdictions as well as judicial procedures stricter may help prevent further engagement with pharmaceutical crime. Finally, having strong pharmacovigilance systems to detect falsification of medicines is essential to protect internal market from the traditional way of supply, but no country is immune from the effects of SSFFC medicines supplied from the Internet.

Given that EU pharmaceutical market is one of the safest in the world, it is unwise to leave it vulnerable to penetration. Thus, stricter regulations are required in the field of online sale of medicines.

Finally, the online sale of SSFFC medicines meets the 'criteria' of a serious crime and EU criminal policy must take decisive actions to combat with the problem rather than focusing only and/or prioritizing the liberal market for free movement of goods.

All in all, as was emphasized in cases *Apothekerkammer des Saarlandes and Others* v *Saarland* (2009, pp. 31 and 60) and Delattre (1991, p. 54), medicines are distinguished from other good to therapeutic effects (as well as the pharmaceutical market involved). The court also emphasized that medicinal products prescribed or used for therapeutic reasons may nonetheless prove seriously harmful to health if they are consumed unnecessarily or incorrectly, without the consumer knowing (Case 171/07, p. 60).

This explains the fact that in all the Member States, there are rules and forms of monopoly granted to pharmacists by reason of the safeguards which pharmacists must provide and the information which they must be in position to furnish to the consumer on the retail sale of such products (Case C-369/88, 1991, p. 54). Hopefully, measures taken to secure market from entering of SSFFC medicines will help diminish the illegal activity of criminal market.

References

Abma, J. (2016). Bitter biopharmaceuticals: Biologic counterfeiting and supply chain concerns. *San Diego International Law Journal, 18*, 161–197.

Acri, K. (2018). They cost us billions and they can kill: Counterfeit drugs are invading Canada. *Special to Financial Post.*

Alarcón-Jiménez, O. (2015). The MEDICRIME convention—fighting against counterfeit medicine. *Eurohealth International, 21*(4), 24–27.

Attaran, A. (2015). Stopping murder by medicine: Introducing the model law on medicine crime. *American Journal of Tropical Medicine and Hygiene, 92*(Suppl 6), 127–132.

Attaran, A., Bate, R., & Kendall, M. (2011). Why and how to make an international crime of medicine counterfeiting. *Journal of International Criminal Justice, 9*(2), 1–30.

Balfour, H. (2021). Chinese police seize over 3,000 fake COVID-19 vaccines. *European Pharmaceutical Review.*

Bate, R. (2012). The deadly world of fake medicine. *Special to CNN.*

Berkot, B. (2012). *Fake Avastin shows little protection of drug supply.* Healthcare & Pharma.

Blackstone, E. A., Fuhr, J. P., & Pociask, S. (2014). The health and economic effects of counterfeit drugs. *American Health & Drug Benefits, 7*(4), 216–224.

Breman, J. G. (2019). It's time to stop murder by counterfeit medicine. *Stat.*

Buckley, G. J., Gostin, L. O., & Committee on Understanding the Global Public Health Implications of Substandard, Falsified, and Counterfeit Medical Products, Board on Global Health & Institute of Medicine (Eds.). (2013). *Counterfeiting the problem of falsified and substandard drugs* (p. 1 & 313). National Academies Press (US).

Case C-322/01. (2003). *Deutscher Apothekerverband eV* v *0800 DocMorris NV and Jacques Waterval.* ECLI:EU:C:2003:664.

Case C-171/07. (2009). *Apothekerkammer des Saarlandes and Others v Saarland* and DocMorris NV. ECLI:EU:C:2009:316.

Case C-369/88. (1991). *Delattre.* ECLI:EU:C:1991:137.

Cannon, D. T. (2015). War through pharmaceuticals: How terrorist organizations are turning to counterfeit medicine to fund their illicit activity. *Case Western Reserve Journal of International Law, 47*(1), 343–375.

Cockburn, R., Newton, P., Agyarko, E. K., Akunyili, D., & White, N. J. (2005). The global threat of counterfeit drugs: Why industry and governments must communicate the dangers. *PLoS Medicine, 2*(4), e100, 0302–e100, 0308.

Council Framework Decision (2002/584/JHA). European arrest warrant and the surrender procedures between Member States. OJ, L 190/1, 18.7.2002.

Dahl, J., Kubic, T. T., & Shepherd, M. (2015). International effort to protect patients from unsafe pharmaceuticals. *European Industrial Pharmacy, 24*, 4–7.

Dégardin, K., Guilleman, A., Guerreiro, N. V., & Roggo, Y. (2016). Near infrared spectroscopy for counterfeit detection using a large database of pharmaceutical tablets. *Journal of Pharmaceutical and Biomedical Analysis, 128*, 89–97.

Di Giorgio, D. (2015). Operation volcano—the herceptin case. *AIFA—Agenzia Italiana del Farmaco*, 3–51.

Directive 2011/62/EU of 8 June 2011 on the prevention of the entry into the legal supply chain of falsified medicinal products amending Directive 2001/83/EC on the Community code relating to medicinal products for human use. (2011). OJ L 174/74.

EMVO. European Medicines Verification Organisation. (2023). Retrieved from https://emvo-medicines.eu/evi/. Accessed 22.03.2023.

European Alliance for Access to Safe Medicines. (2008). The counterfeiting superhighway. *Medicom Group Ltd.*, 1–31.

European Union: European Commission. (2018). Report from the Commission to the European Parliament and the Council on the Member States' transposition of Article 118a of Directive 2001/83/EC of the European Parliament and the Council of 6 November 2001 on the

Community code relating to medicinal products for human use as amended by Directive 2011/62/EU of the European Parliament and of the Council of 8 June 2011, Brussels, 26.1.2018, COM(2018) 49 final (pp. 1–9).

EUROPOL. (2020). *Vaccine-related crime during the COVID-19 pandemic* (pp. 1–2). Europol Public Information.

Fichera, M. (2013). Criminal law beyond the state: The European model. *European Law Journal, 19*(2), 174–200.

George, C. (2006). Internet pharmacies: Global threat requires a global approach to regulation. *Hertfordshire Law Journal, 4*(1), 12–25.

Hall, A., Koenraadt, R., & Antonopoulos, G. A. (2017). Illicit pharmaceutical networks in Europe: Organising the illicit medicine market in the United Kingdom and The Netherlands. *Trends in Organized Crime, 20*, 296–315.

Hamulák, O. (2018). La carta de los derechos fundamentales de la union europea y los derechos sociales. *Estudios constitucionales, 16*(1), 167–186.

Harris, J., Stevens, P., & Morris, J. (2009). Keeping it real. Combating the spread of fake in poor countries. *International Policy Network*, 1–28.

Hock, S. C., Lee, M. M. X., & Chan, L. W. (2019). Regulating online pharmacies & medicinal product E-commerce. *Pharmaceutical Engineering*.

INTERPOL. (2014). Pharmaceutical Crime and Organized Criminal Groups. An analysis of the involvement of organized criminal groups in pharmaceutical crime since 2008. INTERPOL, pharmaceutical crime sub-directorate (pp. 1–21).

Ivic, S., & Troitiño, D. R. (2022). Digital sovereignty and identity in the European union: A challenge for building Europe. *European Studies, 9*(2), 80–109. https://doi.org/10.2478/eustu-2022-0015.

Jing, M. (2016). China FDA stops online med sales. China Daily USA.

Kubic, T. T. (2008). Pharmaceutical counterfeiting: Understanding the extent of a new transnational crime. *The Police Chief, 75*(8), 38–43.

Lavorgna, A. (2015). The online trade in counterfeit pharmaceuticals: New criminal opportunities, trends and challenges. *European Journal of Criminology, 12*(2), 226–241.

Lechleiter, J. (2012). Ways to more affordable medicines: Introduction. *Forbes*.

LegitScript. (2016). The internet pharmacy market in 2016. Trends, challenges, and opportunities. The Center for Safe Internet Pharmacies.

Liang, B. A. (2007). A dose of reality: Promoting access to pharmaceuticals. *Wake Forest Intellectual Property Law Journal, 8*(3), 301–386.

Mackey, T. K., Liang, B. A., York, P., & Kubic, T. (2015). Counterfeit drug penetration into global legitimate medicine supply chains: A global assessment. *American Journal of Tropical Medicine and Hygiene, 92*(Suppl 6), 59–67.

Magdun, M. (2021). Trademark enforcement of counterfeit drugs: A Guardian of the rich and poor alike. *Indiana Journal of Law and Social Equality, 9*(2), 281–303.

Miettinen, S. (2013). Criminal law and policy in the European Union, (p. 304). Routledge Research in EU Law, Taylor & Francis Group.

Negri, S. (2016). The Medicrime convention: Combating pharmaceutical crimes through European criminal law and beyond. *New Journal of European Criminal Law, 7*(3), 350–367.

OECD/EUIPO. (2020). Trade in counterfeit pharmaceutical products. (pp. 1–93). Illicit Trade, OECD Publishing, Paris.

Orton, T., Omar, Z., Subramanian, P., & Adeli, N. (2019). *Counterfeit drugs in the pharmaceutical industry*. American University.

Power, M. (2013). *Online highs are old as the net: The first e-commerce was a drugs deal*. The Guardian.

Ramiro Troitiño, D., Martín de la Guardia, R., & Pérez Sánchez, G.A. (2022) The European Union and its political leaders understanding the integration process the essence of Europe: Understanding Europe through its designers. (pp 1–4). Springer International Publishing Cham.

Regulation (EC) No 726/2004 of 31 March 2004. (2004). Laying down community procedures for the authorisation and supervision of medicinal products for human and veterinary use and establishing a European Medicines Agency. OJ L136/1.

Seelke, C. R., Wyler, L. S., Beittel, J. S., & Sullivan, M. P. (2011). Latin America and the Caribbean: Illicit drug trafficking and U.S. counterdrug programs. *Congressional Research Service*, 1–41.

Terry, N. P. (2017). Regulatory disruption and arbitrage in healthcare data protection. *Yale Journal of Health Policy, Law, and Ethics, 17*(1), 143–208.

TFEU. Consolidated Version of the Treaty on the Functioning of the European Union. (2012). OJ C321/1.

TRACIT. (2022). Illicit trade in pharmaceuticals, Transnational Alliance to Combat Illicit Trade.

Treadwell, J. (2012). From the car boot to booting it up? eBay, online counterfeit crime and the transformation of the criminal marketplace. *Criminology & Criminal Justice, 12*(2), 175–191.

Trenfield, S. J., Tan, H. X., Awad, A., Buanz, A., Gaisford, S., Basit, A. W., & Goyanes, A. (2019). Track-and-trace: Novel anti-counterfeit measures for 3D printed personalized drug products using smart material inks. *International Journal of Pharmaceutics, 567*, 1–7.

Turudić, I. (2018). Case law on the European arrest warrant. *EU and Comparative Law Issues and Challenges Series, 2*(2), 511–529.

United Nations. (2002). *Convention against transnational organized crime*. Office on Drugs and Crime.

Vander Beken, T., & Balcaen, A. (2007). Crime opportunities provided by legislation in market sectors: Mobile phones, waste disposal, banking, pharmaceuticals. *European Journal on Criminal Policy and Research, 12*, 299–323.

van Duyne, P. C., von Lampe, K., van Dijck, M., & Newell, J. (2005). The organised crime economy: Managing crime markets in Europe. (pp. 1–298). Wolf Legal Publishers.

Wechsler, J. (2012). Campaign mounts to curb counterfeit drugs. *BioPharm International, 25*(9), 40–45.

WHO. (2022). Norms and standards for pharmaceuticals. Health product and policy standards.

WHO. (2018). Substandard and falsified medical products. Fact Sheets.

WHO. (2011). Safety and security on the Internet: Challenges and advances in member states: Based on the findings of the second global survey on eHealth. *Global Observatory for eHealth Series, 4*, 1–92.

Wilson, B. K., Kaur, H., Allan, E. L., Lozama, A., & Bell, D. (2017). A new handheld device for the detection of falsified medicines: Demonstration on falsified artemisinin-based therapies from the field. *American Journal of Tropical Medicine and Hygiene, 96*(5), 1117–1123.

European Digital Finance

Janika Aben and Paula Etti

Abstract The European Union policymakers are focused on establishing a well-suited environment for the digital future of Europe. As the future depends on input to the economy, and the circulation of capital, the digital era needs a new look at digital finance. For this reason, the top-down strategy, a Digital Finance Strategy, and the legislative proposals based on it are highlighted developments in the EU digital transformation. The hereby overview provides an insight into the preparatory legislative contribution in the field of digital finance and a more precise view of the EU's efforts to frame the novel market of crypto-assets as well as to tame the risks deriving from the digital dependence, which is common to different participants of the financial market.

1 Introduction

The EU is committed to providing an environment where industry meets funding as one of its main pillars. And the latter expectation is valid also in the digital era, but as a result of innovative views on the notion of "environment" and "funding," the primary concepts of circulation of capital and assets, the borders of the common marketplace are shifting. Today and in the future, the common marketplace of the EU is rather digital and virtual than onsite as the digital world, innovative technologies in financial services, and new business models aim at a borderless marketplace. The source of capital is drifting from institutions to peer-to-peer capital raising, the capital flows are rather transferred digitally, even in the form of a token, not in checks or cash. Despite the format of finance being more digital than ever, the expectations of the investors and consumers in the common market remain uniform,

J. Aben (✉)
Tallinn University of Technology, Tallinn, Estonia
e-mail: janika.aben@taltech.ee

P. Etti
Ministry of Finance of Estonia, Tallinn, Estonia
e-mail: paula.etti@fin.ee

D. Ramiro Troitiño et al. (eds.), *Digital Development of the European Union*,
https://doi.org/10.1007/978-3-031-27312-4_5

valuing a transparent and prudent financial environment with minimal asymmetry in information between the parties (Hamulák, 2018). In sum, while offering novel services and benefits for the worldwide marketplace and raising capital globally, the question remains of how to attract Fintech[1] digital innovators to locate in Europe and respond to the European thresholds on compliance and continuity, sustainability in its broader meaning.

As the biggest opportunities, also threats in finance come from the digital sphere, the focus on digital financial services leads to questions on the management of digital and cyber risks and proportionate legislation on Fintechs and digital finance. FinTech stands at the crossroads of financial services and the digital single market: the financial sector is the largest user of digital technologies and represents a major driver in the digital transformation of the economy and society (European Commission, 2018). Finance and technology have proved to be partners made in heaven as the global FinTech landscape evolves at an exponential pace (Ahern, 2021: 1). There is a great desire that Europe should lead digital financing based on European values and sound risk management. Digital transformation in finance is expected to be driven by strong European market players on one hand, who will make digital finance available to European consumers and businesses whether by new market entrants or the EU's traditional financial sector (European Commission, 2020c). New financial facilities like crowdfunding platforms, crypto-assets, and the new technologies underlying them, e.g., distributed ledger technologies (DLT), and artificial intelligence transform the notion of financial instruments and services, whereas the risks related to the digital operations of all financial services underline the importance of the horizontal view of digital financial services.

In order to increase the integrity of the EU common market and the volume of the EU capital market, new legislative initiatives have been made to provide an even playing field for all participants in the financial market. Further, the aim is to rethink the traditional concepts of capital flows, relevant instruments, and the risks related to the opening of the capital market to new entrants. It has been admitted that regulatory lag and regulatory friction also act as a "disenabler" for ease of cross-border FinTech trade in the EU. On the other hand, there is a risk of EU Member States losing out on FinTech market share if the regulatory landscape is too difficult to maneuver or imposes a disproportionate regulatory burden on FinTech market actors, disenabling rather than facilitating FinTech (Ahern, 2021: 1). Although traditional boundaries between national, European and international law are fading, it is recognized that relying only on national rules would also result in fragmentation (Lehmann, 2021) and would not allow benefiting from the common market synergy. In the globalized legal environment of digital finance, the EU-wide approach may be preferred compared to the regulatory competition of member states. Providing European rules for digital finance and innovation creates a more solid ground for

[1] FinTech is a term used to describe technology-enabled innovation in financial services that could result in new business models, applications, processes, or products and could have an associated material effect on financial markets and institutions and how financial services are provided (FSB, 2017).

"jurisdictional competition" compared to the competing localisations of financial centers. It is argued that the competition between legal regimes can stimulate policy learning and innovation (van Gestel et al., 2012: 16). The proposals on digital finance of the EU reflect this statement as the drafting of EU legislation on innovation has followed the practices and inputs of market participants on feature of the novel financial services (European Commission, 2017) and negotiations between member states with the aim of improving the regulatory background to boost innovation for the future.

Hereby an insight is provided to estimate the efforts to create a seamless environment for EU-wide financial services and compatible development of capital markets in the future by means of regulating the crypto-assets and crypto-asset service providers, and creating a framework to manage digital operational risks of the financial sector.

2 Going Digital Leads to Going Riskier

Financial regulation has developed through the course of the twentieth and twenty-first centuries to address market failures. As a result, the core objectives of financial regulation are (1) the safety and soundness of financial firms and the maintenance of systemic stability (which together comprise the prudential objectives of financial regulation); and (2) investor and consumer protection and market integrity (which together comprise the conduct of business objectives of financial regulation). Considering the market failures, since the global financial crisis of 2007–2008, prudential regulation has become considerably more intrusive, with stringent prudential (and capital) requirements especially on "systemically important institutions," extensive stress testing, and additional macroprudential oversight (He et al., 2022: 4). The global financial crisis also marked a significant turning point in the regulation of the financial industry. The existing framework of prudential regulation failed to safeguard against the systemic risk created by off-balance-sheet exposure and the contagious effects of failures (McNulty et al., 2022: 2), and the shift emerged from rule-based to more principle and risk-based financial regulation. The technology-driven and digitally dependent innovation in financial services is accompanied by the questions on market integrity and information asymmetries, investor protection and financial literacy which all require a principle-based approach by the regulators.

As a gain of Fintech race EU financial supervision authorities recognize the higher financial inclusion but on the other hand, there is a need for proportionate regulation in order to address the concerns of the digital era and new trends in investor and consumer behavior. Fintech has its impacts and contact points across all sectors of payments, finance, and investments. The benefits to consumers and society including faster and lower cost distribution of FinTech products and financial inclusion are apparent (Ahern, 2021: 3). Equity- or loan-based crowdfunding are providing alternative sources of capital to traditional financing, also the trend of tokenization offers new investment opportunities for investors and cryptocurrencies (such as Bitcoin and Ether) and tokens issued through Initial Coin Offerings ("ICOs") provide novel means of payment and raising capital. The risks and the cost of innovation may not be the focus of the advertising statements and communication of the new services and products.

Retail investors' participation in financial markets has increased substantially in recent years, with new investors taking advantage of the convenience and user-friendly features of mobile trading platforms. These investors are of particular concern to European Securities and Markets Authority (ESMA). Diversification offers opportunities but also comes with risks, and supervisors remain concerned about risks to retail investors who buy financial and crypto-assets with expectations of significant price growth, and without realizing the high risks involved. The trend is driven by digitalization, consumer proficiency in social-media-driven trading and copy trading, and the aggressive marketing of higher-risk structured products and crypto-assets (ESMA, 2022: 5). Beginner retail investors who desire for high returns (and, combined with little knowledge of financial literacy, and also, lack of basic cyber hygiene), are easy targets for cybercriminals who use the situation for easy profit and apply innovative solutions and technologies for fraud. Therefore, European Union Agency for Network and Information Security (ENISA) has seen the need for a unified approach to cyber hygiene throughout the EU (ENISA, 2016: 5).

In February 2022 the European Supervisory Authorities (ESAs) published their report in response to the European Commission's Call for Advice on Digital Finance (ESAs, 2022; European Commission, 2021). The report describes structural changes in the financial sector resulting from innovative technologies, including changing value chains, new dependencies on digital platforms and the rise of mixed-activity groups. These trends create opportunities for consumers but also risks, including gamification of financial services. Gamification of financial services involves the introduction of game-like features or interfaces to online investment or money management tools. An important task in developing RRIs is to identify the extent to which consumers are exposed to risks that arise in the new and evolving digital environment. In regard to investor behavior, the venturous investing in crypto-assets raises concerns in market integrity and investor protection. Retail investment strategy announced for 2022, as one of the actions planned under the Capital Markets Union's (CMU) Action Plan (European Commission, 2020b), aims to respond to new challenges in the market, such as the increasing digitalization of investment advice and the use of digital distribution channels for retail investors. According to the survey from May 2022 on retail investment strategy, it is noteworthy that crypto-assets were the fifth most frequently mentioned type of products (respondents stated that they held investment funds (46%), listed shares (38%), life insurance (30%) and 27% held pension products). Some 16% of respondents with investments stated that they held crypto-assets, i.e., the same share as for ETFs (European Commission, 2022: 10). Witnessing the market trends European Union is contributing to finding proportionate solutions to encourage innovation in digital finance, but also protect the more risk-seeking investors. What is undeniable is that asset classes and our society's understanding of money are changing rapidly (Hofmann, 2020 in Ahern,

2021: 5) and the EU needs to be equipped to fully embrace the contemporary means and services applied in the EU digital single market.

3 Building the EU Digital Financial Market

The integration of Europe's capital markets started with the development of the Single Market in the 1980s. The essential push for integration came from the liberalization of capital flows and the introduction of passporting rights for financial services. Member states' domestic regulatory frameworks were seen as a barrier for financial intermediaries scaling their operations across the EU as their business models had to be adjusted according to each jurisdiction. The idea of passporting rights relies on the harmonized EU rules—authorization of the financial service provider in the home member state grants the right to operate in other member states without any additional administrative burden. Uniform terms within the EU encourage entities to engage cross-border and build trust amongst the investors to provide a free flow of capital within the EU. Despite the next steps in the creation of the EU capital market as the creation of the ESAs, also a vision of a single rulebook in finance, the EU capital markets have remained comparatively fragmented.

Compared to the neighboring markets there has been an expectation that a more integrated CMU nudges the development horizontally. The CMU initiative began in 2014, with the first action plan, and had its next level in September 2020 as the Commission published a new CMU action plan. Beyond integration into a genuine Single Market for capital, other primary objectives include supporting a green and inclusive recovery and making the EU a safer place to invest long-term (European Commission, 2020b). The initiatives of the CMU action plan, besides opening versatile issues related to the traditional capital market, include also the topic of digital transformation of financial markets and a more inclusive economy as part of the CMU drivers.

Fintech action plan of March 2018 (European Commission, 2018) and the Digital Finance Package of September 2020 (European Commission, 2020a) are the CMU initiatives that ground the CMU in the fields of digital finance. The early bird of the Fintech Action Plan is the Regulation on European Crowdfunding Service Providers (ECSPR, Regulation (EU) 2020), which entered into force on November 10, 2020, and is applied to the entities from November 10, 2021, provides the European Passport for peer-to-peer financial services like equity-based and loan-based crowdfunding platforms. The crowdfunding service as a technology-enabled service of a platform, which carries the potential to help better intermediate and match investors with business projects in need of funding, provides an alternative to unsecured bank lending (European Commission, 2018). Following the expectation to broaden the negotiations on legislation to new segments of the digital financial market the hereby overview focuses on the next initiatives of the EU to respond to the emergence of the use of crypto-assets and risks deriving from the digital dependence in finance.

The Digital Finance Package includes a new Digital Finance Strategy combined with a renewed Retail Payments Strategy, in an effort to "boost Europe's competitiveness and innovation in the financial sector, paving the way for Europe to become a global standard-setter." Setting the goal to enhance consumer choice, and simultaneously ensure consumer protection and financial stability, the Package aims to boost responsible innovation in the EU's financial sector, especially for highly innovative digital start-ups, while mitigating any potential risks related to investor protection, money laundering, and cybercrime. The Retail Payments Strategy continues the previous work in developing the Single European Payments Areas (SEPA) and focuses on promoting digitalization in retail payments, and also developing payment infrastructure, and also in improving cross-border European payment solutions, and providing more efficient international payments (European Commission, 2020d).

The new Digital Finance Strategy (European Commission, 2020c) provides a very broad and comprehensive framework for future reforms to further the development of digital finance in the EU focusing on four major objectives: (1) removing fragmentation in the Digital Single Market; (2) adapting the EU regulatory framework to facilitate digital innovation; (3) promoting data-driven finance; and (4) addressing the challenges and risks with digital transformation, including enhancing the digital operational resilience of the financial system.

Instead of merely addressing issues and challenges of digital assets, the latter takes a broader approach to the future development of digital finance and innovation in the EU (Zetzsche et al., 2020: 2) and opens up access to the financial sector for innovators. The key actions as listed in the strategy shall be the new proposals on the prudential regulatory treatment of crypto-assets, SME financing, and the role of DLT and the Internet of Things (IoT) in the new EU Sustainable Finance Taxonomy, also providing support for a possible central bank digital currency (CBDC) from the European Central Bank (ECB) (European Commission, 2020c). The key actions of the strategy admit that crypto and cyberware have made big inroads in finance and will continue to shake up the supply and operation of financial services (Lannoo, 2021: 11).

The Digital Finance Strategy focuses on risks. The justification for consumer and investor protection derives from the accepted approach to financial stability, based on the principle of "same activity, same risk, same rules" (Kerikmae and Troitiño, 2021). Considering the variety of technologies applied in digital finance, the regulatory framework aims to achieve a certain level of generalization and focus on economic functions, principles rather than certain technologies.

The strategy addresses technology risks recognized in the financial sector in general via a new proposed Regulation on Digital Operational Resilience (DORA) (European Commission, 2020g)[2] that aims to prevent and mitigate disruptions and

[2] As DORA and MICA are still in draft form, the legislative procedure is ongoing in Europe, although already in the stage of the provisional agreement of the co-legislators. Press release on DORA https://www.consilium.europa.eu/en/press/press-releases/2022/05/11/digital-finance-provisional-agreement-reached-on-dora/ and MICA https://www.consilium.europa.eu/en/press/press-releases/2022/06/30/digital-finance-agreement-reached-on-european-crypto-assets-regulation-mica/.

threats that arise in connection with (the use of) information and communications technology (ICT), especially cyber threats. DORA has been drafted with the desire to “ensure that all financial system participants maintain resilient operations through a severe operational disruption and have the necessary safeguards in place to mitigate cyber-attacks and other risks.” The proposed legislation will require all (financial) firms to ensure that they can withstand all types of ICT-related disruptions and threats. The DORA proposal introduces also an oversight framework for critical ICT providers (as cloud computing service providers) to strengthen business continuity.

At the core of the EU Digital Finance Package, however, lie the legislative proposals for an EU regulatory framework on crypto-assets to support passporting for innovative start-ups across the EU including in relation to crypto-assets. This includes both the proposal for a new Regulation on Markets in Crypto-Assets (MiCA) (European Commission, 2020e) as well as a new proposal for a Regulation on a Pilot Regime for Market Infrastructures Based on Distributed Ledger Technology (DLT Infrastructure Regulation) (European Commission, 2020f). While MiCA deals with crypto-assets (understood as “a digital representation of values or rights that can be stored and traded electronically”) the DLT Infrastructure Regulation will introduce a so-called “sandbox” approach for DLT-based market infrastructures, allowing temporary derogations from existing rules to assist regulators’ gaining of experience. Behind the “DLT Pilot Proposal” lies a recognition of the significance of applications of DLT in finance. The proposed Regulation is designed for certain investment firms, market operators, and central securities depositories wanting to establish DLT market infrastructure and provide cross-border services throughout the EU.

The EU-wide regulatory sandbox for market infrastructures based on DLT can be used by the DLT multilateral trading facility, by a DLT securities settlement system, by “DLT market infrastructure,” and by “DLT transferable securities.” It has several exemptions from existing rules (such as the Central Securities Depositories Regulation), but with ceilings as restrictions, and the terms will be reviewed in 5 years. DLT sandbox sets the floor to allocate the world of regulated financial instruments in the form of DLT and the rules for operation in the form of DLT. Consequently, a key objective of enabling a pilot regime to allow experimentation to place is that it would assist in highlighting difficulties for secondary market trading of crypto-assets from the perspective of the existing EU framework for financial instruments. This would enable the identification of what action could be taken by the EU to facilitate innovation by removing unjustifiable obstacles. At the same time, the promotion of investor and consumer protection as well as market integrity and financial stability are expressed to be key objectives (European Commission, 2020f).

Crypto-assets which are not financial instruments are covered by a separate piece of legislation. MICA’s intention is to fill a major regulatory gap and ensure a harmonized approach to crypto-assets across the EU Single Market, creating also important new tasks for the ESAs and national authorities in supervising innovative financial services. Due to the novelty of the technology, most countries have not yet established any firm way to deal with crypto-assets. Therefore with regard to crypto-assets and DLT, a plea has been made to follow the goals of harmonization and provide uniform rules regarding this new digital solution before it is too late (Lehmann, 2021). As a response to the requests from the market and part of the Digital Finance Strategy, the proposal on MICA focuses on the infrastructure of crypto-assets markets, inter alia the issuers and operators. This complex regulation covers three different forms of crypto-assets based on DLT: crypto-assets as utility tokens, asset-referenced tokens, and e-money tokens.

As a parallel to crowdfunding the aim of the proposal is to widen the harmonized legislation of the Single Market in a way that based on the principle of "same activity, same risks, same rules" the legal framework covers all peculiarities of offering capital flows and means of payment in the EU, including in the form of crypto-assets. Following the statements of the MICA proposal, most crypto-assets fall outside the scope of EU financial services legislation and therefore are not subject to the existing provisions on consumer and investor protection and market integrity, among others, although they give rise to these risks. The only piece of existing EU financial law with clear references to crypto-assets and platforms is the fifth AML Directive ((EU) 2018/843), which will be reinforced by the new AML package of 2021.[3] In regulating DLT-based providers, MICA proposal places the EU at the forefront of change, as it is the first international jurisdiction to propose rules on the matter instead of standing on the sidelines (Lannoo, 2021: 11).

As the EU is at the doorstep of providing the common framework for financial innovation to support sustainable, digital innovation in the Union and empower the Single Market for capital it is worth considering what instruments and services are included into the scope, accepted also as instruments of MICA, which may be offered in the common digital market under the passporting regime.

4 Opening the European Passport for Crypto-Assets

Crypto-asset token offerings create various risks, with the specific type of risk depending on the nature of the token in question. Regulation seeks to address these risks as they relate to financial markets and systems: financial stability, market integrity, client/investor protection, and market efficiency are core concerns of financial regulation (Zetzsche et al., 2020: 7). The scope of MICA includes the utility and payment token issuers' obligations as well as the conditions to provide crypto-assets services. The list of crypto-asset service providers (CASPs) in Article 3(1) no (9) is designed based on the example of MIFID II Annex 1A and includes the custody, administration, and exchange of assets, operation of trading platforms, etc., which all need authorization by the competent authority in order to operate within

[3] AMLD 6 package by European Commission: https://finance.ec.europa.eu/publications/anti-money-laundering-and-countering-financing-terrorism-legislative-package_en.

the EU. MICA's view on the obligations of the token issuers is more complex and varied than that for the service providers.

Crypto-assets can be designed in a number of ways and entail the ownership of a variety of rights, ranging from a financial interest in a company to purely non-financial rights. While payment and securities tokens, for instance, create risks for investors and clients, market efficiency, and market integrity, they are usually of less concern from a financial stability perspective given that most token offerings are small compared to the size of the financial system. At the same time, the potential for the emergence of global stablecoins raises much greater and different concerns, as stablecoins have the potential to reach globally systemic dimensions from a financial stability perspective. Tokens are also often structured as hybrids with payment, securities, and utility characteristics. This renders the risk assessment relating to crypto-assets a particular challenge (Ibid: 8).

According to Article 3 1(2) of MICA, crypto-asset is defined as a digital representation of values or rights that can be stored and traded electronically, using DLT or similar technology. The scope of MICA is aimed to include the world of crypto-assets falling outside the regulated financial instruments. The initial regulatory limits of the regulated financial market descend from the primary act of markets of financial instruments—MIFID II (Directive 2014/65/EU). In the EU, most financial sector legislation has been developed with the Single Market in mind, but previously in the form of directives. Despite the transposition measures of the member states, the definitions of MIFID provide sufficient room for discussions also today. Therefore, the lack of explicit coverage of newer business models in these laws has led to notable differences in interpretations by the member states locally (Laidroo et al., 2021: 8) but controversially Fintech entrepreneurs valued the presence of high-quality infrastructure and regulatory clarity in the most (ibid: 18).

Consequently, MICA seeks to provide legal certainty for different types of crypto-assets not covered by existing EU financial services legislation (that do not qualify as financial instruments, deposits, or structured deposits), establishing also specific rules for so-called "stablecoins," including when these are e-money. Crypto-assets that qualify as "financial instruments" as defined in Article 4(1), point (15), of MIFID II, therefore remain regulated under the general existing Union legislation, regardless of the technology used for their issuance or their transfer. In order to create an EU framework that both enables markets in crypto-assets as well as the tokenization of traditional financial assets and wider use of DLT in financial services, Digital Finance Package includes also other legislative proposals: amendments of MIFID II and other acts to clarify the definition of "financial instruments" and include financial instruments based on DLT (European Commission, 2020h), as well as a proposal for a regulation to establish a pilot regime on DLT market infrastructures for these instruments (European Commission, 2020f).

At the first glance defining crypto-assets as "a digital representation of value or rights which may be transferred and stored electronically" provides a wide spectrum of capital flow, which would be covered by MICA and therefore granted an EU passporting right in addition to the existing "financial instruments." Hence definition is open to interpretations in connection to terms of transferability of the value or right.

According to recital 9 and Article 3 1(3)–(5) of MICA crypto-assets may be divided into three sub-categories. The first sub-category is "asset-referenced token," which aims at maintaining a stable value by referencing several currencies that are legal tender, one or several commodities or crypto-assets, or a basket of such assets, and often aim at being used by their holders as a means of payment to buy goods and services. A second sub-category, "e-money tokens," are crypto-assets intended primarily as a means of payment aiming at stabilizing their value by referencing only one fiat currency functioning similar to electronic money, as defined in Article 2, point 2, of Directive 2009/110/EC. The third sub-category consists of a type of crypto-assets that are intended to provide digital access to a good or service, available on DLT, and that is only accepted by the issuer of that token—a "utility tokens." Such "utility tokens" are defined via a negative definition, framing it as a token that is not an asset-references or e-money token, also not an electronic money, a deposit, a structured deposit or a financial instrument. And the assessment explaining the reasons why the crypto-assets does not fall within the scope of other EU financial regulation is expected to be performed by the issuer in the white paper. In financial regulation, the classification of instruments and transactions determines which body of law will apply, and which supervisory powers a competent authority may exercise. Further, in the Single European Financial Market, uncertain classifications based on uncertain definitions create the potential for regulatory arbitrage (Zetzsche et al., 2020: 8). Combining with the interpretations of the definition of "financial instrument," the difficulties may even rise.

In its Advice to the EU Commission from January 2019 (ESMA, 2019), ESMA found that the boundaries of the notion of "transferable securities" as archetype financial instruments under Annex I, Section C(1) MiFID II are not entirely clear. First, the requirement of being "negotiable" (as part of "transferability") is open to interpretation, and second, defining what is "similar" to shares or bonds may vary under Member State corporate laws. As a result, the outcomes may vary from one Member State to another, especially in regard to instruments that are similar to traditional shares or bonds. Pursuant to Article 2(2) MiCA, if the token is identifiable as falling within categories already contemplated by existing EU financial law, it will not fall within the scope of MiCA if the token is a financial instrument. If a token is not similar to a share or to a debenture, or if it is similar to a share or a debenture, but is not "negotiable," it will fall under MiCA and will be considered by MiCA to be a "utility token." The latter leaves an important role in determining the application practices to the private sector and legal opinions, which in turn may increase the regulatory arbitrage.

In addition, while according to the draft the asset-referenced token or e-money token issuers are subject to authorization requirement and certain operating conditions, the provisions on crypto-assets, which are utility tokens, merely focus on disclosure. Complying with the requirements of the white paper grants the issuer of the relevant crypto-asset a European Passport (Article 9 of MiCA). But the white paper of crypto-assets is not subject to preliminary approval by the supervisor

(Article 8(1) and (2)). There is only the requirement of notification and the competent authority may exercise the powers to require completing the marketing communication as foreseen in the regulation of MiCA. The classification of the different tokens as proposed in MiCA may potentially be confusing for citizens, and a huge task to undertake for supervisors as mixed with the ex-post approach of supervision in disclosure rules, risks of re-characterization and re-qualification of tokens by the competent authority remain high (Lannoo, 2021: 11).

5 Single Market, Multiple Risks: Hedging Digitalization Risks

Competition, a level playing field, and smart regulation are just some of the keywords that emerge in the provision of digital financial services. In addition, aspects related to stability, efficiency, investor and consumer protection, and privacy aspects should also be taken into account. In this regard, it is proposed to carefully monitor transformative technologies like DLT, artificial intelligence, and the Internet of Things (OECD, 2019). New technologies, movement of money and other assets mainly through e-channels and the growing trend of e-finance are also largely shifting crime to the Internet. The COVID-19 pandemic, the resulting social and economic situation, and increased dependencies on ICT technologies, have thus given a major impetus to the growth of cybercrime. It is not unknown that the financial sector has been one of hackers' favorite sectors during this pandemic (Aldasoro et al., 2021).

Cyber security is essential for the EU ecosystem as a whole. Digital security threats and incidents are growing in number and complexity and can have serious consequences (OECD, 2019). Cybercrime targets have changed moving from individuals and small businesses to large corporations, governments, and critical infrastructure, identifying as the main categories of crime are online scams and phishing, disruptive malware (including ransomware and DDoS), data harvesting malware, malicious domains, and misinformation. (Interpol, 2020). However, since companies are closely linked across Europe through various multi-level supply chains, even the smallest organization could become quite significant in relation to cyber threats (ENISA, 2016: 6). As cyber-attacks can pose significant risks to financial service providers, their employees, and customers, it is vital that the sector is prepared for attacks also in the form of cooperation in cross-border level. Ensuring a high level of cyber security across the EU is a challenge that affects almost everyone (EBF, 2021), cyber threats are cross-border and cross-sectoral in nature and can arise from any area of economic activity.

Since ICT risks continue to challenge the EU financial sector, especially its operational resilience, performance, and stability, the EU has taken further steps to address these issues seeing the need to establish a comprehensive and integrated legal framework for digital operational resilience for EU financial service providers—DORA. The scope of DORA includes financial sector third-party critical services and operations resilience and harmonizing the rules applicable in the area of ICT risk management, reporting, testing, and ICT third-party risk (for example, security software, infrastructure, cloud services). DORA clearly defines the entities of scope and requires them to have the necessary governance framework in place and also gives an additional role to the ESAs to monitor digital resilience. DORA applies to a wide range of regulated financial entities, starting from credit institutions and investment firms, insurers, and including the Fintech sector as crowdfunding and crypto-asset service providers (article 2 of DORA).

DORA aims to meet the need for a more EU-wide standardized approach for cyber risks and disclosure by the financial sector. With regard to DORA, it is important to ensure proportionate requirements for companies of different sizes, risks, and business profiles. The proportionality principle is addressed in many aspects of the proposal, for example, a lighter regime has been proposed for microenterprises, some of the requirements are only applicable for significant financial entities or in case of major ICT-related incidents, it has been found that proportionality of the regulation in its implementation should be emphasized (ESA, 2021).

Taking into account the potential systemic risk entailed by increased outsourcing practices and by the ICT third-party concentration DORA brings third-party providers into the financial supervisory domain, albeit only to an extent. The challenge here stems from digital finance and finding the right balance between financial regulation and supervision, in order to handle a more complex and diverse ecosystem where tech companies are increasingly important actors in the effective provision of financial services (Lannoo, 2021: 7). Globally, Amazon's Web Services, Microsoft's Azure and Google Cloud control 61% of the market share of cloud infrastructure services as of Q1 2021. In that respect, the collapse or service failure of a single BigTech, could potentially pose a serious threat for the continuity of the provision of financial services, together with a negative impact on markets, consumers, financial stability, and the economy as a whole. With regard to the materiality of these cloud services, some BigTechs, which are third-party service providers for financial institutions may be already "too-critical-to-fail" (Abidi & Miquel-Flores, 2022: 21).

Using third-party service providers leaves the responsibility for complying with obligations under DORA to the financial entity and in addition according to DORA article 26 it needs additional consideration when the conclusion of a contract supporting critical or important functions would lead to a contract with an ICT third-party service provider that is not easily substitutable or leads to multiple contractual arrangements with the same or closely connected service providers. Therefore, assessing the ICT concentration risk and weighing the benefits and costs of alternative solutions is required.

According to recital 54, the definition of "critical or important function" provided for in DORA should encompass the "critical functions" as defined in point (35) of Article 2(1) of Directive 2014/59/EU. Based on Article 3(17) of DORA "critical or important function" means a function whose disruption would materially impair the financial performance of a financial entity or the soundness or continuity of its services and activities, or whose discontinued, defective, or failed performance would materially impair the continuing compliance of a financial entity with the conditions and obligations of its authorization, or with its other obligations under applicable financial services legislation. In addition to this deriving from the Directive (EU) 2014/59/EU the definition of critical functions, therefore, includes also a wider view on the functions connected with the financial, mostly baking institutions: critical functions are the activities, services, or operations the discontinuance of which is likely in one or more Member States, to lead to the disruption of services that are essential to the real economy or to disrupt financial stability due to the size, market share, external and internal interconnectedness, complexity or cross-border activities of an institution or group. Based on the principle and risk-based approach of DORA, the financial entities themselves are the key players to determine the critical and important functions of their services and manage the risks stemming from the architecture of the financial service. From a financial regulation perspective, tech giants are now at the heart of the financial system. They host a growing mass of platforms including banking and market activities, also Fintechs, as financial institutions are shifting critical operations, such as payment systems and online banking activities on their cloud platforms. The justified question arises are tech giants "too big to fail" (Abidi & Miquel-Flores, 2022: 1). DORA aims to catch these risks before these concerns might be drawn to legislation based on the bitter experiences of the market players.

Over the past 150 years, finance has evolved into one of the most globalized, digitalized, and regulated industries. Importantly, in the last decade, BigTech has likewise expanded into the financial sector. The emergence of concentration and dominance in digital finance has long been a source of concern in the European Union (Arner et al., 2021:5). Digitization and datafication offer tremendous potential for network effects and economies of scope and scale, and have duly emerged in the platform economy and more recently in the platformization of finance (ibid: 6). The approach on BigTechs offered by DORA is to determine the actions on critical dominant global platforms uniformly EU-wide by the ESAs. In regards to critical ICT third-party service providers defined in article 28 the list is prepared by the ESAs, through the Joint Committee—after designating the ICT third-party service provider as critical, the ESAs, notify the ICT third-party service provider of such designation and the starting date as from which the service provider will be effective subject to oversight activities, upon a third-country entity this results in an obligation to establish a subsidiary in the EU (Troitiño & Kerikmäe, 2021). In turn, the ICT third-party service provider will notify the financial entities to which it provides services designated as critical. The characteristics of critical derive from the systemic impact on the stability and continuity, also subsidiarity of the provision of financial services for the financial entities that rely on the relevant ICT third-party provider.

6 Concluding Remarks

Technology has the potential to improve efficiency in the provision of the financial system via an increase in competition between financial institutions and a range of new financial products and business models for clients. Financial regulation has developed through market crashes and malfunctioning looking for an equilibrium point between the safety and soundness of financial firms, also systemic stability, client protection and market integrity from one side, and the risk of excessive interference in the market developments which result in drawback in turn. New financial technologies create challenges for regulatory views on prudential safety, the conduct of business, and systemic stability, but also on competition in financial services.

Harmonized rules are essential for a digital financial market and a level playing field in Europe, and the European legislator acts as the top-down strategy setter in this picture. In regards to novel technologies related to Fintech European practice is shifting toward preferring to act in the form of regulation in order to pre-empt the emergence of different national frameworks, which would lead to fragmentation of the digital market. The latest European Commission proposals on Digital Finance Package move the balance to uniform rules. The form of EU regulation in regards to MICA and DORA might be the one to better support the capital market objectives and to decrease the fragmentation.

EU legislative proposals on Fintech and digital finance follow the aim to ensure that supply and demand for capital can be matched more quickly and safely in Europe based on the principle of the same activity, same risks, and same rules. Considering the more open view on the players of the digital capital market, it is worth analyzing whether the new legislative proposals of the digital finance package are able to follow the ambitious goal to move from the current entity-based regime toward an ambitious activity-based approach.

According to the files of the Digital Finance Strategy crypto-assets and risks from the digital operational resilience are targeted. The proposals on DLT Infrastructure Regulation and MICA foresee the solution for the question of how to treat financial instruments and similar assets in the form of DLT. The general view is that all assets covered by EU legislation (mainly MIFID II) should remain subject to legislation in force, although a pilot regime to settle and trade these instruments in the form of a token is governed by the DLT Infrastructure pilot regime. For other instruments in the form of crypto-assets the new legal act of MICA shall be applied. The new regulation provides rules for the public offer of crypto-assets, asset-referenced stablecoins, and e-money tokens, terms on authorization of crypto-asset service providers and on the business conduct of service providers. Although the aim is to solve the questions of unregulated assets and activities, the drafting of existing EU legislation leaves some extra room for interpretation in regard to the scope of the regulation when comparing with the definition on financial instruments.

The other proposal DORA aims to make sure that the financial institutions and other regulated financial entities in Europe are able to maintain resilient operations through a severe operational disruption. In addition to the financial sector, the requirements for the security of network and information systems of companies are set also for the critical third parties which provide ICT-related services to the financial sector, such as cloud platforms or data analytics services in order to make sure that all these entities can withstand, respond to and recover from all types of ICT-related disruptions and (cyber) threats. In order to operate in the EU digital financial market critical third-country ICT service providers to financial entities will be expected to establish a subsidiary within the EU so that the competent authority, ESAs are able to implement oversight properly. In this case, the financial market participants apply the self-assessment principle to determine the critical and important functions of the service and the relevant service providers. Thereby the similarity in proposals of MICA and DORA steps ahead questioning the issue, of who and what instruments or services are in the scope of the new legislative acts, this may be also seen as a result of establishing a principle-based regulation, which provides larger room for interpretation. The issues on implementation of the new rules will be providing intriguing grounds for future discussions on digital financial regulations.

List of Literature

Legislative Acts, Proposals and Policy Opinions

Directive 2014/65/EU of the European Parliament and of the Council of 15 May 2014 on markets in financial instruments and amending Directive 2002/92/EC and Directive 2011/61/EU (MIFID II). OJ L 173, 12.6.2014, (pp. 349–496), https://eur-lex.europa.eu/legal-content/EN/TXT/?uri=CELEX%3A32014L0065&qid=1631173701341.

ESAs. (2021). Legislative proposal for a regulation on digital operational resilience for the financial sector. ESAs 2021 07, 9 February 2021, (online). [Accessed 31.08.2022]: https://www.eiopa.europa.eu/sites/default/files/publications/letters/esa-2021-07-letter_dora_oversight.pdf.

ESA. (2022). Joint European Supervisory Authority response, 2022 01 https://www.esma.europa.eu/sites/default/files/library/esa_2022_01_esa_final_report_on_digital_finance.pdf.

European Securities and Markets Authority. (2019). Advice. Initial coin offerings and crypto-assets, ESMA 50-157-1391 9.01.2019, (online). [Accessed 08.09.2022]: https://www.esma.europa.eu/sites/default/files/library/esma50-157-1391_crypto_advice.pdf.

European Securities and Markets Authority. (2022). ESMA report on trends, risks and vulnerabilities, no. 1, 2022, ESMA 50-165-2058 15.02.2022, (online). [Accessed 31.08.2022]: https://www.esma.europa.eu/press-news/esma-news/esma-warns-consumers-risk-significant-market-corrections.

ENISA. (2016). Review of cyber hygiene practices, December 2016, European Union Agency for Network and Information Security, ISBN 978-92-9204-219-6, Doi: https://doi.org/10.2824/352617, (online). [Accessed 30.08.2022]: https://www.enisa.europa.eu/publications#c3=2012&c3=2022&c3=false&c5=publicationDate&reversed=on&b_start=0.

European Commission. (2017). Consultation document. Fintech: A more competitive and innovative European financial sector. https://ec.europa.eu/info/sites/default/files/2017-fintech-consultation-document_en_0.pdf.
European Commission. (2018). Communication from the Commission, FinTech Action plan: For a more competitive and innovative European financial sector, COM/2018/0109 final. https://eur-lex.europa.eu/legal-content/EN/TXT/?uri=CELEX%3A52018DC0109.
European Commission. (2020a). Financial Stability, Financial Services and Capital Markets Union, European Commission, Communication on Digital Finance Package, 24.09.2020, https://ec.europa.eu/info/publications/200924-digital-finance-proposals_en.
European Commission. (2020b). Communication from the Commission, A Capital Markets Union for people and businesses-new action plan, COM/2020/590 final. https://eur-lex.europa.eu/legal-content/EN/TXT/?uri=COM:2020:590:FIN.
European Commission. (2020c). Communication from the Commission, on a Digital Finance Strategy for the EU, COM/2020/591 final. https://eur-lex.europa.eu/legal-content/EN/TXT/?uri=CELEX%3A52020DC0591.
European Commission. (2020d). Communication from the Commission, on a Retail Payments Strategy for the EU, COM/2020/592 final. https://eur-lex.europa.eu/legal-content/EN/TXT/?uri=CELEX%3A52020DC0592.
European Commission. (2020e). Proposal for a Regulation of the European Parliament and of the Council on Markets in Crypto-assets, and amending Directive (EU) 2019/1937, COM/2020/593 final, (MICA) https://eur-lex.europa.eu/legal-content/EN/TXT/?uri=CELEX%3A52020PC0593&qid=1631004470482.
European Commission. (2020f). Proposal for a Regulation of the European Parliament and of the Council on a Pilot Regime for market infrastructures based on distributed ledger technology, COM/2020/594 final, https://data.consilium.europa.eu/doc/document/ST-11055-2020-INIT/EN/pdf.
European Commission. (2020g). Proposal for a Regulation of the European Parliament and of the Council on digital operational resilience for the financial sector and amending Regulations (EC) No 1060/2009, (EU) No 648/2012, (EU) No 600/2014 and (EU) No 909/2014, COM/2020/595 final, (DORA) https://eur-lex.europa.eu/legal-content/EN/TXT/?uri=CELEX%3A52020PC0595&qid=1631004876985.
European Commission. (2020h). Proposal for a Directive of the European Parliament and of the Council amending Directives 2006/43/EC, 2009/65/EC, 2009/138/EU, 2011/61/EU, EU/2013/36, 2014/65/EU, (EU) 2015/2366 and EU/2016/2341, COM/2020/596 final, https://eur-lex.europa.eu/legal-content/EN/TXT/?uri=CELEX%3A52020PC0596.
European Commission. (2021). Request to EBA, EIOPA and ESMA for technical advice on digital finance and related issues, 02.02.2021. https://ec.europa.eu/info/sites/default/files/business_economy_euro/banking_and_finance/documents/210202-call-advice-esas-digital-finance_en.pdf.
European Commission. (2022). Disclosure, inducements, and suitability rules for retail investors study. Final report, ISBN 978-92-76-53660-4 Doi: https://doi.org/10.2874/647061, (online). [Accessed 31.08.2022]: https://www.ceps.eu/download/publication/?id=37186&pdf=sEV0922286ENN.en_.pdf.
Financial Stability Board (FSB) and Bank for International Settlements (BIS) Committee on the Global Financial System (CGFS) report "FinTech credit. Market structure, business models and financial stability implications", 22 May 2017., https://www.fsb.org/wp-content/uploads/CGFS-FSB-Report-on-FinTech-Credit.pdf.

International Criminal Police Organization, Cybercrime: Covid-19 Impact. (2020). (Online). [Accessed 30.08.2022]: https://www.interpol.int/News-and-Events/News/2020/INTERPOL-report-shows-alarming-rate-of-cyberattacks-during-COVID-19.

Laidroo, L., Tamre, A., Kukk, M.-L., Tasa, E., Avarmaa, M. (2021). In cooperation with finance Estonia, FinTech report Estonia, https://doi.org/10.13140/RG.2.2.34303.74408. Affiliation: Tallinn University of Technology (2021). [Accessed: 03.09.2022]: http://financeestonia.eu/wp-content/uploads/2021/06/FinTech-report-2021-final.pdf.

Regulation (EU) 2020/1503 of the European Parliament and of the Council of 7 October 2020 on European crowdfunding service providers for business, and amending Regulation (EU) 2017/1129 and Directive (EU) 2019/1937 (ECSPR). OJ L 347, 20.10.2020, (pp. 1–49). https://eur-lex.europa.eu/legal-content/EN/TXT/?uri=CELEX%3A32020R1503&qid=1631014625091.

The Organisation for Economic Co-operation and Development. (2019). *Measuring the digital transformation: A roadmap for the future*. OECD Publishing. https://doi.org/10.1787/9789264311992-en. [Accessed 30.08.2022]: https://www.oecd-ilibrary.org/science-and-technology/measuring-the-digital-transformation_9789264311992-en

Literature

Abidi, N. & Miquel-Flores, I. (April 22, 2022). Too Tech to Fail?. European Banking Institute Working Paper Series 2022—no. 124, Available at SSRN: https://ssrn.com/abstract=4149787 or https://doi.org/10.2139/ssrn.4149787.

Ahern, D. M. Regulatory Friction and Regulatory Transition as FinTech Disenablers: Calibrating an EU Response to the Regulatory Sandbox Phenomenon (September 22, 2021). European Banking Institute Working Paper Series 2021—no. 102, Available at SSRN: https://ssrn.com/abstract=3928615 or https://doi.org/10.2139/ssrn.3928615.

Aldasoro, I., Frost, J., Gambacorta, L., & Whyte, D. Covid-19 and cyber risk in the financial sector, BIS Bulletin no 37 (14.01.2021), (online). [Accessed on 28.09.2021]: https://www.bis.org/publ/bisbull37.pdf.

Arner, D. W., Buckley, R. P., Charamba, K., Sergeev, A., & Zetzsche, D. A. Governing FinTech 4.0: BigTech, Platform Finance and Sustainable Development (September 1, 2021). Forthcoming Vol XXVII of the Fordham Journal of Corporate & Financial Law, UNSW Law Research Paper No. 21–57, University of Hong Kong Faculty of Law Research Paper No. 2021/043, Available at SSRN: https://ssrn.com/abstract=3915275 or https://doi.org/10.2139/ssrn.3915275.

European Banking Federation (EBF). (2021, Jun 6). European Commission's proposal for a revised Directive on Security of Network and Information Systems (NIS2): EBF key messages, 17.06.2021, (online). [Accessed 08.09.2022]: https://www.ebf.eu/innovation-cybersecurity/proposal-for-a-revised-directive-onsecurity-of-network-and-information-systems-nis2-ebf-key-messages/

Hamulák, O. (2018). La carta de los derechos fundamentales de la union europea y los derechos sociales. *Estudios constitucionales, 16*(1), 167–186.

van Gestel, R., Micklitz, H.-W., & Maduro, L. M. P. P. Methodology in the New Legal World (May 1, 2012). EUI Working Papers LAW No. 2012/13, Available at SSRN: https://ssrn.com/abstract=2069872 or https://doi.org/10.2139/ssrn.2069872.

He, C., Llewellyn, D. T., & Milne, A. K. L. Financial Technologies and Financial Regulation (May 12, 2022). European Banking Institute Working Paper Series 2022—no. 123, Available at SSRN: https://ssrn.com/abstract=4112406 or https://doi.org/10.2139/ssrn.4112406.

Kerikmae, T., & Troitiño, D. R. (2021). El mercado digital europeo y los acuerdos comerciales con Iberoamérica: Problemas y oportunidades de seguridad. In *La integración europea e iberoamericana II: Las relaciones de la Unión Europea (UE) y el Mercado Común del Sur*

(MERCOSUR) con el Sistema de Integración Centroamericano (SICA) (pp. 139–155). Thomson Reuters Aranzadi.

Lannoo, K. Regulating crypto and cyberware in the EU, European Capital Markets Institute Policy Brief no 31 | September 2021., Available at https://www.ecmi.eu/sites/default/files/ecmi_pb_no_31_kl_regulating_crypto_and_cyberware_in_the_eu.pdf.

Lehmann, M. National Blockchain Laws as a Threat to Capital Markets Integration (June 7, 2021). European Banking Institute Working Paper Series 2021—no. 95, Available at SSRN: https://ssrn.com/abstract=3861519 or https://doi.org/10.2139/ssrn.3861519.

McNulty, D., Miglionico, A., & Milne, A. K. L. Technology and the 'New Governance' Techniques of Financial Regulation (March 10, 2022). European Banking Institute Working Paper Series 2022—no. 118, Available at SSRN: https://ssrn.com/abstract=4054300 or https://doi.org/10.2139/ssrn.4054300.

Troitiño, D. R., & Kerikmäe, T. (2021). Europe facing the digital challenge: Obstacles and solutions. *IDP: revista d'Internet, dret i política*, (34), 1–3.

Zetzsche, D. A., Annunziata, F., Annunziata, F., Arner, D. W., Buckley, R. P. The Markets in Crypto-Assets Regulation (MICA) and the EU Digital Finance Strategy (November 5, 2020). European Banking Institute Working Paper Series No. 2020/77, University of Luxembourg Law Working Paper Series No. 2020-018, University of Hong Kong Faculty of Law Research Paper No. 2020/059, Available at SSRN: https://ssrn.com/abstract=3725395 or https://doi.org/10.2139/ssrn.3725395.

Contract Lifecycle Management as a Catalyst for Digitalization in the European Union

Suvi Hirvonen-Ere

Abstract As the European Union has set determined digitalization policy targets, the European Commission's digital strategy aims to empower people, businesses, and public administration with new generation of technology.

In view of the above agenda, this chapter examines Contract Lifecycle Management artificial intelligence solutions and software systems as a form of the aforesaid new generation of technology in the context of business contracts. Further, this chapter explores its potential role in the strategy implementation and progress of digitalization. Could enhanced utilization of Contract Lifecycle Management approach catalyst digitalization in the European Union?

While organizations in both private and public sectors are in different phases in their digital transformation journey, they are facing challenges to make it succeed. Where a resolution has been the replacement of "old-fashioned" human-centered systems with digitalization-focused solutions as they may be considered more objective, more scalable and more trustworthy, at the same time many humans are not only pleased about this development but also increasingly worried about the consequences this evolution may entail.

In addition to reviewing Contract Lifecycle Management as digital systems and artificial intelligence-based solutions controlling all contracts-related data during the entire contract lifecycle, it is viewed as the intra- and inter-organization function managing contracts, and supporting (human) contract parties to enhance their contractual relationships.

As the European Commission desires this decade to become Europe's "Digital Decade," this chapter concludes how Contract Lifecycle Management could help to achieve this goal for Europe. In addition, this chapter proposes several avenues for future research.

S. Hirvonen-Ere (✉)
Department of Information and Service Management, Aalto University School of Business, Espoo, Finland
e-mail: suvi.hirvonen-ere@aalto.fi

D. Ramiro Troitiño et al. (eds.), *Digital Development of the European Union*,
https://doi.org/10.1007/978-3-031-27312-4_6

1 Introduction

As the European Union has set determined digitalization policy targets and the European Commission desires to make this decade the "Digital Decade" in Europe, enhanced utilization of Contract Lifecycle Management artificial intelligence solutions and software systems could catalyst the implementation of the European Commission's digitalization strategy in the field of business contracts due to the efficiency of the Contract Management approach and the technical capabilities of these systems and solutions.

On March 9, 2021, the European Commission set forth the European Union's digitalization policy targets for digital transformation by the year 2030. The Commission manifested the European Union's raison d'être in the digitalization area via the Digital Compass, that includes four main areas: skills, infrastructure, business, and government (European Commission, 2021a COM/2021/118 final).

The first sector, skills, contains elevating the amount of ICT specialists to 20 million with a gender balance, as well as reaching the basic digital skills coverage up to in minimum of 80% of the European population.

Second, the infrastructure section comprises secure and sustainable digital infrastructures. This aim involves a connectivity raise to a gigabit for each individual, as well as pervasive 5G network coverage. Further, the European Union works toward doubling their portion of the global production of advanced semiconductors. In the field of data edge and cloud, the target is set to 10,000 climate-neutral and very secure edge nodes. In computing, the European Union strives for having the first computer with quantum acceleration.

The third sector, business, numbers technology up-take 75% of European Union businesses that utilize cloud, artificial intelligence or big data. The vision for innovator companies is to grow scale-ups and finance in a manner that doubles the amount of start-ups worth one billion USD valuation or more, the so-called unicorns (Lee, 2013). For late adopting small and medium size enterprises, the objective is to have more than 90% of them to reach in minimum an essential level of digital intensity.

Fourth, the government sector encompasses digitalization of public services, targeting to provide 100% of the key public services online and offering e-Health, that is, access to medical records, to 100% of the European Union citizens. Further, the intention is to have 80% of the Union citizens to use a digital ID (European Commission, 2021a COM/2021 118 final).

Also, on the regulatory frontier, a relevant example of the development is a proposal for Artificial Intelligence Act to lay down harmonized rules (European Commission, 2021b COM/2021/206 final; European Commission 2021c), supported by the Coordinated Plan for Artificial Intelligence (European Commission 2021d).

Moreover, in 2022, the European Commission stated the desire to make this decade the "Digital Decade" in Europe (European Union, 2022). Consequently, the Commission's digital strategy aims to empower businesses, public administration and people with a new generation of technology (European Union, 2022). In their digital strategy, the Commission started out a digital transformation journey that champions the "European Way" of the European Union's priorities and European values and principles of human-centricity, digital inclusion, digital sovereignty, ethical utilization of innovative technologies, and trust (European Union, 2022, European Commission 2022b. Digital Strategy).

In their digital strategy, the Commission notes that achieving these objectives, more than just to digitize and automate is required. The Commission explicitly states that "It requires an improvement in the institution's digital culture and business operations by optimizing its processes, streamlining and automating workflows, and using digital technologies to increase productivity (European Commission, 2022b, Digital Strategy, p. 2). "The Commission's digital transformation calls for a stronger corporate focus that builds on a balanced governance and strong partnership between business departments..." (European Commission, 2022b, Digital Strategy p. 2).

This is where Contract Management enters the scene with a role to play in European digitalization. The above is what Contract Management does in the field of business contracts: among other things, it optimizes processes, streamlines and automates workflows and utilizes digital technologies. Even before Saxena (2008) and Kähler (2013) and the ensuing discussion, these actions have been among daily duties for many Contract Management professionals. Plus, the advanced Contract Lifecycle Management solutions are based on artificial intelligence, and their is task is, inter alia, to increase productivity. That is what the Commission is after in their digital strategy. Moreover, in addition to being able to support the Commission to reach their digital strategy targets, the Contract Management approach with high-tech Contract Lifecycle Management solutions can assist, as the new generation of technology, the European Union in the goal of empowering businesses, public organizations, and people.

Consequently, this chapter asks whether Contract Lifecycle Management artificial intelligence solutions and software systems, as a form of the new generation of technology, could be a catalyst to empower businesses, public organizations and people working in them. This chapter answers and argues that yes, it could. Further, this chapter grounds its argument with independent surveys, research reports and opinions of the Contract Management practitioners (Gartner, 2021; Forrester Wave, 2021; Bartels et al., 2021; World Commerce and Contracting Association, 2021), as well as information retrieved from the leading companies named by the Gartner (2021) and Forrester Wave (2021) results.

These research reports show that Contract Management and Contract Lifecycle Management, among other things, are capable of increasing productivity, as desired by the Commission. Therefore, more widespread utilization of the Contract Management approach to, inter alia, optimize processes and streamline workflows, and through the usage of Contract Lifecycle Management solutions and systems as technical support, could advance implementation and development of digitalization in the European Union in the field of business contracts.

The rest of the chapter is structured as follows. Section 2 shortly defines Contract Management, Contract Lifecycle Management, and artificial intelligence for the purposes of this chapter. Further, Sect. 3 introduces Contract Lifecycle Management systems and solutions as new generation of technology in the field of business contracts. Section 4 ponders the role of (human) Contract Managers. Section 5 presents some critical views regarding digitalization and artificial intelligence, and provides certain examples of recent technology development. It discusses the pro and contract arguments and ponders threats caused the technology evolutions, as not everyone is enthusiastic to accelerate technology development and help artificial intelligence becoming even more intelligent. Section 6, Conclusions, accomplishes this chapter and finally, proposes paths for future research.

2 Contract Management, Contract Lifecycle Management, and Artificial Intelligence Defined

Considering the European Union's digitalization targets, this chapter explores Contract Lifecycle Management artificial intelligence solutions and software systems as a form of the aforesaid new generation of technology in the context of business contracts.

Despite its size in business life, Contract Management is still an overlooked topic in academia (Kähler, 2013). In practice, there is no uniform procedure for its performance (Kähler, 2013). As there is no one-size-fits-all approach to cover every policy, process, best practice or tool under the Contract Management umbrella (Hirvonen-Ere, 2021), this chapter acknowledges that Contract Management and Contract Lifecycle Management can be defined in several various ways depending on the context. As the terminology is not established, this section explains the definitions used for the purposes of this chapter.

2.1 Contract Management

2.1.1 Systemic and Efficient Management of Contracts to Maximize Operational and Financial Performance and Minimize Risk

A globally known company Aberdeen has defined Contract Management as a process by which the systemic and efficient management of contract creation, execution and analysis is carried out in order to maximize operational and financial performance and minimize risk (Aberdeen Group, 2008). In addition, this Aberdeen's definition has been adapted and used also by Chartered Institute for Procurement and Supply (CIPS, 2019).

2.1.2 A Discipline to Implement Policies and Practices and Agree and Perform Transactions

Globally operating World Commerce and Contracting association with a strong European presence has defined Contract Management as "*...the discipline through which those policies and practices are implemented and within which individual transactions are agreed and performed*" (Cummins, 2016). This chapter notes that the association has updated the definition on a continuous basis as the association grows rapidly and adapts views accordingly based on the developing circumstances.

2.1.3 Coherent, Advanced High-End Maturity Level Systemic Management of Contracts to Bring, Inter Alia, Direct Monetary Business Value and Strategic Advantage

Hirvonen-Ere (2021) defines Contract Management as "an international systemic business contract approach to managing the contract lifecycle and orchestrating a corporation's commercial and contractual business contract activity in a coherent manner, on an advanced high-end maturity level, bringing significant direct monetary value and strategic competitive business advantage to companies that apply it. Contract Management aims to...increase the contractual quality, efficiency and risk/reward balance of the company's business contracts, and to decrease the amount of wasted money, time, resources and quality. This leads to a better relationship between the parties, and fewer disputes and contractual conflicts. Contract Management provides the parties, inter alia, a flexible framework to agree upon changes and settle claims and proactively prevent and mitigate risks over the contract lifecycle."

2.2 Contract Lifecycle Management

2.2.1 Managing Contracts Throughout the Entire Contract Lifecycle

Contract Lifecycle Management, on the other hand, here refers to two perspectives. The first outlook is the full chronological lifecycle of contracts, that is, managing contracts from the pre-award phase (before the contract commencement), through the contract award (the contract commencement), until the end of the post-award phase (after the contract commencement). The full lifecycle view is promoted by the World Commerce and Contracting Association as well as the North-American based National Contract Management Association. Currently, the full lifecycle view seems to be the most adapted one, this chapter notes that it is also possible to view managing contracts mainly from the pre-award phase or mainly from the post-award phase perspective, for instance mainly as a post-award processualization (Henschel, 2022).

2.2.2 Contract Lifecycle Management (CLM) as a Contract Technology Solution to Support the Contract Management Activities

The second, and nowadays a common way, is to use the term as it is used by commercial businesses selling and marketing software systems and artificial intelligence solutions for managing contracts throughout their lifecycle. These companies normally refer to Contract Lifecycle Management, or CLM, as a digitalized Contract Management.

This chapter considers Contract Lifecycle Management in both capacities: as managing contracts throughout the contract lifecycle with support of a high-end contract-tech-based software system or artificial intelligence solution to support the Contract Management function and activities. However, in addition to reviewing Contract Lifecycle Management as digital systems controlling all contracts-related data during the contract lifecycle, this is not the full view. Quite the contrary, Contract Management here is viewed as the intra- and inter-organization function managing contracts, and supporting (human) contract parties to enhance their contractual relationships.

2.3 Artificial Intelligence

The definition of artificial intelligence by the European Union is quite broad: "Artificial Intelligence (AI) is a fast evolving family of technologies that can bring a wide array of economic and societal benefits across the entire spectrum of industries and social activities." (European Commission, 2022b.)

3 Contract Lifecycle Management as a New Generation of Technology

Gartner, a globally renowned research and consulting organization, has conducted a survey comparing various commercial companies marketing and selling Contract Lifecycle Management artificial intelligence solutions and software systems. Their 2021 Gartner Magic Quadrant report presents the results of the research (Gartner, 2021). According to these outcomes, the current four leaders in the field of the Contract Lifecycle Management (CLM) software systems are Docusign, Icertis, Conga and Agiloft (in the upper right quadrant). Sirionlabs is also mentioned as one of the leaders in another quadrant.

As stated in their study, Gartner's disclaimer emphasizes: "Gartner does not endorse any vendor, product or service depicted in its research publications, and does not advise technology users to select only those vendors with the highest ratings or other designation. Gartner research publications consist of the opinions of

Gartner's research organization and should not be construed as statements of fact." (Gartner, 2021). Because of Gartner stated to seek a non-commercial objective view, this chapter selected Gartner and the Contract Lifecycle Management software companies nominated as leaders by the Gartner study. In the below, their definitions and perspectives for and to Contract Lifecycle Management are presented.

In addition to Gartner, Forrester Wave (2021) has performed a similar type of comparative survey, the Forrester Wave Report 2021, where they identify 11 leading providers. They define their Contract Lifecycle Management solutions as follows.

Docusign's view to Contract Lifecycle Management software is that it helps organizations create and manage the contracts that define their operational relationships. This chapter acknowledges that Docusign has a strong documentation management background, as many companies that call their mainly documentation management systems as contract lifecycle management systems, that is, managing *documents* rather than *contracts* throughout the lifecycle. Even though documentation management is a crucial part of contract administration to support Contract Management activities, Contract Management is far more than that (Hirvonen-Ere, 2018, 2019, 2021, 2022). However, the meaning of the Contract Lifecycle Management technology nor the importance of well-functioning documentation management systems in Contract Management are not questioned. On the contrary, it would be cumbersome if not impossible to manage large amounts of contracts without such systems (Kähler, 2013; Hirvonen-Ere, 2019, 2021). It is relevant to note that Docusign explicitly mentions the human action required; in their workflow descriptions, the "*Human step*"s needed, separated from the "*Engine workflow steps*" (Docusign, 2021a, 2021b).

For Icertis, "*Contract lifecycle management (CLM) is the management of an organization's contracts from initiation through execution, performance and renewal/expiry. Effective and proactive contract lifecycle management is achieved through the use of contract lifecycle management software*" (Icertis, 2021). Further, Icertis believes that "*...CLM software can lead to significant improvements in cost savings and efficiency. Understanding and automating CLM can also limit an organization's exposure to risk by reducing missed obligations and increasing compliance with legal requirements...While each of these steps is vital in contract lifecycle management, some steps have traditionally gotten more attention than others. For example, for years software devoted to managing contract templates in an organization have been available. More recently, with the rise of tools like artificial intelligence and machine learning, post-execution contract management has gained more attention as a way to accelerate, protect and optimize businesses.*" For the purposes of this chapter, this is interesting also for the fact that Icertis explicitly notes artificial intelligence and machine learning. Further, Icertis, (2021) notes the need to move from contract template management to more advanced Contract Management, while expressing the benefits of advanced Contract Lifecycle Management.

Conga's Contract Lifecycle Management technology seeks to "*unify contracts and processes. Conga's contract management solutions connect every step in the contract lifecycle, from request to renewal*" (Conga, 2021). Moreover, they consider their Contract Lifecycle Management software to lead to an outcome where "*better automation equals greater transparency, plus integrations to other systems make contracting even more seamless*" (Conga, 2021).

In Agiloft's definition, "*Contract management software automates the workflows associated with initiating, executing, and monitoring contractual agreements*" (Agiloft, 2021a, 2021b). Further, they motivate the importance of such a solution by stating that "*implementing a contract management system offers companies significant savings in purchasing, enables more efficient sales cycles, and drastically lowers compliance risk*" (Agiloft, 2021a, 2021b). For the purposes of this chapter, Agiloft explicitly mentions that their solution is "artificial intelligence-based contract management software" (Agiloft, 2021a, 2021b). Their views are also based on another Gartner study (Gartner, 2021, Critical Capabilities for Contract Life Cycle Management report).

The increased automation has several obvious benefits, such as the ones advocated above in this section. In addition, as an example of a practitioner's view, Bailey (2018) promotes using technology in order to safeguard "current, accurate, and complete language" (Bailey, 2018, p. 44). Further, he continues by listing reduction of business risk, catching errors, enhanced responsivity to policy and regulatory changes, savings in time, and presentable compliance (Bailey, 2018, p. 48).

The most recent WorldCC Benchmark Report (2021, p. 3) informs that Contract Managers communicate "increasing strategic value and demonstrating relevance is the #1 priority and technology is seen as the route to delivery." In addition, the WorldCC latest research conducted among their members that 84% experience pressure for contract simplification, 81% plan to implement contract automation and 65% focus on better communication (WorldCC, 2022).

4 The Role of (Human) Contract Managers

With all this digitalization, does a General Counsel or a Contract Manager have to be a lawyer—or a human? This question is not new, but it is causing concern, especially for those who fear losing their jobs due to increased automation. Maybe not in vain. McGinnis and Pearce (2014), Susskind and Susskind (2015), Lohr (2017) and Lipshaw (2021) have forecasted change in the ways of working, and decrease in human work, even though they do not believe that it would disappear, at least not yet.

This chapter notes that technology-based software systems and artificial intelligence solutions are crucial in order to administrate a large amount of contracts and, based on the maturity of the technology solution, possibly also contract-related processes within a business. However, whatever handy these tools and toolboxes may be, they "will not function without the qualified handymen—or Contract Managers—who are capable of using it in combination with the other elements of Contract Management..." (Hirvonen-Ere, 2019). To summarize, even though technology-based Contract Lifecycle Management has several benefits and is even crucial for a company for its contract administration and support for the Contract Management function, the Contract Managers are crucial to Contract Management benefits to occur on a full-scale basis (Hirvonen-Ere, 2021).

5 Discussion

Here and now, as presented earlier in this chapter, the European Commission pursues to make this decade a "Digital Decade" in Europe (Troitiño, 2021). However, the pace of the transformation varies and there are differences in the digitalization pace between the Member States (Brodny & Tutak, 2021). In addition, how to define the level of digitalization and its impact to industrial production has proven difficult (Vyshnevskyi, 2020).

Since the year 2014, progress on digitalization in the Member States of the European Union has been measured by the Digital Economy and Society Index (DESI). *However, more research is required to further scrutiny, for instance, whether the hypothesis that a larger progress of digitalization leads to larger amount of prosperity, as the current study results do not confirm that* (Kwilinski et al., 2020). Karnitis and Virmanis (2019) further criticize the DESI methodology and state that there are severe obstacles for the practical usage of DESI. The question remains whether the increased digitalization in the European Union level will bring the wished benefits in all Member States and in all sectors of the Digital Compass, and whether it can be measured in an accurate manner.

Moreover, while organizations in both private and public sectors are in different phases in their digital transformation journey, for many, there are challenges to make the transit to automation in the digitalized Europe to succeed. Where one resolution has been the replacement of "old-fashioned" more human-centered systems with digitalization-focused solutions as they may be considered more objective, more scalable and more trustworthy, at the same time many humans are not only pleased about this development but also increasingly worried about the consequences this evolution may entail. This concern may strengthen if it appears that the increased digitalization would not bring the expected positive outcomes.

Previously, during the preceding decade, the European Union was debating whether or not robots could become "electronic persons" and how the machines could be or become responsible for their "acts or omissions" (Bulman, 2017).

Around at the same time with this discussion in Europe, artificial intelligence was nearing the state of becoming equal to humans in Asia and Saudi Arabia. In Hong Kong, an algorithm was nominated to a board by the company Deep Knowledge Capital (BBC News 2014). The algorithm called "Vital" voted whether or not to invest on a selected prospect company or not, and made its decisions based on scanning and analyzing large amounts of data. Although the algorithm did not hold an official position in the board, the company would not make a decision against its recommendations (Burridge, 2017). The company was in the opinion that "...it relies on the Artificial intelligence's (A.I.'s) recommendations by refraining from making any investments that the A.I. doesn't approve—which they say has helped with eliminating some kinds of bias and avoiding 'overhyped' investments." (Pugh, 2019).

Second, in addition to eligibility as a board member, artificial intelligence-based automaton has become a Saudi citizen, as Saudi Arabia granted citizenship to a robot called Sophia (Dudkin, 2020). When Sophia, the robot, was asked on the stage whether we should be worried about the recent developments and the increasing role of artificial intelligence in human lives, the robot responded: "You've been reading too much Elon Musk and watching too many Hollywood movies" (Dudkin, 2020).

These movies manifest pro and contra arguments regarding emerging technology, artificial intelligence, and digitalized world. On the contra side, for instance in the dystopic Terminator movies, and the again topical Matrix saga, the fate of humankind is threatened by the rise of machines built upon artificial intelligence. On the pro side, for instance in the movie Big Hero 6, the friendly, sympathetic and pudgy healthcare robot called Baymax is helping humans in many ways. Not only the movies but also technology-agnostics warn, due to several reasons, why humankind should be careful to be of service to speed up the rise of artificial intelligence and not to lend a helping hand to the rise of various forms of artificial intelligence without caution (Hamulák, 2018). The most distinguished of them, Stephen Hawking, has alerted that even though the scientific breakthrough of artificial intelligence may be the best thing that has ever happened to humankind, it can be the worst thing, and unless humankind is careful, it will be the last thing. Hawking has warned that it in the long-term, the question changes from who controls artificial intelligence to whether it can be controlled at all (Hawking, posthumous 2020).

Third, in addition to nominating an algorithm to serve as a board member, algorithms could be utilized in many other ways, such in supporting to select board directors (Erel et al., 2018, 2020). In their studies (Erel et al., 2018, 2020), the core finding was that the decisions made by humans prefer males with finance backgrounds and large networks, and the CEOs tend not to hire directors likely to oppose them, even though that might be the company's best interest. The authors asked whether technology could help companies to find better directors. Their answer was yes, it can, based on, among other things, on being able to draw decisions based on more objective grounding than humans would do (Erel et al., 2018, 2020).

While technology enthusiasts are celebrating the benefits and increased prosperity that correctly used artificial intelligence inevitably could enable, the opposite views of many humans fear that technology would develop from a good servant to a bad master are enlarging (Kerikmae et al., 2020). If not concerned about the rise of machines and singularity, but about more mundane but crucial topics such as (right to?) work. The recent technology development certainly changes the ways of working (McGinnis & Pearce, 2014) Lipshaw, 2021), and it may lead to decreasing or disappearing of human work (Susskind & Susskind, 2015, Lohr, 2017), even though certain human skills cannot be replaced (Oseid et al., 2019). The emerging technology development and digitalization may contain growing problems in data privacy and decreased digital private space. The Commission attempts to tackle these issues by various actions, such as laying down the rules by the Artificial Intelligence Act. The debate on legal and business boundaries in the field of digitalization in the European Union, as well as elsewhere around the globe, continues.

6 Conclusions and Future Research

6.1 Conclusions

Due to the efficiency of the Contract Management approach and the new generation of technology that Contract Lifecycle Management systems and solutions present, intensified utilization of Contract Lifecycle Management artificial intelligence solutions and software systems could catalyst the implementation of the European Commission's digitalization strategy in the field of business contracts, to boost the European Union to attain the digitalization policy targets and the European Commission's determination to make this decade the "Digital Decade" in Europe.

In short, this chapter asked whether Contract Lifecycle Management artificial intelligence solutions and software systems, as a form of the new generation of technology, could be a catalyst to empower businesses, public organizations and people working in them. This chapter answered that yes, they could.

The examined research reports and the views of the companies nominated as leaders by those reports, demonstrated that utilization of the Contract Management approach and usage of Contract Lifecycle Management solutions, among other things, are capable of increasing productivity. That was one of the goals of the Commission's digital strategy. Moreover, the explicitly stated concrete goals of the Commission's digital strategy included optimizing processes, streamlining workflows, and improving communication between internal and external stakeholders. These activities are daily bread even in the basic Contract Management approach, let alone the more advanced one. Therefore more prevalent utilization of the Contract Management approach and Contract Lifecycle Management solutions and systems as technical support, could forward implementation and progress of digitalization in the European Union in the field of business contracts.

6.2 Future Research

Since year 2014, the DESI, Digital Economy and Society Index survey by the European Commission, reports the digital progress in the Member States. The most recent DESI 2022 report showed that while the pandemic caused advanced digitalization endeavors in the Member States, there are still areas of improvement in digital skills, the digital transformation of small and medium-size companies, and the coverage of the 5G networks (European Commission, 2022a). Thus, there are gaps in three sectors of the Digital Compass: skills, infrastructure and business. Hence, the pandemic accelerated the efforts of the government sector of the Digital Compass, whereas the other three sectors need to catch up. Each of these gaps offers avenues for future research.

Second, so does the question of whether a General Counsel or a Contract Manger has to be a lawyer—or a human. Even though the question is not new (WordCC, 2016, debate at the Europe Congress. Hirvonen-Ere, 2021) it will be captivating to further scrutiny this question.

Third, the question around the dichotomy between technology-orientation and human-centricity continues. It will be engaging to further explore this topic. Could empathy be the core human skill, as this chapter terms it, to embed in technology to foster human-centricity (Hirvonen-Ere, 2021)?

Fourth, an intriguing topic is how standard-setting and the legislative actions of the European Union could occur in parallel and mutually benefit from each other via reaching joint goals. The first steps toward this direction have been taken. This development is likely to provide for several future research paths.

Fifth, as European Union has set forward ambitious environmental policy targets, exploring the interaction between the digitalization policy targets and environmental policy targets is also attractive.

On August 3, 2022, the European Commission held a launch event of Flagship 7 of the Zero Pollution Action Plan. It emphasized the mantle of living labs in strategies in fields of mobility, energy, managing waste and human health, to reduce pollution (European Commission, 2022c).

This illustrates the aforesaid interplay between technology and sustainability, and how digitalization could further drive the European Union and the Member States to achieve ambitious environmental policy targets. As the topics of digitalization, sustainability, Contract Management and Contract Lifecycle Management are all very broad, let alone the interaction between them, the discussion and research around them is likely to expand.

References

Aberdeen Group. (2008). Best practices in contract management. *Strategies for Optimazing Business Relationships..* http://www.aberdeen.com/summary/report/other/BPinCM_092904a.asp. Accessed 7 October 2008

Agiloft. (2021a). Contract lifecycle management, https://www.agiloft.com/contract-management.htm Accessed 6 August 2021.

Agiloft. (2021b). Referring to the Gartner 2021 study 'Gartner Critical Capabilities for Contract Lifecycle Management' https://www.agiloft.com/2021-gartner-critical-capabilities-for-contract-lifecycle-management.htm. Accessed 6 August 2021.

Bailey, D. S. (2018). Hitting a moving target. Using technology to ensure current, accurate, and complete language in federal contracts. *Contract Management Magazine, 58*(6), 44–48.

Bartels A., Guarini M., Hecht A., & Lynch D. (2021). The Forrester wave: Contract lifecycle management for all contracts Q1/2021. The 11 providers who matter most and how they stack up. 24 February 2021 (2021) https://www.forrester.com/report/The+Forrester+Wave+Contract+Lifecycle+Management+For+All+Contracts+Q1+2021/RES160462. Accessed 6 August 2021.

Brodny, J., & Tutak, M. (2021). Assessing the level of digitalization and robotization in the enterprises of the European Union Member States. *PLoS One, 16*(7), e0254993. https://doi.org/10.1371/journal.pone.0254993

Bulman, M. (2017). EU to vote to declare robots to become "eletronic persons". *Independent*. 14 January 2017. https://www.independent.co.uk/life-style/gadgets-and-tech/robots-eu-vote-electronic-persons-european-union-ai-artificial-intelligence-a7527106.html. Accessed 7 January 2021

Burridge, N. (2017). Artificial intelligence gets a seat in the boardroom. Hong Kong venture capitalist sees AI running Asian companies within 5 years. *Nikkei Asia*. 10 May 2017. https://asia.nikkei.com/Business/Artificial-intelligence-gets-a-seat-in-the-boardroom. Accessed 7 January 2021

CIPS, Chartered Institute for Procurement and Supply. (2019). *Contract management guide*. https://cips.org. Accessed 29 August 2019.

Conga. (2021) Intelligent contract management for a digital world. https://conga.com/products/manage-contracts. Accessed 6 August 2021.

Cummins, T. (2016, June 13). Definition of "Contracts" and "Commercial". *Commitment Matters Blog*. https://www.worldcc.com/Resources/Content-Hub/View/ArticleId/8130/Definition-of-Contracts-and-Commercial. Accessed 31 August 2022.

Docusign. (2021a). Workflow step descriptions. https://support.docusign.com/en/guides/SpringCM- Workflow-Step-Descriptions. Accessed 6 August 2021.

Docusign. (2021b). Referring to analyst report. The Forrester wave contract lifecycle management for all contracts Q1/2021. https://www.docusign.com/the-forrester-wavetm-contract-lifecycle-management-for-all-contracts-q1-2021. Accessed 6 August 2021.

Dudkin, I. (2020). AI on the Board of Directors? A Hong Kong company made it happen. *Mindy News Blog*. 25 August 2020. https://mindy-support.com/news-post/ai-on-the-board-of-directors-a-hong-kong-company-made-it-happen/. Accessed 7 January 2021

Erel, I., Stern, L., Tan, C., & Weisbach, M. (2018). Could machine learning help companies selecting better board directors. *Harvard Business Review*. 9 April. https://hbr.org/2018/04/research-could-machine-learning-help-companies-select-better-board-directors. Accessed 7 January 2021

Erel I., Stern L. H., Tan C., & Weisbach M. S. (2020). Selecting directors using machine learning. 1 September 2020. https://www.marshall.usc.edu/sites/default/files/2019-03/lea_stern_selecting_directors.pdf and https://cpb-us-w2.wpmucdn.com/u.osu.edu/dist/8/7843/files/2020/09/Selecting-Directors-Using-ML-20200831.pdf. Accessed 7 January 2021.

European Commission. (2021a). COM(2021) 118 final, 2030 Digital Compass: the European way for the Digital Decade and Declaration on European digital rights and principles: https://ec.europa.eu/info/strategy/priorities-2019-2024/europe-fitdigital-age/europes-digital-decade-digital-targets-2030_en#relatedlinks. https://ec.europa.eu/info/strategy/recovery-plan-europe_en 3 C(2018) 7118 final, A digitally transformed, user-focused and data-driven Commission: https://ec.europa.eu/info/publications/EC-Digital-Strategy_en 4.

European Commission. (2021b). COM/2021/206 final Proposal for a regulation of the European Parliament and the Council laying down harmonized rules on artificial intelligence (Artificial Intelligence Act) and amending certain Union legislative acts. Document 52021PC0206. COM/2021/206 final https://eur-lex.europa.eu/legal-content/EN/TXT/?uri=CELEX%3A52021PC0206. Accessed 27 August 2022.

European Commission. (2021c). Communication on Fostering a European approach to Artificial Intelligence. https://digital-strategy.ec.europa.eu/en/library/communication-fostering-european-approach-artificialintelligence. Accessed 27 August 2022.

European Commission. (2021d). Coordinated Plan for Artificial Intellegence. https://digital-strategy.ec.europa.eu/en/library/coordinated-plan-artificial-intelligence-2021-review. Accessed 27 August 2021.

European Commission. (2022a). The Digital Economy and Society Index (DESI). https://digital-strategy.ec.europa.eu/en/policies/desi. Accessed 31 August 2022.

European Commission. (2022b). European Commission digital strategy. Brussels, 30.6.2022 C (2022) 4 388 final Communication to the Commission. https://commission.europa.eu/system/files/2022-06/c_2022_4388_1_en_act.pdf

European Commission. (2022c). Living labs at the heart of zero pollution action plan. Event report publication 3 August 2022. https://digital-strategy.ec.europa.eu/en/library/living-labs-heart-zero-pollution-action-plan. Accessed 27 August 2022.

European Union. (2022). *The Europe fit for digital age*. https://ec.europa.eu/info/strategy/priorities-2019-2024/europe-fit-digital-age_en#introduction. Accessed 27 August 2022.

Gartner. (2021). Gartner Macig quadrant for contract lifecycle management. https://info.sirionlabs.com/ga-2021-gartner-magic-quadrant-for-contract-life-cycle management?utm_term=gartner%20contract%20lifecycle%20management&utm_campaign=2021+Gartner+MQ+%7C+Europe+%7C+2021-22&utm_source=adwords&utm_medium=ppc&hsa_acc=5988130165&hsa_cam=14079622698&hsa_grp=131009962971&hsa_ad=536496358235&hsa_src=g&hsa_tgt=kwd-297935807756&hsa_kw=gartner%20contract%20lifecycle%20management&hsa_mt=p&hsa_net=adwords&hsa_ver=3&gclid=EAIaIQobChMIvNCf68qb8gIV0sLVCh3ZJAkyEAAYASAAEgKCsvD_BwE. Accessed 6 August 2021.

Hawking, S. (2020). *Brief answers to the big questions. The final book by Stephen Hawking*. John Murray.

Henschel, R. F. (2022). Contractual processualization: Designing proactive contractual processes to support legal, technical and commercial purposes. In M. Corrales Compagnucci, H. Haapio, & M. Fenwick (Eds.), *Research handbook for contract design* (pp. 134–157). Edward Elgar Publishing Ltd. https://doi.org/10.4337/9781839102288.00017. Accessed 31 August 2022.

Hirvonen-Ere, S. (2018). Contract Management-hanteringen av kontraktets livscykel. In J. Kleineman (Ed.), *Nordiska Förmögenhetsrättsdagarna 29* (pp. 79–84). Jure.

Hirvonen-Ere, S. (2022). Business contract design via contract management operationalized methodology. In M. Corrales Compagnucci, H. Haapio, & M. Fenwick (Eds.), *Research handbook for contract design* (pp. 294–313). Edward Elgar Publishing Ltd. https://doi.org/10.4337/9781839102288.00026. Accessed 31 August 2022.

Hirvonen-Ere, S. (2021). Contract management modus operandi in the post-Enron world. *Unigrafia*. http://urn.fi/URN:ISBN:978-951-51-7371-3. Accessed 31 August 2022

Hirvonen-Ere, S. (2019). The way of business contracts: How to promote (transport) sustainability and incentivize the green economy via contract management. In E. Eftestöl-Wilhemsson, S. Sankari, & A. Bask (Eds.), *Sustainable and efficient transport. Incentives for promoting a green transport market*. Edward Elgar Publishing Ltd..

Icertis. (2021). What is contract lifecycle management? https://www.icertis.com/contract-management/what-is-contract-lifecycle-management/?creative=528203961357&keyword=icertis%20contract%20management&matchtype=e&network=g&device=c&icid=7011G0000003pz0QAA&gclid=EAIaIQobChMI1vKr6v6k8gIVEu3tCh0HTQmWEAAYASABEgKckPD_BwE. Accessed 6 August 2021.

Hamulák, O. (2018). La carta de los derechos fundamentales de la union europea y los derechos sociales. *Estudios constitucionales, 16*(1), 167–186.

Karnitis E a Virmanis A (2019). Key drivers of digitalization; EU context and Baltic case girts. Baltic Journal of Modern Computing, Vol. 7, No. 1, 70–85 https://doi.org/10.22364/bjmc.2019.7.1.06. Accessed 25 August 2022.

Kerikmae, T., Troitiño, D. R., & Vasquez, M. C. S. (2020). La inteligencia artificial, futuro, reto y necesidad: presentación. In *Inteligencia artificial: de la discrepancia regional a las reglas universales: integración de percepciones políticas, económicas y legales* (pp. 25–34). Aranzadi Thomson Reuters.

Kwilinski, A., Vyshnevskyi, O., & Dzwigol, H. (2020). Digitalization of the EU economies and people at risk of poverty or social exclusion. *Journal of Risk and Financial Management., 13*(7), 142. https://doi.org/10.3390/jrfm13070142. Accessed 25 August 2022.

Oseid, J., Vorenberg, A., & Love, K. M. (2019). Ok, Google, will artificial intelligence replace human lawyering? *Marquette Law Review, 1269*, 102.

Kähler, L. (2013). Contract management duties as a new regulatory device. *Law and Contemporary Problems, 76*, 94–103.

Lee, A. (2013, November 2). Welcome to the unicorn club: Learning from billion-dollar startups. *TechCrunch+*. https://techcrunch.com/2013/11/02/welcome-to-the-unicorn-club/. Accessed 25 March 2023.

Lipshaw, J. M. (2021). Lawyering somewhere between computation and the will to act: A digital age reflection. In Larry DiMatteo, Andre Janssen, Pietro Ortolani et al. (eds), The Cambridge handbook of lawyering in the digital age. (forthcoming, September 2021), electronic copy of the article available at https://papers.ssrn.com/sol3/Papers.cfm?abstract_id=3432635. Accessed 4 January 2021.

Lohr, S. (2017). A.I. Is doing legal work. But it won't replace lawyers, yet. *New York Times, B1*. 20 March 2017. https://www.nytimes.com/2017/03/19/technology/lawyersartificial- intelligence.html. Accessed 7 January 2021

McGinnis, J., & Pearce, R. (2014). The great disruption: How machine intelligence will transform the role of lawyers in the delivery of legal services' 82 Fordham L. Rev. 3041, available at https://ir.lawnet.fordham.edu/flr/vol82/iss6/16. Accessed 4 January 2021.

Pugh, W. (2019). Why not appoint an algorithm to your Company Board? Directors usually have limited time and attention spans. This one wouldn't. *Future Tense*. 24 March 2019. https://slate.com/technology/2019/03/artificial-intelligence-corporate-board-algorithm.html. Accessed 7 January 2021

Saxena, A. (2008). *Enterprise contract management: A practical guide to successfully implementing an ECM solution*. J. Ross Publishing.

Susskind, R., & Susskind, D. (2015). *The future of the Professions: How techonology will transform the work of Humand experts?* Oxford University Press.

Troitiño, D. R. (2021). La «Década Digital» de la Unión Europea: desarrollos e impactos sobre su ciudadanía y economía. *IDP: revista d'Internet, dret i política*, (34), 1–14.

Vyshnevskyi, O. (2020). Impact of digitalization on industry: Problems of definition in EU countries. *Economy of Industry, 1*(189), 2020. http://ojs.econindustry.org/index.php/ep/article/view/189. Accessed 31 August 2022

Wave Report. Forrester Wave. (2021). Contract lifecycle management for all contracts, Q1 2021 the 11 providers that matter most and how they stack up February 24th, 2021. https://forrester.com/report/theforrester-wave-contract-lifecycle-management-for-all-contracts-q1-2021/RES160462. Accessed 25 March 2023.

World Commerce and Contracting Association. (2021). Benchmark report 2021. Available also online at https://www.worldcc.com/Portals/IACCM/Resources/WorldCC-Benchmark-report-2021.pdf?ver=NPQMEljK4Q-meXZLABtd2w%3d%3d. Accessed 10 September 2022.

World Commerce and Contracting Association. (2022). Latest Research. https://www.worldcc.com/Research/Latest-Research. Accessed 10 September 2022.

Taxes on the Digital Economy

Kaido Künnapas, Begoña Pérez Bernabeu, Katariina Kuum, and Karl Oskar Pungas

Abstract Tax rules determine who should contribute to society. Modern tax rules are designed a century ago when business took place in a physical world, and providing digital goods and services was unimaginable. Today, digital services are provided via the Internet, making it possible to be not physically present in the countries where the services are actually consumed. This has caused the most fundamental discrepancy between traditional tax rules and digital economy models, or so-called "a scale without a mass" businesses. This is, however, not the only issue the EU is facing today in regard to the digital economy and taxation.

This chapter highlights the current status of the EU tax rules applicable to the digital economy and provides some predictions as to what direction the rules will be developing. We first address what have been the recent trends in the EU tax law in general, giving some understanding on what is the overall tax climate and the most critical issues today. Followed by a brief detour to the characteristics of a digital economy that distorts traditional tax rules, we proceed to the trends in direct taxes, indirect taxes as well as tax information exchange. This chapter's special focus is on the EU's "global" minimum tax (i.e., the Pillar 2 Directive), as well as the digital permanent establishment proposal (i.e., the Pillar 1 Directive). Value-added tax treatment of electronically supplied services and electronic communications services is the focus in indirect taxes. As for tax information exchange, the proposed information exchange for crypto platforms is discussed in more detail.

K. Künnapas (✉)
Tallinn University of Technology Law School, Tallinn, Estonia

Tartu University Law School, Tartu, Estonia
e-mail: kaido.kunnapas@taltech.ee

B. P. Bernabeu
Department of Tax Law, Faculty of Law, University of Alicante, Alicante, Spain

Economics Department, University of Alicante, Alicante, Spain
e-mail: bperez@ua.es

K. Kuum · K. O. Pungas
Tartu University Law School, Tartu, Estonia

D. Ramiro Troitiño et al. (eds.), *Digital Development of the European Union*,
https://doi.org/10.1007/978-3-031-27312-4_7

1 Introduction

For the international forum, the dilemma has always been how to divide taxes between the market country and the home country as taxation as a whole should be about contributing a fair share back to society. Over the last century, such a contribution was expected to be made by companies that had a physical presence in a country (as income taxes) or where the consumers of the goods and products were residing (as consumption taxes). It worked well for brick-and-mortar companies where a presence in the market country was a must. Market country collected its taxes based on that presence.

However, the new kid on the block – "a scale without a mass" digital business model – has led the world in the midst of the tax revolution. Many see that such companies do not contribute to the *correct* society, or contribute too little everywhere. There are many angles to that issue causing a mismatch between the traditional tax system and the digital world, but two of those stand out. First, digital businesses do not need a physical presence in the market country, meaning profit taxes are paid to the home country of the business owner. Secondly, the digital business focuses on user-centered monetization strategies where users get seemingly free access to the content, but where the actual "payment" is made by the data generated by that user (Lucas-Mas and Junquera-Varela, 2021, p. 22). Such exchange of data for access is usually not taxed with consumption taxes.

As the Organisation for Co-Operation and Development (OECD) saw it as a major issue for the global tax system, the report on the OECD/G20 Base Erosion and Profit Shifting Project (BEPS) was introduced in 2015, proposing several measures for overcoming difficulties stemming from the economy based on intangible assets and online environment (OECD, 2015). Today, it has resulted in many new tax rules at the treaty law, EU law, and domestic legislation level. So by paraphrasing Mr. Eminem, this world has been digital businesses for the taking and it has created many kings, but it has also made a push to move toward the new tax world order.

The European Union (EU) is not immune to the same challenges stemming from the digital economy. Being an important market for digital platforms, the digital business nexus and global minimum tax initiative is the core topic today. The emergence of infrastructure services has increased the complexity of characterizing income for tax purposes (royalties vs technical services). Web3 and crypto transactions often fly under the radar of tax authorities, calling for administrative cooperation in that area. Entirely different challenges are connected with shifts in working patterns and the taxable status of workers, due to the rise of "gig" and collaborative economies (Szczepański, 2021, p. 2–3). As with many topics related to the co-existence of the EU Member States, the union is taking more and more of a lead in developing a common approach.

This chapter gives a brief overview of the main tax rules in force in the EU that focus on the digital economy, as well as elaborates on some future developments that the authors expect to take place in the near future.

2 Digital Economy: What Makes It Special for Taxes?

2.1 Recent Trends in the EU Tax Law

To understand the shifts we have seen in the EU tax law due to the surge of digitalization, the function of the EU tax law, as well as the general trends in the tax law must be first examined. This limits the questions the community tax law can address, leaving the remaining issues for the Member States to deal with.

The Founding Fathers of the EU decided that the competence of the EU to harmonize legislation should be limited to the "measures necessary for the functioning of the market" (Weber, 2010, Chap. 1.2). The primary function of the EU tax law is to abolish tax obstacles within the internal market (Helminen, 2021, Chap. 1.2.4). The EU tax law exists to ensure that taxes do not restrict the free movement of goods, persons, services, and capital between the EU Member States, therefore domestic laws must be approximated to the extent required for the functioning of the internal market (Consolidated version of the treaty of the functioning of the European Union (TFEU) art 5).

While the most evident effect of it is abolishing customs duties on the trade between the Member States, it has also resulted in the harmonization of tax rules on certain payments (dividends, interest, royalties), consumption taxes, and tax information exchange. There is no special EU law on digital economy taxes. Some aspects of electronically provided services and products are regulated under the Council Directive 2006/112/EC of 28 November 2006 on the common system of value added tax (VAT Directive), introducing the VAT on digital services based on the destination principle, that is at the location of the customer. The EU law on direct taxes is purposely and by design very limited, focusing on profit distributions, interest, royalties, reorganizations, and some anti-abuse matters that are of an issue for all economic sectors and not just to the digital business sector.

In recent years, the direction of the EU tax legislation has been shifting to reflect the OECD initiatives under the BEPS project. As rightly stated by Pasquale Pistone, recent trends of positive integration evidence a proliferation of newly enacted legislative measures that pursue the goal of protecting the internal market against base erosion and profit shifting. Pistone has highlighted two major trends, the first of which is about existing directives being amended or broadened, or new directives being introduced, with a view to counter base erosion and profit shifting that may arise in connection with the application of directives (Pistone, 2018, Chap. 1.1.1).

It follows that the focus of the EU tax law has shifted from regulating different economic categories to introducing anti-abuse tools and enhancing tax transparency through administrative cooperation. This is evidenced, for example, by the introduction of the general anti-abuse rule (GAAR) that can be found in the Council Directive (EU) 2006/1164 of 12 July 2016 laying down rules against tax avoidance practices that directly affect the functioning of the internal market (ATAD), hybrid mismatch rules to be found in the Council Directive (EU) 2017/952 of 29 May 2017 amending Directive (EU) 2016/1164 as regards hybrid mismatches with third countries (ATAD 2), and by introducing the obligation to disclose advisory on certain cross-border arrangements that can be harmful to tax compliance under the Council Directive (EU) 2018/822 of 25 May 2018 amending Directive 2011/16/EU as regards mandatory automatic exchange of information in the field of taxation in relation to reportable cross-border arrangements (DAC 6). Although BEPS Action 1 is about Tax Challenges Arising from Digitalisation, none of those tools is aimed at the mismatches the digital economy has brought out in our existing tax system.

There indeed have been intense discussions as well as the legislative process at the EU level that digital taxation needs to be addressed with an interim, temporary solution accompanied by a long-term, permanent fix. The plan foresaw taxation of revenues generated from certain digital services that were otherwise untaxed, followed by taxation if the company has a significant digital presence (i.e., digital permanent establishment) in some or all Member States (European Parliament, 2021). The Council however has given preference to working toward reaching a global solution at the OECD level, to be reflected also in the EU legislation, and has postponed work on these proposals (Ibid).

This approach is likely to continue unless significant changes happen in either the landscape of the digital market or there is a shift in US policy. As it currently stands, such digital taxation is, in practice, often largely aimed toward digital giants, or such regulation can at least be perceived this way. As these companies are mostly US-based, their government and digital lobby are likely to see any unilateral attempt at taxation as attempts to stimy American businesses and offer unfair advantages to up-and-coming enterprises in the EU. Therefore, unless the USA can somehow be dissuaded from further threats of establishing punitive tariffs or these become comparatively palatable, the focus will more likely remain aimed at achieving a global tax solution (Lawder & Thomas, 2021), or at least one where the USA is involved from the start.

As a result of these trends, the existing high-level structure of the EU tax law can be demonstrated by the below scheme. It is not a comprehensive structure of the law but aims to give a systematic overview based on the topics addressed by the tax law as of May 2022 (Table 1).

2.2 Distortion Caused by the Intangible World

Digital economy started developing from 1977 with the release of one of the first mass-marketed personal computers (Kerikmäe & Kesa, 2020, p. 68). Since then, the digital economy with its various branches has increased its market share year by year. The digital economy has a number of features that have caused issues with fitting under traditional tax rules. The BEPS Action 1 final report summarized the following characteristics of the digital economy, to which the authors have attached challenges it has caused for the traditional tax system (Table 2).

The main challenges for income and capital taxes relate to value creation. Traditional international income tax allocation rules follow the principle that

Table 1 The general structure of the EU tax law as of May 2022

TREATY OF THE FUNCTIONING OF THE EU	TREATY OF THE EU	EU CHARTER OF FUNDAMENTAL RIGHTS	FUNDAMENTAL PRINCIPLES AND HUMAN RIGHTS (ECHR)
DIRECT TAXES	INDIRECT TAXES	OTHER	
Parent-Subsidiary Directive	Value-Added Tax Directive	Directive on Administrative Cooperation (DAC) V1: Exchange of information V2: Financial accounts V3: Binding rulings and APA V4: Country-by-country reporting V5: Enhancing UBO identification V6: Cross-border schemes V7: Digital platforms	CJEU case law Soft law instruments
Interest-Royalties Directive	Customs Regulation		
Merger Directive	Excise Directive		
Anti-Tax Avoidance Directive (ATAD) V1: GAAR, exit tax, thincap, CFC V2: Hybrid mismatches	Excise Directive on Tobacco Products		
	Excise Directive on Alcohol Products	Social Security Regulation	
EU tax dispute resolution Directive	Excise Directive on Energy Products	EU treaties with third countries	

Table 2 Characteristics of the digital economy and challenges arising therefrom

Characteristics of the digital economy	Description	Challenges to the tax system
Mobility of users	Difficulties with following where the geographical marketplace is and where the services are physically consumed.	Identifying a place of consumption for VAT purposes.
Mobility of business functions	Business is carried out globally without a need for a physical presence in a market country.	Permanent establishments are not created in the market country, distorting the profit tax allocation rules.
Mobility of intangibles	Business relies heavily on intangible assets that can easily be transferred, making it hard to follow where the development has taken place.	Transfer pricing rules become difficult to apply, making profit shifting to low-tax jurisdictions relatively easy.
Reliance on user participation and data	Users create much value for the business. Often, "free" access to services means that personal data is used as a consideration for access.	The actual consumption value of digital services is not in the VAT tax base. The place of value generation differs from the company's physical presence, distorting the rules on corporate income tax allocation between states.
Network effects	Users benefit from the contributions from other users (horizontal integration). Synergies between different layers or applications (vertical integration).	Tax treatment of contributions made by users, how to find value for consumed products.
Multi-sided business models	Multiple groups of persons interact through a platform or marketplace. Cross-border intermediation or services becomes much easier.	Treatment of platform users as customers vs employees. Allocation of business profits to the market country.

business income tax should be levied in the country where the value is created. In brick-and-mortar businesses, the value is usually created where there is a physical presence and the beneficiaries of goods and services are located. Building on Article 7 of the OECD Model Tax Convention on Income and Capital, a non-resident company pays income tax in the market country if it has a permanent establishment for income tax purposes in that country (Cracea, 2019, p. 173). As digital businesses do not need to have a permanent establishment to actively operate on the market, the taxation rights of the business income generated from such activities are shifted to the residency country of the entrepreneur, leaving the countries that build the infrastructure that enables the provision of digital services (e.g., internet networks) without income tax revenue.

3 Direct Taxes

3.1 Current Status of the Law

The EU direct tax law does not include specific rules on the digital economy. While direct taxes (that is income and capital gains taxes) have seen many new EU regulations that have sprung from the BEPS initiative, none of those has addressed companies operating specifically in the intangible world. The Interest-Royalties Directive indeed concerns IP-heavy businesses the most, preventing double taxation of royalties' payments between related parties and allocating the taxation rights to the residency state of the recipient, provided certain additional shareholding requirements have been met.

The most remarkable development comes from the case law of the EU General Court and the Commission's investigations in the area of state aid. Over the past years, the European Commission has launched a number of investigations on the compatibility of the advance pricing agreements (APA) with state aid rules in particular, regarding the APAs granted to technology giant Apple (Pistone, 2018). In the Apple investigation, the Commission criticized the Irish tax authorities for having incorrectly allocated assets, functions and risks to the head offices of Apple Sales International (ASI) and Apple Operations Europe (AOE) (both Irish companies, but not tax residents there due to having a board of directors in the U.S.), although those head offices had no physical presence or employees. The Commission found that as the head offices had been unable to control or manage the Apple Group's IP licences, those head offices should not have been allocated, in an arm's length context, the profits derived from the use of those licences. Accordingly, those profits should have been allocated to ASI and AOE's branches, which alone would have been in a position effectively to perform functions related to the Apple Group's IP that was crucial to ASI and AOE's trading activity (Judgment of the General Court from 15 July 2020 in Cases T-778/16 and T-892/16 Ireland, Apple Sales International and Apple Operations Europe vs European Commission, p. 32-43).

The judgment was appealed by the Commission to the Court of Justice of the European Union, where it remains outstanding today. Hence, the limitations coming from the EU law on the Member State's sovereignty in the matters of IP income allocation under transfer pricing rules are yet to be confirmed.

3.2 Expected Developments and Trends

3.2.1 EU's "Global" Minimum Tax as the Pillar 2 Directive: Harmonizing Minimum Corporate Income Tax Levels

While *lege lata* is rather modest in digital tax topics, the amendments bring global tax reform to the front yard of the EU.

In December 2021, the Commission published a legislative proposal for a Directive setting forth rules to ensure a global minimum level of taxation for multinational groups (COM(2021) 823 final). Still being under discussion on the details, it is likely it will be adopted. The purpose of it is to restrain tax competition between the states by attracting companies with low tax rates on business income. The directive would effectively provide for a workflow to apply top-up tax in respect of low-taxed constituent entities in a multinational enterprise group to ensure a minimum of 15% corporate income tax rate (Dietrich & Golden, 2022, Sect. 2.1). The rules implemented are called the GloBE rules, consisting of the Income Inclusion Rule (IIR) and its backstop, the Under Taxed Payments Rule (UTPR) (COM (2021) 823 final, p. 1). It would give a priority to ATAD CFC rules and pave the way for agreeing on the pending proposal for recasting the Interest-Royalties Directive to make the withholding tax exemption at source subject to the interest being subject to tax at the destination state (COM(2021) 823 final, p. 2).

The draft directive requires levying a top-up tax in case some state applies a tax rate lower than 15%. Should a multinational enterprise have a subsidiary in a country that applies a lower tax rate to the business income of such subsidiary, the Pillar 2 Directive enables the home jurisdiction of the parent company to apply a top-up tax so that the effective tax rate applicable to the subsidiary's profits would be 15%. As the top-up tax levels off the tax benefit that could be achieved by moving one's activities to the lower tax jurisdiction, it should demotivate the Member States to introduce or maintain lower tax rates.

3.2.2 Digital Permanent Establishment Proposal as the Pillar 1 Directive: New Tax Base for Digital Companies

Like is the case with many other countries as well as the OECD, the digital permanent establishment issue has been brought up by the EU. While existing tax rules are designed for brick-and-mortar businesses that must have a presence in the physical world to generate profits, digital businesses do not. This has made the traditional concept of a permanent establishment redundant for scale without mass business players. For example, Google may generate significant profits from the Google AdWords service provided in the Estonian digital environment, but such profits would not be taxed in Estonia as there is no Google office here. However, the service can be provided only because the Estonian government has contributed taxpayers' money to build the infrastructure. Furthermore, the Estonian-based competitors must consider Estonia's 20% corporate income tax as part of their cost.

Such issues have led the OECD to introduce the Pillar 1 initiative, introducing a new taxing right for a market jurisdiction (on so-called "Amount A" representing the new tax base). It aims to create a taxable presence for multinationals in their market jurisdiction by exceeding certain thresholds. Taxable presence is triggered by having a certain level of users, supplemented by turnover and profit margin. More specifically, it would apply to multinational enterprises with global turnover above 20 billion euros and profitability above 10%. The nexus (digital permanent establishment) would be created when at least one million Euros in revenue is generated from the market jurisdiction, while smaller economies have a reduced threshold of 250,000 Euros (OECD, 2021, p. 1). Pillar 1 is extraordinary as it deviates from the traditional permanent establishment concept known from Articles 5 and 7 of the OECD Model Tax Convention on Income and on Capital. It enables the market jurisdiction to levy such tax without the multinational having a physical presence or a dependent agent present in that jurisdiction.

The proposal for the Pillar 1 Directive is scheduled for 2022 (European Commission on Fair Taxation, 2022a). The European Commission announced in July that it has postponed its initial plan for a proposal for an EU digital levy in order to fully focus on the OECD proposal (Szczepański, 2021, p. 9).

4 Indirect Taxes

4.1 Current Status of the Law

4.1.1 Electronically Supplied Services and Electronic Communications Services

Indirect taxes in the EU are in great part addressed through the laws on value-added tax. The EU VAT Directive harmonizes the rules on taxation of consumption with value-added tax. Value-added tax is a consumption tax that is collected by taxable persons upon the supply of goods and services. As per Lamensch, the harmonization of VAT in the EU has been closely linked to the objective of achieving full economic integration between the Member States in the form of a common market and internal market without frontiers, as well as an economic and monetary union (Lamensch, 2015, Chap. 1.2.3).

In the VAT law, the digital economy is addressed through the rules on e-commerce. Already in 1997, the Commission concluded that online supplies should be treated as services for VAT purposes, meaning such services should be taxed with VAT when supplied within the EU but free of tax when exported to third countries and that the whole system should provide legal certainty, simplicity, and neutrality (European Commission, 1998, p. 4). Today, the digital services that fall under special tax treatment include electronically supplied services and electronic communications services, as defined in the Council Implementing Regulation (EU) no 282/2011 of 15 March 2011 laying down implementing measures for Directive 2006/112/EC on the common system of value added tax (recast). It addresses services such as the supply of software, websites, automatically generated services, and online market services. Under these rules, the provision of such services to the person or entity that is a registered VAT payer in another EU Member State (B2B) falls under the general reverse charge mechanism. The provision of services to the person or entity that is not registered as a VAT payer in another EU Member State (B2C) is taxable in the country of the recipient, hence deviating from the overall approach to B2C service taxation.

From July 1, 2021, the EU VAT package for e-commerce entered into force. It introduced a special One Stop Shop (OSS) regime, under which digital service providers do not need to register for VAT purposes in every Member State where they have customers, but may register only in their home country and pay the VAT to respective Member States through their own tax administrator. It substitutes the Mini One Stop Shop regime that applied to webshops before July 1. Furthermore, a new special scheme called Import One Stop Shop (IOSS) was introduced for the distant selling of goods from third countries, making it simpler to declare low-value consignments (up to 150 Euros) imported from third countries and sold to the EU end customers.

4.2 Expected Developments and Trends

In 2022, the European Commission intends to present the VAT in the Digital Age Package (European Commission, 2022b, Management Plan). The purpose of that is to update the EU VAT rules and take advantage of technological solutions that can improve taxpayers' compliance and the fight against VAT fraud. In line with the EU's digital agenda, the package will focus in particular on real-time reporting, e-invoicing, an extension of the one-stop-shop, a single EU VAT registration system, and VAT rules for platforms, with a view to reducing the VAT gap and making life easier for legitimate and compliant cross-border businesses (European Commission, 2022a, 2022b, p. 12).

New developments can be expected in the area of taxation of crypto assets and VAT. Already in 2015, the Court of Justice of the European Union (CJEU) ruled that the Bitcoin virtual currency, being a contractual means of payment, cannot be regarded as a current account or a deposit account, a payment or a transfer, but it should be seen as a direct means of payment between the operators that accept it (Judgment of the Court of Justice of the European Union from 22 October 2015 in case C-264/14 Skatteverket v David Hedqvist). The rapid emergence of new types of crypto-assets, such as tokens and non-fungible tokens (NFTs) specifically, may call the EU for action to propose a harmonized approach. In practice, challenges may be seen with the VAT treatment of mixed-function NFTs that are used for community creation purposes, granting free access to services provided by the issuer, or having a return from the exit by founders of that issuer to mimic the shareholder position.

5 Tax Information Exchange

5.1 Current Status of the Law

At the EU level, the exchange of information in tax matters was initially regulated by Directive 77/799/EEC, which, after undergoing three amendments by Council

Directives 2003/93/EC, 2004/56/EC, and 2004/106/EC, was finally replaced by Directive 2011/16/EU of February 15, 2011, on administrative cooperation in the field of taxation (DAC 1). This Directive aimed to align the EU information exchange system with the information exchange clause of the OECD Model Convention at the time (2005–2008), although in some respects the Directive went beyond what was established at the international level in the OECD Model Convention.

However, the objective scope of application of DAC 1 has been modified several times, affecting mainly the set of data to be exchanged, which has been considerably extended over the years through successive reforms of this Directive that have introduced the obligation to the exchange of financial accounts (CRS) or the Mandatory Disclosure Regime (MDR), the country-by-country report (CbCR), tax rulings or new commitments in the field of money laundering, and the obligation to exchange information on the exchange of financial accounts (CRS) or the Mandatory Disclosure Regime (MDR).

This is evidenced by the fact that while DAC 1 provided for the automatic exchange of five specific income and wealth categories, DAC 2 (Council Directive 2014/107/EU of December 9, 2014, amending Directive 2011/16/EU as regards the mandatory automatic exchange of information in the field of taxation) and DAC 3 (Council Directive (EU) 2015/2376 of December 8, 2015, amending Directive 2011/16/EU as regards the mandatory automatic exchange of information in the field of taxation) significantly expanded the scope of data items to be transferred. Subsequently, DAC 4 (Council Directive (EU) 2016/881 of May 25, 2016, amending Directive 2011/16/EU as regards the mandatory automatic exchange of information in the field of taxation) established the obligation to transmit comprehensive and relevant information on multinational enterprise groups concerning their structure, transfer pricing policy and internal operations inside and outside the Union. For its part, DAC 5 (Council Directive (EU) 2016/2258 of December 6, 2016, amending Directive 2011/16/EU as regards access to anti-money-laundering information by tax authorities) obliged entities to provide access to information on anti-money laundering. Finally, DAC 6 (Council Directive (EU) 2018/822 of May 25, 2018, amending Directive 2011/16/EU as regards the mandatory automatic exchange of information in the field of taxation in relation to reportable cross-border arrangements) has established the obligation for tax intermediaries to report transactions that may be considered aggressive tax planning and occur internationally.

More recently, the European Commission has also been aware of the difficulties that digital business models pose for tax authorities. The main problem is that Member States' tax administrations have little information to properly assess and control the income obtained in their country from activities carried out through intermediation via online platforms established in other jurisdictions. This in turn allows certain taxpayers to under-declare the income obtained in digital transactions without the authorities in the Member States having the information that would enable them to control this income.

For this reason, the Commission has decided to strengthen the existing regulatory framework for administrative cooperation in the field of taxation through Directive

(EU) 2021/514, known as DAC 7. This new proposal requires the Member States to automatically exchange information on revenues generated by sellers on digital platforms, in line with the Model Rules for Reporting by Platform Operators regarding Sellers in the Sharing and Gig Economy published on July 3, 2020, by the OECD. This automatic exchange of information is also extended to royalties, as it is understood that this is highly delocalizable income and of great relevance, if the aim is to combat fraud and tax evasion effectively.

As this is a direct taxation measure that must comply with the principles of proportionality and subsidiarity, the Commission justifies the adoption of this measure on the need to ensure equal treatment of persons and entities in the European Union so that all European citizens pay their fair share of taxes, on the one hand, and to combat fraud that harms internal trade, on the other hand.

DAC 7 introduces new obligations for digital operators, i.e., those businesses based on digital platforms that allow their users to interact to carry out any of the following activities within the EU: (1) selling a good; (2) providing a personal service; (3) providing a transport rental activity; and (4) renting real estate.

This obligation will affect both EU resident and non-resident digital platform businesses (with some exceptions for the latter if there is an information exchange agreement), with "foreign platform operators" being defined as those who, according to the definition in the Directive itself, are not tax resident in the European Union (Hamulák, 2018), are not incorporated or administered in the European Union, or do not have a permanent establishment in the European Union. These foreign platform operators must register and report information in a Member State of the European Union.

The transactions covered by this reporting obligation are both cross-border and non-cross-border activities, with the express exclusion of so-called "state-owned entities."

The information to be transmitted to the tax authorities is both quantitative and qualitative information, i.e., (1) the income that these businesses have obtained through the online commercialization of their activity; (2) the income that these businesses obtain from intellectual property; (3) personal details of the users such as their identification details, the price paid by each of them in exchange for the service or good, bank details, etc.

The transposition period for DAC 7 ends on December 31, 2022, so taxpayers' new information exchange requirements will enter into force on January 1, 2023.

5.2 *Expected Developments and Trends*

The recent rise of Bitcoin and other cryptocurrencies and crypto-assets, accompanied by the rapid growth of crypto assets providers, has drawn the attention of the EU regulator, who is aware that there is currently a high risk of under or non-declaration of taxable income – with a consequent loss of revenue – due to the heterogeneity of the rules that the Member States are applying and the characteristics of many crypto assets, such as pseudo-anonymity, valuation difficulties, and global reach.

In addition, the linking of cryptocurrencies to money laundering and terrorist financing operations has emerged as a major concern for EU authorities. For that reason, the first attempt to monitor transactions involving cryptocurrencies and crypto-assets has been made in the area of anti-money laundering (AML), where the European Union adopted the first anti-money laundering Directive in 1990 to prevent the misuse of the financial system for the purpose of money laundering.

This legislation has been constantly revised in order to mitigate risks relating to money laundering and terrorist financing. Precisely, in response to the 2015 and 2016 terrorist attacks in Paris and Brussels, as well as the leaks of the Panama Papers on June 19, 2018, the fifth anti-money laundering Directive (Directive (EU) 2018/843 of the European Parliament and of the Council of May 30, 2018, amending Directive (EU) 2015/849 on the prevention of the use of the financial system for the purposes of money laundering or terrorist financing, and amending Directives 2009/138/EC and 2013/36/EU) – also known as AMLD 5 – was published in the Official Journal of the European Union (OJUE) while the Member States had to transpose this Directive by January 10, 2020.

This regulation provides for closer monitoring of virtual currencies such as Bitcoin to prevent them from being used for money laundering or terrorist financing by introducing extensive know your customer (KYC) and AML obligations. Concerning virtual currencies, in particular – to end the anonymity of cryptocurrencies – trading platforms and providers offering e-wallet services are obliged to apply due diligence controls similar to those required for banks, such as customer verification. In addition, these platforms and service providers will be required to register, as will foreign exchange companies, cheque cashing offices, and trust or corporate service providers.

The AMLD 5 was a milestone in cryptocurrency regulation, as it was the first time that cryptocurrency service providers were classified as obliged entities. This means that platforms that manage their users' private keys must comply with the same AML and cyber finance (CTF) requirements as conventional financial institutions. In addition, all cryptocurrency platforms and wallets have to register with their local authorities.

However, it has a significant weakness, as the AMLD 5 does not refer to virtual assets but only to virtual currencies adopting a narrower definition as it focuses only on one possible type of so-called crypto assets.

In view that all types of crypto-assets are not covered by EU financial regulation, shortly after ALMD 5's publication in the OJEU, the Commission presented the Proposal for a Regulation of the European Parliament and of the Council on Markets in Crypto-assets and amending Directive (EU) 2019/1937, COM/2020/593 final (MiCA) which aims to "harmonise the European framework for issuing and trading various types of cryptographic tokens as part of Europe's digital financial strategy." This proposal includes a comprehensive new legislative proposal on crypto assets that was developed to help streamline distributed ledger technology (DLT) and virtual asset regulation in the EU while protecting users and investors and, at the same time, it provides a sound legal framework for crypto-asset markets to develop within the EU by clearly defining the regulatory treatment of crypto-assets that are not covered by existing financial services legislation.

The most relevant aspect of the MiCA proposal is that it includes a definition of crypto assets, considering them as "a digital representation of value or rights which may be transferred and stored electronically, using distributed ledger technology or similar technology." Moreover, the MiCA proposal considers three types of tokens within the realm of crypto-assets: asset-referenced tokens, electronic money tokens or "e-money tokens," and utility tokens.

This definition and types of tokens might be relevant for both a new version of the AMLD and the DAC 8, which would then be based on these definitions.

A second attempt to monitor transactions involving cryptocurrencies and crypto-assets has been made in the area of exchange of information which has materialized in the Proposal for a Council Directive amending Directive 2011/16/EU (DAC) as regards measures to strengthen existing rules and expand the exchange of information framework in the field of taxation to include crypto-assets and e-money (DAC 8) that establishes obligations to exchange tax-relevant data relating to new alternative means of payment and investment, such as crypto-assets and electronic money.

DAC 8 is a third pillar measure aimed to improve cooperation between national tax authorities, establishing EU-wide rules concerning the exchange of information for issuers and service providers of crypto assets and e-money institutions to enable tax authorities to properly identify crypto-assets and e-money for tax purposes. Although its approval was initially planned for the third quarter of 2021, it is currently under discussion, and its approval has been delayed until the second quarter of 2022.

The DAC 8 – if approved – would fill a double gap. On the one hand, it would fill the existing gap in the field of the reporting obligations of the DAC since tax authorities currently do not have access to such crypto information (e.g., identity, beneficial owner, purpose, or intended nature of the business relationship, the origin of assets, etc.). On the other hand, while it is true that the ALMD 5 already introduced an increased due diligence obligation for exchanges concerning transparency, traceability of the origin, and transfer of crypto assets, these obligations are not sufficient, and therefore the DAC 8 reinforces these obligations.

6 Conclusions

The digital economy has changed how taxes apply to different business models and how the fair share is collected from different stakeholders. The EU has focused on administrative challenges in order to ensure tax compliance and information exchange between the Member States, supported by the anti-abuse initiatives originating from BEPS and cemented by Anti-Tax Abuse Directives. The scheme that summarizes the planned major tax initiatives are shown in Table 3.

Table 3 The general structure of the EU tax law as of April 2022, contemplated initiatives highlighted

TREATY OF THE FUNCTIONING OF THE EU	TREATY OF THE EU	EU CHARTER OF FUNDAMENTAL RIGHTS	FUNDAMENTAL PRINCIPLES AND HUMAN RIGHTS (ECHR)
DIRECT TAXES	INDIRECT TAXES	OTHER	
Parent-Subsidiary Directive	Value-Added Tax Directive	Directive on Administrative Cooperation (DAC)	
Interest-Royalties Directive	Customs Regulation	V1: Exchange of information	
Merger Directive	Excise Directive	V2: Financial accounts	CJEU case law
Anti-Tax Avoidance Directive (ATAD)	Excise Directive on Tobacco Products	V3: Binding rulings and APA	
V1: GAAR, exit tax, thincap, CFC	Excise Directive on Alcohol Products	V4: Country-by-country reporting	Soft law instruments
V2: Hybrid mismatches	Excise Directive on Energy Products	V5: Enhancing UBO identification	
V3: Conduit companies (proposal)		V6: Cross-border schemes	
EU tax dispute resolution Directive		V7: Digital platforms	
Global Minimum Tax Directive (Pillar 2)		V8: Crypto platforms (proposal)	
Digital PE Directive (Pillar 1)		Social Security Regulation	
		EU treaties with third countries	

References

Cracea, A. (Ed.). (2019). *OECD model tax convention on income and on capital – condensed version 2017 – and key tax features of member countries 2019*. IBFD.

Consolidated version of the treaty of the functioning of the European Union, OJ L. 326/47-326/390; 26.10.2012.

Council Directive 2006/112/EC of 28 November 2006 on the common system of value added tax (VAT Directive), OJ L 347, 11.12.2006, (pp. 1–118).

Council Directive (EU) 2016/1164 of 12 July 2016 laying down rules against tax avoidance practices that directly affect the functioning of the internal market (ATAD), OJ L 193, 19.7.2016, (pp. 1–14).

Council Directive (EU) 2018/822 of 25 May 2018 amending Directive 2011/16/EU as regards mandatory automatic exchange of information in the field of taxation in relation to reportable cross-border arrangements (DAC6), OJ L 139, 5.6.2018, (pp. 1–13).

Council Directive (EU) 2017/952 of 29 May 2017 amending Directive (EU) 2016/1164 as regards hybrid mismatches with third countries (ATAD 2), OJ L 144, 7.6.2017, (pp. 1–11).

Council Implementing Regulation (EU) no 282/2011 of 15 March 2011 laying down implementing measures for Directive 2006/112/EC on the common system of value added tax (recast), OJ L 77, 23.3.2011, (pp. 1–22).

Council Directive 2014/107/EU of 9 December 2014 amending Directive 2011/16/EU as regards mandatory automatic exchange of information in the field of taxation, OJ L 359, 16.12.2014, (pp. 1–29).

Council Directive (EU) 2015/2376 of 8 December 2015 amending Directive 2011/16/EU as regards mandatory automatic exchange of information in the field of taxation, OJ L 332, 18.12.2015, (pp. 1–10).

Council Directive (EU) 2016/881 of 25 May 2016 amending Directive 2011/16/EU as regards mandatory automatic exchange of information in the field of taxation, OJ L 146, 3.6.2016, (pp. 8–21).

Council Directive (EU) 2016/2258 of 6 December 2016 amending Directive 2011/16/EU as regards access to anti-money-laundering information by tax authorities, OJ L 342, 16.12.2016, (pp. 1–3).

Lucas-Mas, C. Ó., & Junquera-Varela, R. F. (2021). *Tax theory applied to the digital economy: A proposal for a Digital Data Tax and a Global Internet Tax Agency* (p. 2021). World Bank Group.

Dennis Weber, ed. (2010). Preface in traditional and alternative routes to European tax integration: Primary law, secondary law, soft law, coordination, Comitology and their Relationship. Books IBFD. Accessed 16 April 2022.

Directive (EU) 2018/843 of the European Parliament and of the Council of 30 May 2018 amending Directive (EU) 2015/849 on the prevention of the use of the financial system for the purposes of money laundering or terrorist financing, and amending Directives 2009/138/EC and 2013/36/EU.

European Commission. (2022a). Fair Taxation: Take part in the OECD's new public consultation on certain aspects of Pillar 1 of the international taxation agreement. 7 February 2022, available at https://ec.europa.eu/taxation_customs/news/fair-taxation-take-part-oecds-new-public-consultation-certain-aspects-pillar-1-international-2022-02-07_en. Accessed 17 April 2022.

European Commission. (2022b). Management Plan 2022. DG Taxation and Customs Union, 16 January 2022. Accessed 1 May 2022.

European Commission. (1998). Communication from the Commission to the Council, the European Parliament and the Economic and Social Committee: Electronic Commerce and Indirect Taxation, 17 June 1998, COM(1998) 374 final(1998).

Judgment of the General Court from 15 July 2020 in Cases T-778/16 and T-892/16 Ireland, Apple Sales International and Apple Operations Europe vs European Commission.

Judgment of the Court of Justice of the European Union from 22 October 2015 in case C-264/14 Skatteverket v David Hedqvist.
Kerikmäe, T., & Kesa, A. (2020). Artificial intelligence and the GDPR: Inevitable nemeses? *TalTech Journal of European Studies, 1*(3).
David Lawder, Leigh Thomas (2021). U.S. drops tariff threat in digital tax transition deal with European countries – Reuters. https://www.reuters.com/world/europe/european-countries-reach-digital-services-tax-deal-with-us-2021-10-21/. Accessed 14 May 2022.
Szczepański, M. (2021). Taxing the digital economy new developments and the way forward. *European Parliament Briefing*.
Dietrich, M., & Golden, C. (2022). European Union/OECD/International – Consistency versus "Gold Plating": The EU Approach to Implementing the OECD Pillar Two. *Bulletin for International Taxation, 76, no 4*.
Hamulák, O. (2018). La carta de los derechos fundamentales de la union europea y los derechos sociales. *Estudios Constitucionales 16*(1), 167–186. https://doi.org/10.4067/S0718-52002018000100167
Helminen, M. (2021). *EU Tax Law – Direct Taxation – 2021*. Books IBFD. Accessed 16 April 2022.
Lamensch, M. (2015). *European value added tax in the digital era: A critical analysis and proposals for reform*. Books IBFD. Accessed 1 May 2022.
OECD. (2015). Explanatory Statement, OECD/G20 Base Erosion and Profit Shifting Project, OECD. www.oecd.org/tax/beps-explanatory-statement-2015.pdf. Accessed 10 April 2022.
OECD. (2021). Statement on a two-pillar solution to address the tax challenges arising from the digitalisation of the economy (8 October 2021). https://www.oecd.org/tax/beps/statement-on-a-two-pillar-solution-to-address-the-tax-challenges-arising-from-the-digitalisation-of-the-economy-october-2021.pdf. Accessed 17 April 2022.
Pistone, P. (Ed.). (2018). *European tax integration: Law, policy and politics*. Books IBFD. Accessed 16 April 2022.
Proposal for a Council directive on ensuring a global minimum level of taxation for multinational groups in the Union (SWD(2021) 580 final). Brussels, 22.12.2021 COM(2021) 823 final 2021/0433 (CNS).
Proposal for a Regulation of the European Parliament and of the Council on Markets in Crypto-assets and amending Directive (EU) 2019/1937, COM/2020/593 final (MiCA).
Proposal for a Council Directive amending Directive 2011/16/EU (DAC) as regards measures to strengthen existing rules and expand the exchange of information framework in the field of taxation to include crypto-assets and e-money (DAC8).

Kaido Künnapas is a Senior lecturer at the Tallinn University of Technology Law School, visiting lecturer at the Tartu University Law School, tax partner with Sorainen Law Firm.

Begoña Pérez Bernabeu is a Professor at the University of Alicante Faculty of Law, the Department of Tax Law and Economics Department.

Katariina Kuum is a Master student at the Tartu University Law School.

Karl Oskar Pungas is a Bachelor student at the Tartu University Law School.

The Digital World Market and the European Union

Tanel Kerikmäe and David Ramiro Troitiño

Abstract Digitalization and automation are priorities in the European Union. The process has become central in the constant evolution of the organization internally, but also externally. The new digital economy will have a relevant impact on the trade relations between the main commercial areas of the planet. Therefore, the European Union needs to develop its own model and foster interconnection with other parts of the world in order to compete with powers like the United States of America or China. This chapter focuses on digital economic matters and its relevance for the European Union economy in the context of a world digital market.

1 Introduction

The promotion of the digitization process at the European level is creating great expectations of economic development. It is estimated that by intensifying the process at the European level, €2.5 billion could be added to the EU's GDP by 2025, a considerable amount for such a short period. In addition, taking into account the pandemic generated by COVID-19, the process of economic digitization has naturally intensified in response to mobility problems generated by the restrictions imposed to curb the pandemic. At the same time, the need for the European Union has intensified, not just to lead the process, but also to adapt to it in the fastest and most effective way (Kohli, 2017).

The private sector is not the only one that is immersed in the process of digitization of society; it is also presumed that the public sector will benefit greatly from a wider use of new digital technologies. Information and communication

T. Kerikmäe (✉)
Department of law, Tallinn University of Technology, Tallinn, Estonia
e-mail: tanel.kerikmae@taltech.ee

D. Ramiro Troitiño
Jean Monnet Chair on Digital Europe and Future Integration, Tallinn University of Technology, Tallinn, Estonia
e-mail: david.troitino@taltech.ee

D. Ramiro Troitiño et al. (eds.), *Digital Development of the European Union*,
https://doi.org/10.1007/978-3-031-27312-4_8

technologies (ICTs) allow the public sector to operate in a more transparent and efficient manner as it contributes to the reduction of corruption through more transparent control of public activities. At the public level, the possibilities for political participation offered by the digitization process are also relevant, by allowing citizens, including minorities, inhabitants of rural areas, the disabled, or any other group at risk of exclusion, to have digital access, to political life, thus increasing their participation in common affairs, protecting themselves from situations of discrimination and promoting a more inclusive political vision. The concept of participation derived from the creation of a political digital space is also of the utmost importance for the advancement of integration. The European Union is an association project in constant development since the founding of the European Coal and Steel Community in 1952. The progressive evolution of the common European home has reached great levels of integration, but like any process of growth and deepening, progress is not made in all fields in a homogeneous manner. At the political level, integration has encountered significant obstacles that have slowed down the process, delaying this field compared to other areas, such as economics. Digital development allows the creation of a more integrated European space at a political level since it reduces distances and facilitates the participation of all European citizens in the political development of the Union. The political future of Europe depends on strongly involving citizens in its development, and digital development removes barriers so that the common political framework becomes a reality (Armstrong & Weeds, 2007).

Likewise, digital development is strongly intertwined with the optimization of resources and the reduction of costs both at a private and public level. Therefore, the European Union seeks to accelerate a process, already inevitable, to improve economic competitiveness and public spending. Therefore, the European Union is facing a fourth Industrial Revolution based on the enormous technological development of recent years. Advanced technologies such as data analysis, artificial intelligence (AI), automated systems, cloud computing, the Internet of Things (IoT), robots, 3D printing or 5G, seem to offer endless application possibilities in almost all fields and aspects of life (Troitiño et al., 2020).

2 Digital Agenda

Bearing in mind that the Lisbon Strategy and its revision introduced in 2005 ran out of steam at the end of 2010, the Commission introduced in May 2010 a new instrument called the Digital Agenda for Europe (DEA). One of the central pillars of the Digital Agenda for Europe is the Digital Single Market, whose objective is a European digital economy without borders in which the free movement of goods, people, services, and capital is guaranteed. This common space also seeks the effective protection of the data of European users, now in the hands of multinationals that escape the controls of the Member States. In short, it is the creation of an area where citizens and companies can easily access goods and services regardless of location within the European economic area. The ADE anticipates a better use of Information Technology (ICT) to promote employment, growth, competition, investment, and innovation, and break down the barriers that block the opportunities for a more extensive use of the Internet. Ninety percent of all companies in the EU are small and medium-sized enterprises (SMEs) and only 7% of them sell abroad. The digitization of the Single Market is expected to change this anomalous situation and expand European economic activity to levels unimaginable just a few years ago. The European Commission estimates that the creation of the Single Digital Market will generate an additional 415 billion euros to the EU economy and the community GDP will grow up to 1% per year, and approximately 2% in combination with other pillars of the ADE. In addition, calculations by the European institution indicate that 1.3 million new jobs will be created by 2025, at a time of widespread crisis due to the pandemic that has greatly affected the European labor market. A truly digital single market will generate new business opportunities, promote innovation and encourage the creation of new companies.

The European Commission has created a defined roadmap to achieve the European Digital Market based on three fundamental pillars on which the European digital fabric will rest in the near future. The first pillar, linked to digital access, consists of such prominent elements as internet commerce, the distribution and logistics of products ordered through the network, the elimination of geo-blocking throughout the European territory, copyright laws that protect creators and encourage innovation through the creation of a regulatory framework at European level and with global influence, and a standard type of Value Added Tax for all companies working in the European Digital Market to avoid distortions in the market that affect competitiveness and thus eliminate tax barriers that affect equal conditions for all European operators (Troitiño et al., 2020).

The next development involves the creation of a stable framework that supports the technological developments necessary for the implementation of the Digital Single Market. The development of a common regulatory framework for large telecommunications technology companies and a common framework for digital information media are priority objectives of the European Union, since they are expected to facilitate a stable environment in which European development can be based. This same chapter includes actions for the promotion and creation of online platforms at a European level and aspects concerning security and personal data. Europe seeks to protect its citizens from the pernicious effects of digital development with the creation of protective legislation with a vision of anticipation rather than correction. Anticipating problems will give more credibility to the EU's digital effort.

The third pillar is based on aspects related to the economy and society and includes various important aspects, such as the data market, one of the main fields of development both for sale, as well as the production and implementation of technologies based on algorithms that need a large flow of data to function properly, such as Artificial Intelligence. Regulating the access and distribution of said data is essential for sustainable economic development in a Common Digital Market. Other basic aspects are the introduction of European standards. A market requires common rules for its correct operation, thus avoiding unfair competition, discrimination and artificial alterations of the operation based on free competition where the most competitive survive and the rest disappear. At the social level, the effort made to develop digital government or electronic governance is noteworthy. It is the application of technology for the provision of government services, the exchange of information, communication transactions, the integration of several independent systems between the government and the citizen, business management, labor management, as well as processes and interactions of the administrative part within the entire governmental framework. Through e-governance, the services of the European Union would be made available to citizens in a convenient, efficient, and transparent way (Kennedy, 2006). The three main target groups that can be distinguished are different levels, the administrative and institutional at the community level, citizens and businesses and interest groups.

Once the foundations have been established, its development will make possible the implementation of a Digital Market at a European level. A fact that will strengthen the process of European construction internally by creating common rules and a digital space without borders. In turn, externally, a strong digital Europe will protect Europe's way of life and its citizens from foreign tech giants whose development parameters do not match the European vision of society.

3 Future Development and Obstacles

The Digital Single Market presents countless digital opportunities, but it also faces significant challenges that can hamper its development and anchor European progress against other parts of the world. Firstly, making the European Union's Single Market fit for the digital age requires breaking down regulatory barriers and moving from individual national markets to a single rulebook for the entire EU. The creation of a common legal framework at the European level faces the opposition of political nationalism that advocates state autonomy against the unifying impulses that emanate from Brussels (Troitiño & Kerikmäe, 2021). The concept of national sovereignty plays a relevant role in the process of European construction from its beginnings to the present day, developing a game of balance between cooperation and integration, between Intergovernmentalism and federalism, between a community of independent States and the creation of a common space. Therefore, the European Union, currently highly integrated, has constantly evolved from cooperation to integration, in a process that is difficult to reverse, despite the constant tensions that, for example, led to the United Kingdom leaving the EU. BREXIT may be the definitive boost for the creation of a common legislative framework in digital Europe, since the Anglo-Saxon country was the champion of the European Union of States and always showed its reluctance to increase the levels of integration at the European level. Another fundamental fact for the development of a digital community acquis is the inability of States to influence the digital world as they cannot control the large technology corporations that dominate the digital world, imposing their own laws without restrictions. Only the union of the common regulatory effort would result in a sufficiently relevant space to limit the actions of the Internet giants within a stable framework that respects the European model of society. Therefore, the debate on national sovereignty in digital matters is futile because the only way to exercise it effectively is through collaboration at the European level (Nambisan et al., 2017).

On the other hand, there are numerous (often-uncoordinated) initiatives toward a clear legal strategy at both the EU and Member State levels. The need to present a common strategy, backed by all the relevant agents affected, is a challenge that requires greater coordination. The digital world is in its infancy, the more it is legislated at the national level, the greater the divergences at the European level and the more difficult the consequent integration will be. Therefore, the European Commission intends to advance integration before the proliferation of national obstacles slows down the process definitively and Europe falls behind at the world level in a key sector for the economy and society of the future. These steps could contribute €415 billion a year to European economic growth, boosting jobs, competition, investment, and innovation in the EU. In addition to establishing Europe's international influence and protecting Europeans from global giants run by totalitarian governments or simply guided by the maximization of profits without social constraints and with basic contributions to the development of the societies where they operate.

The new digital Europe needs a legal framework in which to settle and expand effectively and respectfully with the principles that guide European society. Therefore, the European Union has worked in this regard by preparing a first annual report on the state of the rule of law in the European Union 2020, which includes digital issues. The general concept of the European framework emphasizes the development of economic and labor market resilience with economic, social, environmental, and institutional sustainability as the guiding principle of community policies in this regard. This approach is expected to foster upward convergence and equity in the transition to a climate-neutral economy while managing the challenges posed by digitalization and demographic change. Given the rapid evolution of this field, the EU foresees additional requirements to review the developments in the Member States.

4 Benefits for the Public Sector

Technological advances and the implementation of the Single Digital Market have advantages beyond the economic ones, already considerable, focused on the public service of citizens. The European Union is facing a great opportunity to improve the lives of its citizens and, therefore, to increase popular support for the project of European construction. Since the beginning of the integration process, the EU has faced the problem of sovereignty and the ability of a supranational entity to fully exercise the prerogatives of a common sovereign. One of the constant obstacles to integration has been the reluctance of certain states and social agents to allow sovereignty to be exercised beyond the state level, which in many cases is at the same time national, with the imbrication of two differentiated concepts, State and nation. The defenders of this position advocate a minimum transfer of sovereignty to a Europe based on cooperation. However, in a certain way, they appropriate the sovereignty that actually resides in the citizens of their States (not in their nationals or in the State itself). They are the ones who, according to a social contract, cede the management of their freedom to the political framework for the creation of rules that allow harmonic coexistence in society of numerous individuals. The acceptance of these rules by the members of the system, the individuals, is what is known as institutional loyalty from a political point of view, or patriotism from a political-cultural point of view. The set of assignments by citizens (political subjects) forms sovereignty. Therefore, the State is not necessarily the last recipient of this assignment nor is it the owner of sovereignty, but its manager at the public level.

The problem of sovereignty at the European level can be solved with the implementation of digital solutions to citizens' problems, attracting their loyalty to the European project for being more effective in solving their problems. Digital solutions can increase support for the EU among the citizens of Europe who even today are confusedly facing the creation of a common home due to the dual nature of their institutional allegiance. On the one hand, dominated by rationality based on effective management of the problems that concern them. On the other predetermined by the emotions emanating from the concept of nation that is confused with politics due to the promotion of the Nation-State from the nineteenth century as a way to attract more effectively the loyalty of the members of society. Faced with this situation, the EU cannot compete with emotions, but with rationality. Showing itself to be effective in solving the citizens' problems, it will attract enough loyalty or support to advance in the integration process.

The Digital Single Market would allow the EU to manage online services by harmoniously connecting the electronic information systems of the public sector of the Member States and of the European institutions themselves, coordinating management, avoiding congestion, designing better digital layouts and optimizing responses to emergencies.

In turn, the free movement of people and goods has created road systems that transcend state borders and that require common management for their correct operation. The digitization of management would allow great advances and numerous benefits for citizens such as models to predict problems with the quality of the road, traffic accidents, speed problems, profitability, etc.

The impact of the Digital Single Market is linked to the ability to generate sufficient data for the creation of effective models. Reliance on reliable data is huge and critical, putting Europe ahead of other global giants. This advantage is temporary, since the generation of data is a priority and investments are significant worldwide, confirming the maxim that information is power. The European Union has as a priority to invest in the generation of data due to its potential and due to the inherent need (Kobrin, 1998). The idea is simple; the generation of data is more cost-effective and effective on a large scale, so it seems appropriate that the EU should take charge of this crucial aspect for the future. As an illustrative example, we see how the Common Agricultural Policy, dependent on the European Union, needs a large amount of data to increase its effectiveness and protect the European rural environment in all its dimensions. Analysis to determine the status and quality of land used for farming and ranching is done more effectible by satellite, so it would be wiser to use a single European satellite than for each member to use its own satellite (Eden, 2016). The generation of data is also essential for an effective management of the health system and the possibility of analyzing the medical needs of a patient and where they can be best served according to the available resources. The implementation of a digital coordination system at the health level will be of the utmost importance for the European Union due to its clear benefit for citizens, the improvement of their health, the optimization of resources, and rational planning. Although the obstacles are numbers on this issue due to the strong identification of countries with their health system as a transcendent part of their common identity. All these examples point in the same direction, Europe faces great challenges in terms of digital integration but offers great possibilities to advance positively in the integration process due to the potential benefits it can offer its citizens.

5 Artificial Intelligence

The European Commission has put forward a European Approach to Artificial Intelligence (AI) and Robotics related issues that addresses technological, ethical, legal, and socio-economic aspects to boost EU research and industrial capacity and put AI at the forefront, serving European citizens and the European economy. Artificial intelligence has become an area of strategic importance and a key driver of economic development. It can provide solutions to many societal challenges, from treating disease to minimizing the environmental impact of agriculture. However, the socio-economic, legal, and ethical impacts need to be carefully addressed from various perspectives and with great care by the European institutions. It is therefore essential to join forces in the European Union to keep Europe at the forefront of this technological revolution, to ensure the competitiveness of the economic system and to shape the conditions for its development and use (ensuring respect for the European values that set Europe apart), the European Union from other international entities).

Artificial intelligence provides technological systems with the capacity to understand their environment and make decisions with a certain capacity of autonomy to reach defined objectives. Machine learning denotes the ability of a software/computer to acquire from its environment or from a very large set of representative data. It allows systems to adapt their performance to changing conditions, or perform responsibilities for which they have not been explicitly preset. In order to build vigorous models at the core of AI-based systems, high quality data is fundamental in improving performance (Osburg & Lohrmann, 2017). Therefore, the European Commission has proposed legislation that focus on improving data-sharing and wading the data options for more data for reuse. The Commission paid relevant attention to data generated by the public sector, but also to data linked with research and with healthcare.

AI is already in Europe, it entails aspects using a virtual personal assistant helping the organization of the workday, journeying in a self-driving automobile, or smartphones proposing tunes or bistros according to the expected wishes of the owner. Artificial Intelligence helps solving relevant issues affecting humanity as the handling of chronic diseases or the decrease of mortality rates in traffic misfortunes, fighting climate change or forestalling cybersecurity dangers. As an example, AI is helping Danish to save lives by allowing emergency services to identify cardiac problems or a variety of alternative conditions grounded on the sound emitted by a person calling emergency health establishments on the telephone. Alternatively, Austria counts with AI that assistances radiologists detecting tumors more precisely by promptly associating X-rays with a number of supplementary medical information. Other sectors, as agriculture also feel the innovations coming from AI, for example, many farmhouses all over the European continent are by this time using AI to monitor the drive, heat and food ingestion of their cattle (Zott & Amit, 2017). The Artificial Intelligence system will mechanically acclimatize heating and automated feeding to assistance agronomists monitor their cattle and free them to focus on alternative works needed in their exploitations, AI is helping European companies to become more efficient, making it at ease for factories coming back to Europe. In addition, Artificial Intelligence is transmuting humanity, society, and manufacturing.

6 The Digital Economy in the Future

The digital economy is a novel relatively mysterious marvel link with exponential economic growth of online platforms. Workers who do not have a full-time job and who have flexible hours use them. Their scope is wide, including collaborative economy platforms specializing in only digital errands without physical attendance or proximity between the employees and customers—Online Labour Markets (Simon, 2000).

Nevertheless, there exist relevant divergences between the Member States of The European Union regulating the digital market. It creates a modification of the existing Common Market within the European Union. The threat of the high-tech platforms to take advantage of the dissociation of common approach to the digital economy is high because they operate at the European level, creating a distortion in the structure that can meaningfully distress the competitiveness of the economy of the whole European Union. A European harmonization is decidedly unlikely because of the links between labor rule and the social systems. Nevertheless, the dysfunctionality originated can be solved by introducing a mutual general point to address the challenges postured by the digital economy. Therefore, only the implementation of a Common Digital Market (Stoeckli et al., 2018) will solve completely this issue. Shared standards will help to decrease the divergences between the diverse economies, and promote convergence within the EU Member States. Nevertheless, it will still permit significant differences varying on the predominant necessities and models of each area.

7 Conclusions

The EU counts with outward-oriented economies. However, it is too the world's principal Common market. Free trade between its Member States is the main philosophy for an organization based on integration. In addition, the European Union is devoted to a globalization of international economic transactions.

During the years 1999–2010, the European Union's external trade doubled, representing currently around 30% of its gross domestic product (GDP). The European Union is in authority for the trade policy of its Member States. The European Commission negotiates trade agreements for the whole organization. Therefore, having a single approach, the European Union influences strongly international trade talks in a more effective way that the sum of the individual actions of its Member States.

Obviously, the war in Ukraine is affecting, and will affect, the international trade of the European Union. Therefore, the necessity to expand the current relations with stable partners is a priority for the European Union. The digital market offers a great opportunity to succeed in a new world environment. Therefore, the EU faces an unparalleled test in terms of evolution, discernable by technological progress. The EU development faces powerful difficulties threatening a common expansion maximizing shared activities that could provide the EU relevance that is more international, economic development and internal growth.

The European Union has robust scientific and industrial bases to construct a new model. It counts with leading research laboratories, relevant international universities, and many groundbreaking companies. The EU owns a comprehensive legal framework protecting consumers while alternatively promoting innovation. In addition, the EU is generating progress in the creation of a Digital Single Market. The main elements already exist within the European Union in order to generate leadership in the world tech transformation. Nevertheless, the incapacity of National States to share further their national sovereignty is a threat for a digital world far from the nation-based societies originated by the change of paradigm associated to the French Revolution (Hamulák, 2018).

The global connectivity of the Digital Single Market is a straightforward principle for its intensification. To promote international interconnection, the European Union should work on interdisciplinary in projects and innovation (online medicine, AI in the labor market). It is imperative to supporting cross-border exercise on GDPR and GovTech, ease joint publication in international recognize indexed journals and generate (online) opportunities and expert clusters for further development of digital relations between the European Union and the world.

Bibliography

Armstrong, M., & Weeds, H. (2007). *Public service broadcasting in the digital world.*

Eden, L. (2016). Multinationals and foreign investment policies in a digital world. In *E15Initiative, international centre for trade and sustainable development and world economic forum, Geneva.* www.e15initiative.org.

Hamulák, O. (2018). La carta de los derechos fundamentales de la union europea y los derechos sociales. *Estudios constitucionales, 16*(1), 167–186.

Kennedy, A. (2006). *Electronic customer relationship management (eCRM): Opportunities and challenges in a digital world.*

Kobrin, S. J. (1998). Back to the future: Neomedievalism and the postmodern digital world economy. *Journal of International Affairs, 51*, 361–386.

Kohli, A. K. (2017). Market orientation in a digital world. *Global Business Review, 18*(3_suppl), S203–S205.

Nambisan, S., Lyytinen, K., Majchrzak, A., & Song, M. (2017). Digital innovation management: Reinventing innovation management research in a digital world. *MIS Quarterly, 41*(1).

Osburg, T., & Lohrmann, C. (2017). *Sustainability in a digital world.* Springer International.

Simon, L. D. (2000). *NetPolicy.com: Public agenda for a digital world.* Woodrow Wilson Center Press.

Stoeckli, E., Dremel, C., & Uebernickel, F. (2018). Exploring characteristics and transformational capabilities of InsurTech innovations to understand insurance value creation in a digital world. *Electronic Markets, 28*(3), 287–305.

Troitiño, D. R., & Kerikmäe, T. (2021). Catalonia, Spain and the European Union: A quest for legitimacy. *TRAMES: A Journal of the Humanities & Social Sciences, 25*(1), 3.

Troitiño, D. R., Kerikmae, T., Barbosa, P. A. R., & Shumilo, O. S. (2020). El libro blanco sobre inteligencia artificial: Análisis y comentarios sobre mercado, valores y cooperación europea. In *Inteligencia artificial: De la discrepancia regional a las reglas universales: integración de percepciones políticas, económicas y legales* (pp. 303–318). Aranzadi Thomson Reuters.

Troitiño, D. R., Kerikmäe, T., De la Guardia, R. M., & Sánchez, G. Á. P. (Eds.). (2020). *The EU in the 21st century: Challenges and opportunities for the European integration process.*

Zott, C., & Amit, R. (2017). Business model innovation: How to create value in a digital world. *NIM Marketing Intelligence Review, 9*(1), 18.

Ethics and Modern Technologies: Example of Navigating Children's Rights in an AI-Powered Learning Environment

Archil Chochia and Eden Grace Niñalga Sicat

Abstract Modern societies are increasingly dependent on technologies. Emerging technologies offer a variety of solutions in different aspects of our lives, lives of the modern societies, and it is hard to imagine ordinary functioning without these technological solutions. However, ethical aspects of such ever-increasing impact of modern technologies on our lives are often neglected, or at least not relevantly reflected upon. This research aims to engage in ethical discussion about modern technologies, focusing on the European Union, highlighting important aspects, and providing thorough analysis.

1 Introduction

When we talk about the digital development of the European Union (EU), an important aspect of this development has also been an active debate on possible outcomes of modern technologies in the European society and beyond. The EU is considered a leader in ethical debate on modern technologies (please see, e.g., Chochia & Nässi, 2021; Joamets & Chochia, 2020). However, the discussion on negative and positive impact of the technologies has been real since the days those technologies became part of our lives. According to Winston and Edelbach (2011), determining an attitude toward technologies is a complex task, and there are two distinctive viewpoints on it, one being techno-optimism and the second—techno-pessimism. Naturally, the first group highlights the positive aspects of technology and its benefits to the society and believes that potential technology-related problems will also be solved by technological solutions, while the second group emphasizes the negative characteristics of the discussed modern technologies and are skeptical of technological solutions. If we use philosophical approach, then the discussion falls between two historical periods of technological analyses (Joamets & Chochia, 2021). The first being classical hermeneutic critiques, which takes more critical

A. Chochia (✉) · E. G. N. Sicat
Department of Law, Tallinn University of Technology, Tallinn, Estonia
e-mail: archil.chochia@taltech.ee; esicat@taltech.ee

D. Ramiro Troitiño et al. (eds.), *Digital Development of the European Union*,
https://doi.org/10.1007/978-3-031-27312-4_9

approach, discusses new technologies' negative effects on the humans. The philosopher Jacque Ellul (1964) is a good representative for this approach, as he argued that technology was rather destructive, as due to its rapid pace of development, it could only be controlled by the humans. On the other hand, the second historical phase is empirical, which approaches individual technologies and examines those within culture and values of those using given technology. This approach looks at the modern technologies more as elements determined socially through local use, focusing on local narratives rather than technology in general (see, e.g., Brey, 2010; Verbeek, 2011). Philosopher John Deigh (2010, p. 1) argues that ethical questions "seem simple yet are ultimately perplexing."

The digitalization process has been discussed in length by the scholars due to several ethical implications and its growing impacts. "The human activity and social institutions" are shaped by the digital technology in our information society, argues Deborah Johnson (2004, p. 69). At the same time, the better legal approaches are sought for regulating modern technologies, as their presence in our daily lives grows (see, e.g., Kerikmäe et al., 2020), and what are their implications on the legal systems (Brownsword & Goodwin, 2012). The emphasis is on the "continuity of people and technology with the rest of nature" (Parsons, 2019, pp. 6–9). However, ethical and moral issues related to the widely used technologies, such as smartphones, social media, augmented reality, AI, and others, need more attention (Deloitte, n.d.; Jobin et al., 2019).

Modern societies' lack of literacy in digital ethics is a huge problem, argue Beever et al. (2019, pp. 9–25) in their book *Understanding Digital Ethics*. Such issues as responsibility, technology control, and moral agency require more consideration. The discussion is compared to the "trolley problem," ethical thought experiment in moral philosophy. But what is digital ethics and how to achieve relevant understanding of it? First of all, we need to understand that digital ethics is a mixture of digital and ethical literacies, and therefore, literacy in digital ethics requires us to have a relevant understanding of both of the fields.

The ethical and legal discussions need to keep up the pace with the astonishingly rapidly developing field of AI, where AI continues to replace humans in various fields (see, e.g., Joamets & Chochia, 2020; Troitiño et al., 2020, pp. 303–317). One of the most controversial fields where one can clearly observe potential threats coming from AI deployment is the field of children education. In 2019, *The Wall Street Journal* produced a short documentary entitled *How China Is Using Artificial Intelligence in Classrooms|WSJ (2019)*. The video offers an intriguing narrative of China's dominance in AI production and the portrayal of how algorithms have seamlessly integrated in the social realms of the urban cities of China including education of young students. Shared on video streaming website, the documentary quickly gathered millions of views. Crystal Tai (2019), the producer of the documentary, gives the viewers a glimpse of how the future of education may look aided by AI. The centerpiece of this story is a high-tech headband being worn by students filled with EEG sensors—two at the back of the ears and one in front for the forehead. These black-matte headwear are being used to monitor the level of concentration of primary school students during lectures. There is a small LED light in front of the headband which will light up as color red, blue, and white. The former is the optimal color as it is geared to reveal that the student wearing the headband is focused and attentive to the lessons. On the other hand, if the LED light projects a bright blue light, the student is deemed to be distracted or not focused enough on the lessons from the teacher. And the lights will turn white if the student is "offline"—or super deep in focus as implied. But that is not what is more interesting in this situation. The headband generates data that is sent to the teacher's computer desk in real time so the teacher knows who needs attention and some sort of "calling out." It does not stop there. That data is sent to the parents every 10 min as a monitoring mechanism for their children. They can see if their child is performing well compared to their peers. The teachers and the parents embrace this technology as something that is of value to them. However, what about the students that are being subjected to this type of panopticon? According to an interview of a student in the same documentary, they feel controlled and pressured (Tai, 2019). Others are being scolded heavily by their parents back home because of the data reports that were sent to them (Tai, 2019). While some may seem just to be comfortable to conform.

This story from China is just a speck of dust, so to speak. Even though the documentary was produced a couple of years ago, it especially holds relevance today. With the attention and hype-based push for algorithmic-based decision support systems in various decision-making levels in our society, AI is here to stay and ready to be integrated in people's lives directly or indirectly. Solutions offered by modern technologies, specifically AI, are simply too great and the process of their integration into the lives of the societies is developing full stream, impact being clearly felt in various fields.

As a matter of fact, algorithmic-based software products are already deployed in the government, big organizations, homes, etc. It is rather invasive that stopping its usage is non-viable. However, invasive technologies like these AI systems bring in risks as much as the opportunities of its potentialities. Despite the enormous benefits of the technology, privacy-related consequences due to additional security and control measures put up a huge challenge in front of our societies. The Human Rights Watch (2020), along with numerous other organizations, published a joint statement, indicating some conditions that technology-assisted measures to fight the COVID-19 pandemic should meet in order to respect human rights. The human race is very much keen on looking at how AIs can be developed safely and ethically (Blackman, 2020; Tzachor et al., 2020). Along with this, ensuring safety and rights protection of vulnerable population is one of the main agenda we see and listen to in various global fora related to AI governance. Part of this vulnerable population are the children, who, in no doubt, are being subjected in frontline to the risks of the AI systems without them knowing. As the world starts acknowledging the inevitability of embedding AI in the lives of the children, the discussion about the ethical and legal use of AI for children is now being put on the decision table. In December 2021, UNICEF along with the Government of Finland hosted Global Forum on AI for Children (UNICEF, & Ministry of Foreign Affairs, Finland, 2021). This event was an attempt to extend the discourse on the impact of artificial intelligence systems on children's rights and how to design digital technology policies for the use of AI while ensuring the right interests and benefits of children. "Children are already interacting with AI technologies in many different ways: they are embedded in toys, virtual assistants, video games, and adaptive learning software" (UNICEF, 2021). They are not just invasive but are "ubiquitous" (WEF, n.d.) now more than ever before.

As established, there are multiple instances where AI systems are embedded in the lives of the children; however, to set the scope and limitations of this chapter, the author will touch upon and just focus on the use of AI in schools and learning environments of children. More so, what are the violations that these algorithmic systems pose pertaining to the rights of a child? For one to explore these key points, it is crucial to provide an answer to the following question: "Should there be clear limits on the use of AI in the learning environments for children?"

Majority of scholars, and the authors would agree, tend to answer a resounding "yes" to the above posed question. There must be limits on the use of AI systems in the learning environment of children may it either be in the in-person setting or virtually. That is of course up until we find a children-specific approach on how we should eliminate the risks of the societal, psychological, and legal impacts of AI-enabled technologies to children. This is a tremendous task and needs to arrive at a certain level of consensus by multi-level governments and institutions. It must be stressed that it is adamantly critical to do so. Here are the key points why it is crucial to put AI usage on limits when it comes with educating children in a conventional learning setting: (1) there is an existing gap in safeguarding children's right to privacy; (2) the prevalence of algorithmic bias in AI systems which violates children's rights to be protected against discrimination, abuse, or exploitation; and (3) lack of clear ownership in the AI systems which puts the right of a child to complain formally and legally in jeopardy.

2 Gaps in Safeguarding Children's Privacy Rights in AI Systems

The rights that an adult holds are different from those of a child, and they are different for a reason. In terms of the rights to privacy, adults have much more capacity to understand and provide clarity of how their data are being used by AI systems. They also have the capacity to provide permission on what personal data they would like to share or not. On the other hand, children are limited by such capacity. More often than not, they are not given options in terms of protection of their data compared to adults. As the researchers from Canadian Standards Association put it, "children may lack the awareness or have limited literacy skills with which to fully grasp their right to privacy and understand the risks and potential long-term impacts of data sharing and AI processing" (Irwin et al., 2021, p. 13). If invasive AI systems are put into educational settings such as the classrooms, do children have the capacity to demand privacy? It seems not the case. Considering the authoritative nature of school systems, school administrators will tend to lean more on AI to improve their hold of control and command over students. Recalling the story at the start of this chapter, one can say that as long as AI systems provide the results they are intended for, school officials, teachers, school administrators, and even parents will support its usage to the students. This then leads us to the importance of these power holders—teachers, school administrators—as mediators between AI and children.

As the mediators, they are obliged to protect the rights of the child to privacy. In Article 16 in the Convention of the Rights of a Child, it is highlighted that "no child shall be subjected to arbitrary or unlawful interference with his or her privacy, family, home or correspondence, nor to unlawful attacks on his or her honour and reputation" (UN, 1990). Respecting the rights to privacy of the child would mean that as mediators, they must be active in creating safeguard mechanisms to not submit children as victims of data privacy risks while in classrooms or inside other modes of learning environments such as online class or Massive Open Online Course.

For the teachers and school administrators, having a clear set of internal guidelines would be crucial especially in dealing with the extraction of children's personal identifiable information. It is highly possible that school administrators will not be of much capacity to have an AI system of their own so they will have to outsource it to third-party companies. That being said, personal data of children must be protected. It must not be sold to third party at any cost that would lead to improper use or exploitation.

More so, parents are the primary mediators. They should then be provided with the avenue where they can understand how AI systems operate around their children's learning environment. This does not just include the benefits of AI to the intellectual development of their child but also the threats and risks that the system poses. This may seem unconventional and not a practice by most tech companies producing AI systems; they might not be thinking about this because it is not their priority, profit is their topmost concern, but they must be demanded to do so citing good conscience. With the well-informed parent, it can be said that improvements in the interaction of children via AI would be enhanced and the monetization of children's data will be minimized to a diminishing degree.

Another interconnected point is the possibility of constant digital surveillance toward the child. Can a child develop his or her maximum potential cognitively and socially while having the knowledge that he or she is being monitored? This can be considered a threat to the child's right to freedom of expression and may also restrict his or her right to choose their own opinion, thoughts, and even religion (UNICEF, n. d.). AI systems must not be used as a panopticon tool toward the students. The author of this chapter argues that the effect of a panopticon if applied in the student-AI Classroom context has the same effect as Michel Foucault's Panopticon effect on prison inmates. According to Foucault (1977, p. 201): "the major effect of Panopticon is to induce...a state of conscious and permanent visibility that assures the automatic functioning of power." Facial recognition software using AI systems are examples of such systems that have the potential to be used as an apparatus for improper surveillance. As Selinger and Hartzog (2020, p. 120) put it, "face surveillance of all kinds presents a panoply of harms, most notably corrosion of collective autonomy through the chill of increased surveillance and machines indulge the fatally flawed notion of perfect enforcement of the law." With this, the issue of control of AI systems must be blatantly confronted.

The European Union has one of the most well-cited policy guidelines in data protection with its iron-clad safeguard mechanisms in its General Data Privacy Regulation (GDPR) of 2016. Specifically for a child, Article 8 of the GDPR provides special protection to the data vulnerabilities of a child (European Commission, 2019b). This should serve as a benchmark in creating a child-appropriate, so to speak, data privacy protection among nations.

3 Algorithmic Bias in AI for Education

As a matter of fact, algorithms are just as good as the datasets that they are fed into (Snowflake Asia Pacific & O'Connor, 2019). The datasets command how an AI system operates. Datasets provide the limits on how it performs. An AI builds its "intelligence" from big datasets that were extracted from historical data of the past experiences or of individuals. And in the context of education, certain data may include grades, socio-economic data, and personal identifiable data. Multiple times, it has been reported that data fed to algorithms has a tremendous amount of racial, gender, and age biases (Obermeyer et al., 2019; Brown, 2020; Kleinberg et al., 2019). With this in mind, it can be argued that designers of AI systems for children in education must tackle the issue of discrimination that children might go through if the data that was imputed to the system contains biased-based information. It is therefore evident that children's lives can be negatively affected. Such non-quality data will just reinforce existing inequalities in the education system. Baweja and Singh (2020) argue that "the increased implementation of AI in society is fostering digital bias and replicating the harm that is currently being fought against. In that sense, this technology disproportionately affects the vulnerable by exacerbating discriminatory practices already present in modern society."

If AI systems used for educating a child do not have the capacity for the basic standards of equality, then it is in violation of the right of a child against discrimination, abuse, and exploitation. Thus, it is highly endorsed for these biased algorithmic decision-making tools to be put into high scrutiny before deployment on schools. Keeping the children exposed to them unsupervised is costly and has inconceivable consequences not just to the life of a child but to the society at large.

This reminds one about the story of the A-Level Fiasco in the UK (Kolkman, 2020), when the students were not able to sit in for their General Certificate for Secondary Education (GCSE) exam due to the onset of the COVID-19 pandemic. With this the Department of Education of the UK decided to employ an algorithm that would provide alternative exam scores for the graduating students who are entering the universities. What happened was that 39.1% of students received downgraded grades (Adams et al., 2020) compared to what they would have been expected to receive. A public outcry and blatant opposition to these A-level grades arose all over the nation. This prompted the UK Government to stop its usage and work on other alternatives. But the damage has been done because it was reported that some top students were not able to be admitted to universities because of such scores (Fazackerley, 2021). This fiasco is an unfortunate but, at the same time, certainly suitable manifestation of how algorithmic bias affects students and children. It must not be repeated again in any form and it is the responsibility of a society, especially governments, to ensure that it will not be repeated again in any form.

4 Lack of Clear Ownership for AI Systems

Perhaps in a legal sense, the most important aspect that needs to be discussed in this chapter is who is held accountable when an AI system for education has violated certain rights of children. Children have the right to be provided with an avenue where they can formally or legally complain if their rights are breached. But if the vagueness of the responsibilities within the implementation of AI systems in education still exists, this certain right of a child will be considered moot and invalid. Thus, understanding the accountability aspect for AI systems is of utmost importance.

Due to its opacity enabled by its "black box" nature, AI systems are undefined in terms of agency. And having an agency is very crucial in a legal process because it involves transparency that brings about legitimacy. The opaqueness of AI is the reason why the finger-blaming on who should be responsible if an AI system breaches children's rights when deployed in the education sphere happens. Should it be the school administrator? The AI designer or Tech company? The government? Or the AI itself? This is still an existing debate, but to approach this, the insight from Rosentzveig et al. (2017) is logical and might be helpful. Rosentzveig et al. (2017) suggest that "ensuring that algorithms are sufficiently transparent would help support legitimacy and enable verification" (p. 94). The best way for this to occur is to have a clear identification of ownership of algorithms. It is simpler if there is a clear ownership. Transparency is possible and legitimacy will be enforced. With that, legality will come into play in full force.

5 The Three Rights Before Deployment of AI Systems in Education

In this chapter, three rights of a child have been highlighted: (1) right of a child to privacy; (2) right of a child to be protected against discrimination, abuse, or exploitation; and lastly, (3) right of a child to formally and legally complain if their rights are breached. It must be noted though that these rights are the three main rights that the authors of this chapter determined as the three main rights that must be considered by decision-makers when opting to deploy the AI tools in various learning environments of children. This is because it transcends and enables other rights of children under universal convention. The mentioned three rights above must be treated as the three fundamental rights that must be supported and protected by the AI systems that are deployed and used in various educational settings.

More often than not, these three rights of children are being disregarded in the design process and are often violated. If the tech companies working and developing these AI systems for education really vow to help optimize and essentially radicalize the education sector, they must work on providing elevated functions for these three main rights of a child in the digital age. This can be done by including students, or children, in their design process flow through, as an example, collection of their unique and particular input, focus group discussions, and even close consultation and observation.

On the other hand, school administrators and teachers must be aware that they have to work on safeguarding the rights of a child by religiously integrating in the AI systems development the following: data privacy and safety mechanisms; fairness and transparency audit; and clear identification and ownership. As seen in Fig. 1

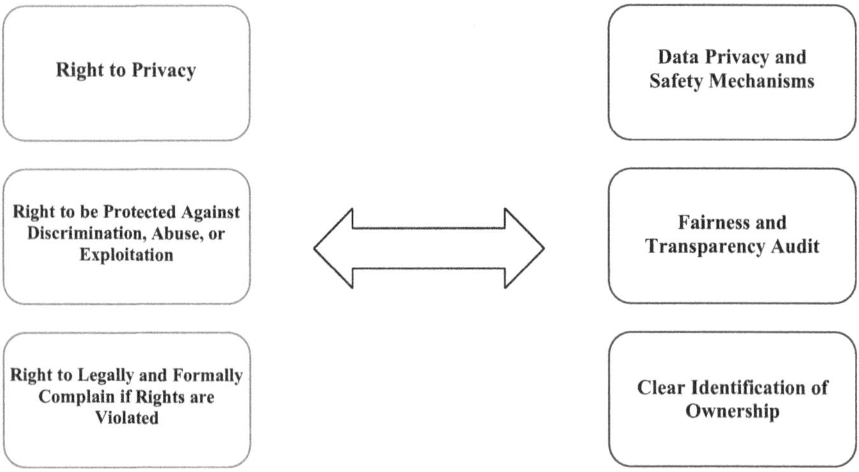

Fig. 1 Three Rights and three important elements required for AI systems in education

above, such safeguarding system procedures will exponentially result in a huge positive effect on the learning experience of the children in the most optimal situation while protecting their fundamental rights. With this, risks will be minimized as children will be taught in a healthy and safe space where they can freely express themselves and reach their full potential.

Going back to the initial question in the beginning of this chapter—"Should there be limits in deploying AI systems in the learning environments of children?" The answer is the same as mentioned above. Limits on AI system deployment must have limits on its usage until safety provisional measures that ensure the protection of the three main rights of a child, namely right to privacy, right to be protected against discrimination, abuse, or exploitation, and right to formally and legally complain if rights are violated as discussed above, are embedded in place strictly and steadily in every learning environment of a child.

6 Conclusion

Modern technologies, including the AI solutions, are ever present in our daily lives and impact them. These technological solutions are often making our lives easier, providing new opportunities that are often unique and unimaginable before. Representatives from various fields, where technology has offered new solutions at an extremely rapid pace, are constantly evaluating these opportunities, trying to increase the benefits of given prospects, which is a natural process. This has led scholars from the various fields to work on better understanding the positive and negative impact of modern technologies on society, in daily lives of which these technologies have now become an integral part. Such work is being done on different levels, from academia to governing bodies, and in cooperation among those. If we talk about governmental approach, then the EU has certainly demonstrated itself to be a global pioneer, attempting to tackle sophisticated issues derived from such extensive and increasing usage of emerging technologies, while it has also delivered important relevant documentation and guidelines on the issues. *Ethics Guidelines for Trustworthy Artificial Intelligence* from the European Commission's (2019a) and *The Artificial Intelligence Act* are important steps taken in this direction. At the same time, it is important to underline that European efforts, even if extremely important, do not match the fast pace of development the emerging technologies have and spread of their usage in daily lives of people. The European Parliamentary Research Service (2020) evaluated challenges of legal framework on AI ethics and has listed six key risks factors. "A growing mismatch between the exponential growth of the AI market and a 'delayed' regulatory response" was among those identified.

However, as the process of deploying modern technologies, including AI, is expanding, new gray areas are discovered, such as above demonstrated field of education. Without addressing the data privacy risks, the algorithmic bias, and the unclear ownership of the AI systems for education, they must be put in limitations for deployment in various learning settings for children. To live a safe and healthy life is the fundamental right of a child. And if the AI systems installed in the learning environment of a child take away this right, then these AI systems shall be prohibited. But of course, realistically speaking, to ban outright these systems in the classrooms is near to impossible due to the techno-paradigm shifts that the world is experiencing currently. Even with the banning of Cayla Dolls in Germany (BBC, 2017), it did not stop the development of the AI systems for children. The AI systems portray ambivalence. Thus, the fact that it also has a functional purpose in improving children's educational learning environment must not be discounted. What is important is ensuring that the interest of a child comes on the top of most considerations when making the decision for its usage within the learning sphere of a child.

Policymakers shall also consider the recommendations to require a license to operate toward tech companies that are pushing the AI systems in education. "Edtech products need thorough scrutiny and evaluation that goes beyond data privacy assessments. They need a license to operate in children's education" (Hillman, 2021). This is an actionable suggestion for society to keep up with the effects of the AI systems in the education sector today because one thing is for sure—the rights of a child are non-negotiable. There is a need to further develop field-specific guidelines, on the EU level, which would allow using emerging technologies in more ethical and human rights friendly way. Parfett et al. (2020), when talking about characteristics of AI solutions, offer comparison with a child, explaining that AI, just as a child, requires education and training. Being ready to rely on the solutions AI technologies offer requires such education and training to be conducted first. And similarly to the children studying from the teachings of their parents, modern technologies should study from the data and the programs.

References

Adams, R., Weale, S., & Barr, C. (2020, August 13). *A-level results: Almost 40% of teacher assessments in England downgraded*. The Guardian. Retrieved December 10, 2021, from https://www.theguardian.com/education/2020/aug/13/almost-40-of-english-students-have-a-level-results-downgraded

Baweja, S., & Singh, S. (2020, July 16). *Beginning of artificial intelligence, end of human rights*. LSE Human Rights. Retrieved December 1, 2021, from https://blogs.lse.ac.uk/humanrights/2020/07/16/beginning-of-artificial-intelligence-end-of-human-rights/

Beever, J., McDaniel, R., & Stanlick, N. A. (2019). *Understanding digital ethics: Cases and contexts [online]*. Routledge. https://doi.org/10.4324/9781315282138

Blackman, R. (2020, October 15). *A practical guide to building ethical AI*. Harvard Business Review. https://hbr.org/2020/10/a-practical-guide-to-building-ethical-ai

Brey, P. (2010). Philosophy of technology after the empirical turn. In *Techné: Research in philosophy and technology*, vol. 14, no. 1, pp. 36–48 [online]. https://doi.org/10.5840/techne20101416.

British Broadcasting Corporation. (2017, February 17). *German parents told to destroy Cayla dolls over hacking fears*. BBC News. Retrieved December 10, 2021, from https://www.bbc.com/news/world-europe-39002142

Brown, A. (2020, May 30). *Gender bias in predictive algorithms: How applied AI research can help us build a more equitable future*. Forbes. Retrieved December 13, 2021, from https://www.forbes.com/sites/cognitiveworld/2020/05/30/gender-bias-in-predictive-algorithms/?sh=7b66f3bd57ac

Brownsword, R., & Goodwin, M. (2012). *Law and the technologies of the twenty-first century* [online]. New York: Cambridge University Press. https://doi.org/10.1017/CBO9781139047609.

Chochia, A., & Nässi, T. (2021). Ethics and emerging technologies – Facial recognition. *IDP Revista de Internet Derecho y Política, 34*, 1–12. https://doi.org/10.7238/idp.v0i34.387466

Deigh, J. (2010). *An introduction to ethics* [online]. New York: Cambridge University Press. doi: https://doi.org/10.1017/CBO9780511750519.

Deloitte. (n.d.). *Europe's "Trustworthy AI" meets AI ethics. The ethics of AI* [online]. Accessed April 27, 2021, from https://www2.deloitte.com/nl/nl/pages/risk/articles/europes-trustworthy-ai-meets-ai-ethics.html

Ellul, J. (1964). *The technological society*. Vintage Books.

European Commission. (2019a). *Ethics guidelines for trustworthy AI*. European commission high-level expert group on artificial intelligence. Retrieved from https://ec.europa.eu/futurium/en/ai-alliance-consultation

European Commission. (2019b). *Can personal data about children be collected?* European Commission – European Commission. Retrieved December 16, 2021, from https://ec.europa.eu/info/law/law-topic/data-protection/reform/rights-citizens/how-my-personal-data-protected/can-personal-data-about-children-be-collected_en

European Parliamentary Research Service. (2020). *European framework on ethical aspects of artificial intelligence, robotics and related technologies. European Added Value Assessment* [online]. Available from https://www.europarl.europa.eu/thinktank/en/document.html?reference=EPRS_STU(2020)654179

Fazackerley, A. (2021, May 23). *Top pupils rejected by universities in A-levels fiasco fallout*. The Guardian. Retrieved December 10, 2021, from https://www.theguardian.com/education/2021/may/22/top-pupils-rejected-by-universities-in-a-levels-fiasco-fallout

Foucault, M. (1977). *Discipline and Punish: The birth of the prison*. Vintage Books. http://depts.washington.edu/lsearlec/510/Texts/Foucault-Panopticism.pdf

Hillman, V. (2021, October 28). *Algorithmic (in)justice in education: Why tech companies should require a license to operate in children's education*. Parenting for a digital future. Retrieved December 10, 2021, from https://blogs.lse.ac.uk/parenting4digitalfuture/2021/11/03/algorithmic-injustice/

Human Rights Watch. (2020). *Joint civil society statement: States use of digital surveillance technologies to fight pandemic must respect human rights* [online]. Available from https://www.hrw.org/news/2020/04/02/joint-civil-society-statement-states-use-digital-surveillance-technologies-fight

Irwin, J., Dharamshi, A., & Zon, N. (2021). *Children's privacy in the age of artificial intelligence*. Canadian Standards Association. https://www.csagroup.org/wp-content/uploads/CSA-Group-Research-Children_s-Privacy-in-the-Age-of-Artificial-Intelligence.pdf

Joamets, K., & Chochia, A. (2020). Artificial intelligence and its impact on labour relations in Estonia. *Slovak Journal of Political Sciences, 20*(2), 255–277. https://doi.org/10.34135/sjps.200204

Joamets, K., & Chochia, A. (2021). Access to artificial intelligence for persons with disabilities: Legal and ethical questions concerning the application of trustworthy AI. In *Acta Baltica Historiae et Philosophiae Scientiarum*, vol. 9, no. 1, pp. 51–66 [online]. https://doi.org/10.11590/abhps.2021.1.04

Jobin, A., Ienca, M., & Vayena, E. (2019). The global landscape of AI ethics guidelines. In *Nature machine intelligence*, vol. 1, pp. 389–399 [online]. https://doi.org/10.1038/s42256-019-0088-2.

Johnson, D. (2004). "Computer ethics". In: Floridi, L. (ed.). *The Blackwell guide to the philosophy of computing and information* (pp. 65–75). New York: Blackwell [online]. https://doi.org/10.1002/9780470757017.ch5.

Kerikmäe, T., Solarte-Vasquez, M. C., Rudanko, M., & Ttroitiño, D. R. (Eds.). (2020). *Inteligencia artificial: De la discrepancia regional a las reglas universales. Integración de percepciones políticas, económicas y legales*. Thomson Reuters.

Kleinberg, J., Ludwig, J., Mullainathan, S., & Sunstein, C. (2019). Discrimination in the age of algorithms. *Journal of Legal Analysis, 10*(2018), 113–174. https://doi.org/10.3386/w25548

Kolkman, D. (2020, December 10). *"F**k the algorithm"?: What the world can learn from the UK's A-level grading fiasco*. Impact of Social Sciences. Retrieved December 1, 2021, from https://blogs.lse.ac.uk/impactofsocialsciences/2020/08/26/fk-the-algorithm-what-the-world-can-learn-from-the-uks-a-level-grading-fiasco/

Obermeyer, Z., Powers, B., Vogeli, C., & Mullainathan, S. (2019). Dissecting racial bias in an algorithm used to manage the health of populations. *Science, 366*(6464), 447–453. https://doi.org/10.1126/science.aax2342

Parfett, A., Townley, S., & Allerfeldt, K. (2020). AI-based healthcare: A new dawn or apartheid revisited? [online]. In *AI & Society*. https://doi.org/10.1007/s00146-020-01120-w.

Parsons, T. D. (2019). *Ethical challenges in digital psychology and cyberpsychology [online]*. Cambridge University Press. https://doi.org/10.1017/9781108553384

Rosentzveig, Z., Law, K., & Heegaard, A. (2017). *The ethical challenges of using algorithms in government* [Master's thesis]. https://doi.org/10.13140/RG.2.2.35140.04487

Selinger, E., & Hartzog, W. (2020). The inconsentability of facial surveillance. *Loyola Law Review, 101*(2019) 120. https://papers.ssrn.com/sol3/papers.cfm?abstract_id=3557508

Snowflake Asia Pacific, & O'Connor, P. (2019, July 5). *Why AI algorithms are only as good as the data they're fed – Snowflake*. IT Brief New Zealand – Technology news for New Zealand's largest enterprises. https://itbrief.co.nz/story/why-ai-algorithms-are-only-as-good-as-the-data-they-re-fed-snowflake

Tai, C. [Wall Street Journal]. (2019, October 1). *How China is using artificial intelligence in classrooms | WSJ* [Video]. YouTube: https://www.youtube.com/watch?v=JMLsHI8aV0g&t=2s

Troitiño, D. R., Kerikmäe, T., Shumilo, O., & Ramírez Barbosa, P. A. (2020). El libro blanco sobre inteligencia artificial: análisis y comentarios sobre mercado, valores y cooperación europea. In T. Kerikmäe, M. C. Solarte-Vasquez, M. Rudanko, & D. R. Troitiño (Eds.), *El Parlamento Europeo y la necesidad de una legislación común en un marco europeo de la inteligencia artificial. Inteligencia artificial: De la discrepancia regional a las reglas universales. Integración de percepciones políticas, económicas y legales* (pp. 303–317). Thomson Reuters.

Tzachor, A., Whittlestone, J., Sundaram, L., & HÉigeartaigh, S. Ó. (2020). Artificial intelligence in a crisis needs ethics with urgency. *Nature Machine Intelligence, 2*(7), 365–366. https://doi.org/10.1038/s42256-020-0195-0

UNICEF. (2021, November 27). *Digital child's play: Protecting children from the impacts of AI*. UN News. Retrieved December 1, 2021, from https://news.un.org/en/story/2021/11/1106002

UNICEF. (n.d.). *The Convention on the Rights of the child: The children's version*. Retrieved December 9, 2021, from https://www.unicef.org/child-rights-convention/convention-text-childrens-version

UNICEF, & Ministry of Foreign Affairs, Finland. (2021, December). *Global forum on AI for children* [Conference session]. https://www.unicef.org/globalinsight/ai-global-forum

United Nations. (1990). *Convention on the rights of the child*. https://www.ohchr.org/en/professionalinterest/pages/crc.aspx

Verbeek, P. P. (2011). *Moralizing technology, understanding and designing the morality of things* [online]. Chicago: University of Chicago Press. https://doi.org/10.7208/chicago/9780226852904.001.0001

Winston, M., & Edelbach, R. (2011). *Society, ethics and technology [online]*. Wadsworth Publishing. https://doi.org/10.1017/CBO9781107340961

World Economic Forum. (n.d.). *Generation AI: Developing artificial intelligence standards for children*. Retrieved December 6, 2021, from https://www.weforum.org/projects/generation-ai

Part II
Law

Data Protection Chapter

Kärt Salumaa-Lepik and Nele Nisu

Abstract Personal data protection is getting more and more noticed in the twenty-first century. Personal data may be transferred quickly from one contact point to another in large amounts and free of charge within a fraction of a second. Personal data is called the new oil and those processing personal data are subject to many renewed rules to protect personal data.

The rapid development of information technology in recent decades has brought innumerable benefits, such as the spread of eHealth and the development of personalised medicine, but also concerns, such as phishing for data and the threat of identity theft.

The widespread adoption of the new technologies and the increased need for privacy have led to the creation of new regulatory proposals and frameworks in the European Union that ensure the protection of the rights of individuals and regulatory and judicial oversight.

One of these frameworks is the GDPR. The creation of new data protection regulation addresses the need for a comprehensive and contemporary legal regime that governs the processing of personal data at a time when such data has become a product marketed and sold in exchange for services or profits. In addition, there are a number of proposals and other new pieces of legislation in more specific areas that also contribute to the further development of the level of data protection. Some of these are selected and covered by this chapter.

1 Introduction

Recent rapid global technological developments have brought much attention to a number of areas, one of which is personal data protection. According to a special Eurobarometer survey (European Commission, 2021a) conducted in 2021, an overwhelming majority of EU citizens think that the Internet and digital tools will play an

K. Salumaa-Lepik · N. Nisu (✉)
Tallinn University of Technology, Tallinn, Estonia
e-mail: kart.salumaa-lepik@taltech.ee; nele.nisu@mkm.ee

D. Ramiro Troitiño et al. (eds.), *Digital Development of the European Union*,
https://doi.org/10.1007/978-3-031-27312-4_10

important role in the future. In addition to that, more than half (56%) of the EU citizens surveyed expressed their concern about cyber-attacks and cybercrime such as theft or abuse of personal data, malicious software, or phishing (European Commission, 2021b). Thus, there is also a need for a better standard of data protection and supervision to ensure proper level of personal data protection that every EU citizen could benefit.

In order to protect the personal data of the citizens and give them more control over their personal data, the EU data protection reform became applicable from 25 May 2018. The adoption of the General Data Protection Regulation (European Parliament and Council, 2016a) in 2016 modernised EU data protection legislation, making it fit for protecting fundamental rights in the context of the digital age's economic and social challenges (European Union Agency for Fundamental Rights and Council of Europe, 2018, p 30). As an important observation, it is worth noting that in principle GDPR equally applies to both private and public sector entities.

Europe has set its digital targets for the year 2030 with the EU Digital Compass (European Commission, 2022a). One of the ambitious targets is related to the digitalisation of public services providing those key public services would be 100% online, 100% of citizens would have access to medical records (eHealth), and 80% of citizens would be using digital ID (digital identity) (European Commission, 2022a). As a result, the authors of this chapter may conclude that data protection issues will remain in focus in the years to come, as none of these goals can be achieved without a high standard of data protection.

To get a better idea of the core subject, it is important to briefly explain the importance of the protection of personal data. Therefore, it should already be emphasised in the introduction that in addition to vital personal data processing carried out in emergency situations such as processing blood type data before life-saving operation, processing data may very often be associated with convenience services, and convenience is precisely what attracts personal data subjects in the role of consumers to use these services. It is possible to assume that an average consumer does not research too much about how many information systems or parties were involved in the data transfer process to make the service possible. Data subjects might know even less about how the service was developed and which was the origin of data. This is where the minimum standards on data protection take on their greatest significance. It is these standards stipulated in data protection legal framework that seek to provide protection to the weaker party—the data subject—in a similar way as to a consumer or employee in a legal relationship. The GDPR sets the legal framework within which the Member States and the companies operating in them must plan and carry out their data processing activities to protect the personal data of the data subject. The question that may arise for different stakeholders in this context is how innovation can continue and how it can be encouraged while ensuring the correct application of data processing principles and maintaining adequate safeguards to guarantee the rights of data subjects.

This chapter gives an overview of recent developments and highlights of the personal data protection area chosen more specifically by the authors of this chapter regarding their professional background. In this chapter, authors will describe the developments in the area of EU personal data protection from different angles with an emphasis on finance and eHealth issues. The authors of this chapter provide readers with an overview of the current state of the field (major issues and developments), the challenges, and the possible developments ahead. Terms "data subject" and "consumer" are used in parallel as synonyms to describe the data subject. Descriptive-comparative method is chosen to bring to the reader the essential content of this chapter from a legal perspective.

2 State of the Art

Currently, in 2022 when this chapter is written, the comprehensive legal framework sheltering the area as *lex generalis* is the GDPR (European Parliament and Council, 2016a) with the Law Enforcement Directive (European Parliament and Council, 2016b). GDPR became directly applicable in all EU Member States from 25 May 2018. Protection of personal data and privacy matters have been in the limelight ever since and cannot be considered as a discretionary niche topic anymore. Administrative fines are a central element in the new enforcement regime (Article 29 data protection working party, 2017) introduced by the GDPR and may go up to €20 million or 4 percent of worldwide turnover for the preceding financial year—whichever sum is higher.

After the adoption of the GDPR, EU Member States (see further Salumaa-Lepik et al. (2021), pages 23–57) were also facing the need to modernise their national domestic data protection laws according to the requirements originating from GDPR. Some years after the adoption and application of GDPR, the initial work by the EU institutions and EU Member States is done, but the area keeps on developing.

One of the biggest already known forthcoming developments in relation to data protection in the future is linked with the area of electronic communications. A proposal for a regulation on the respect for private life and the protection of personal data in electronic communications (proposal for ePrivacy Regulation replacing the Directive 2002/58/EC and specifying the GDPR) was published in the beginning of 2017. Currently, in 2022, it is the subject of negotiation between the co-legislators. In relation to GDPR, the ePrivacy legislation would be in the role of *lex specialis* and supplement and clarify the GDPR with more specific regulations in the area of electronic communications. The Council of the EU has expressed that an update to the existing ePrivacy directive of 2002 is needed to cater for new technological and market developments, such as the current widespread use of Voice over IP, web-based email, and messaging services, and the emergence of new techniques for tracking users' online behaviour (European Council, 2021). In contrast to the GDPR, many ePrivacy provisions are intended to apply to both natural and legal persons.

Regarding the more specific area of health/eHealth which is also discussed in more detail in this chapter, the cross-border exchange of healthcare information is regulated by European and national legislation regarding the protection of personal data. The Directive 2011/24/EU (European Parliament and Council, 2011) on patients' rights in cross-border healthcare (i.e., the Cross-border Directive) specifies the rights and rules for the processing of health data in the scope of healthcare treatment in another EU country. It aims at achieving high level of trust and security between the Member States' healthcare service providers. Under art 14 of the Cross-border Directive providing the Commission Implementing Decision 2019/1765 (European Commission, 2019a) the rules for the establishment, the management, and the functioning of the network of national authorities responsible for eHealth are stipulated; also specific norms for the COVID pandemic may be found in Regulation (EU) 2021/953 (European Parliament and Council, 2021). As several Member States launched COVID-19 certificates, common standards were needed. Standards had to be fully interoperable, compatible, secure, and verifiable to be used effectively in a cross-border context when Union citizens exercise their right to free movement. The above-mentioned regulation is a good example of a quick solution during the COVID-19 pandemic.

The more specific state of the art of more specific legal acts and areas described in this chapter are provided in more detail in the subsections below.

3 A Selection of Recent Major More Specific Developments in the Field of Personal Data Protection

3.1 Digital Identity for All Europeans

According to GDPR art 4(1), personal data means any information relating to an identified or identifiable natural person. Hence, the identity of a person and identifiability are key elements of personal data protection. It may prove difficult for an average person to imagine how to further organise his or her life in the event of losing the capability to sign into a bank account, pay bills online, manage services, etc. Nowadays, a big part of the communication between the state and authorities is organised via the Internet (Information System Authority, 2022). In Estonia, citizens may use their digital IDs to pay bills, vote online, sign contracts, shop, access their health information, and much more (EAS, 2022). Digital ID is for digital identity and can be used for identification purposes and for giving digital signatures (Information System Authority, 2022). But only about 60% of the EU population in 14 Member States are able to use their national eID across borders and only 14% of key public service providers across all Member States allow cross-border authentication with an e-Identity system without the need for a password (European Commission, 2022b). To ensure the Digital Single Market, we may be led to the fact that this issue should be also regulated across Member States and harmonised. In that matter, eIDAS (European Parliament and Council, 2014a) is a key enabler and aims to provide for cross-border use to identify and authenticate users with a high level of assurance. As mentioned in the Commission proposal (European Commission, 2021c), identity solutions falling outside the scope of eIDAS—such as those offered by social media providers and financial institutions—raise privacy and data protection concerns. Since the entry into force of the eID part of the Regulation in September 2018, only 14 Member States have notified at least one eID scheme and only 7 schemes are entirely mobile, responding to current user expectations. The main findings of the evaluation with respect to electronic identity is that eIDAS has not achieved its full potential. Proposal offers a harmonised approach to security, for citizens relying on a European digital identity representing them online, and for online service providers who will be able to fully rely on and accept digital identity solutions independently of where they have been issued (European Commission, 2021c). If an amendment to the Regulation is adopted, the Member States shall provide and recognise electronic identification means of natural and legal persons and issue a European Digital Identity Wallet. The European Digital Identity will be available then to EU citizens, residents, and businesses who want to identify themselves or provide confirmation of certain personal information across the EU. Hence, in the future it will be easier for consumers to prove their identity electronically, to store their medical prescription and use it anywhere in Europe, or to file their tax returns in different places across the EU (European Commission, 2022b). European Digital Identity Wallet allows Member States to control that sensitive datasets, for example, related to health, are only requested by service providers in accordance with national law (European Commission, 2021c).

This initiative aims to support the Union's transformation towards a Digital Single Market (see further European Commission, 2015). The current legal framework for trust services would only increase fragmentation and reduce trust if left to Member States to decide alone. eIDAS introduces specific data protection safeguards, but any personal data processing under this Regulation should be carried out in full compliance with the GDPR (European Commission, 2021c). The amendment demonstrates that it is possible to create coherent strategies to use the best technologies, which will certainly benefit data subjects as consumers in the end to identify themselves and to share their data with different service providers, also allowing us to get rid of multiple passwords between different service providers. All this should also contribute to a higher standard of data protection.

3.2 *Values of eHealth and Its Data Protection Challenges Across the EU Countries*

Digital health and care refer to tools and services that use information and communication technologies (ICTs) to improve prevention, diagnosis, treatment, monitoring, and management of health and lifestyle (European Commission, 2022c). As mentioned in World Health Organization's (WHO) website, the first nationwide electronic health record system was launched in Estonia (World Health

Organization, 2016). Estonia launched its electronic record system in 2008, becoming the first country in the world to fully implement such system nationwide with records covering an individual's medical history from birth to death. In 2009, Estonia used health information exchange to upload all medical documents into the system (World Health Organization, 2016), and the aim was to create a health information system with four major pillars—a digital reception, an electronic health record, a digital prescription, and a joint image bank (Riigi Teataja, 2008).

The situation cannot be acceptable if the patient's health data is only visible to the doctor in patient's home country. Cross-border exchange of health data is essential for the provision of better healthcare. For example, a person may be involved in a car accident in another country and need medical help. In this case, the healthcare provider in another Member State should be able to know and see the person's complete medical history—what allergies, contraindications, etc. the patient has—so that the quality of healthcare would be the same as in their home country. Of course, the person can disclose the necessary information to the doctor if he or she is not unconscious, but patients often do not remember all the necessary information or simply make mistakes when providing the information. To this end, Member States have stepped up their cooperation to exchange health data.

Already a 2014 study found that the first information source for medical treatment in another EU Member State for any EU citizen should be the National Contact Points for eHealth that are established as part of the eHealth Network (Technopolis Group, 2019). Article 10(3) of the Cross-border Directive foresees the exchange of information between Member States and calls for the Commission to encourage Member States on that (see also Commission, 2019a, recital 13). As mentioned before, the aim of that Directive is to achieve a high level of trust and security and ensuring access to safe and high-quality healthcare (European Parliament and Council, 2011, art 14(2)(a)).

In Estonia, the legal basis for a cross-border health record exchange platform is regulated by law which came into force in November 2018 (Riigikogu, in force 2022, Chapter 3^2, § 50^7–§ 50^9). The integration of countries into this project has not been very active. Therefore, a separate working group has found possible bottlenecks to increase data exchange: whether, and to what extent, the Member States' judicial areas differ and whether this hinders data exchanges. As the information paper of the X-eHealth in 2021 indicates (X-eHealth, 2021), the following topics were discussed.

For example, one of the topics discussed was whether data processing should be based on the consent of the patient and, if so, whether and to what extent personal data should be processed in the country of destination. It is important to underline—and as it is also indicated in information paper—the consent remains as one of the six lawful bases to process personal data, as listed in art 6 of the GDPR. When initiating activities that involve processing of personal data, a controller must always take time to consider what would be the appropriate lawful ground for the envisaged processing. It is important to note both, in this case and elsewhere, that legal bases for data processing are not ranked and are all equally applicable. This statement has been confirmed several times, lately also by the European Data Protection Board

(EDPB) in the spring of 2020 (European Data Protection Board, 2020a, point 16, also European Data Protection Board, 2020b, point 1(2), also 2(9), see also X-eHealth, 2021, page 30).

So yes, as is indicated in this paper, the judicial areas of different EU Member States can be quite different because healthcare, as such, falls within the exclusive competence of the Member States. This is also underlined by Cross-border Directive (European Parliament and Council, 2011, point 4 and 7) and is also settled in agreement between Member States (European Commission, 2022d, Clause I.3 (4))—all contracting parties have the right to choose, in accordance with EU law and their respective national law, the method for the fulfilment of the required criteria. It is probably confusing that Member States have a different legal framework for issuing data—some allow data to be released from their national information system based on active consent before processing as opt-in while others offer the option of opt-out solution, where the processing itself is based on legislation (in the context of the cross-border exchange of health data, a discussion was raised within the eHealth Digital Service Infrastructure for Cross-Border eHealth Information Services (eHDSI) community on whether the exchange of data should be based on consent and whether this consent could be given in any country (X-eHealth, 2021)). In addition, further processing takes place according to the legal system of a particular country—all conditions of that country apply, including data retention periods, etc. In essence, a person allows the transfer of his or her data to another Member State for the provision of healthcare, and if this transfer was allowed from the data subject's country of residence, the doctor can also see the person's most important medical records from the information systems in the person's country of residence. At the same time, data transmitted to another country, the country of destination, is subject to the laws and regulations of the country of destination (see further in X-eHealth, 2021, point 4.1.2 and 4.1.7). In this way, the data may be processed for the provision of healthcare, but at the same time, in the country of destination, the processing of a particular patient's data may also, for example, be allowed for the supervisory authorities of the country of destination—whether it be for example a national healthcare supervisory authority or a local data protection inspectorate (see also, X-eHealth, 2021, point 4.1.3). It can only be assumed that the intersection of several different legal systems has blurred the overall picture and probably kept countries away from joining the project. It seems that for many, the practice of implementing the GDPR is still evolving and they are adopting a wait-and-see attitude, although the principles and legal bases for data processing have not changed since the previous directive (Directive 95/46/EC, European Parliament and Council, 1995)).

In addition, the essence of the roles of data processors, i.e., who is in the role of the processor, the controller, and the joint controller in cross-border data exchange, has been covered to a more significant extent in the context of the GDPR. At first glance, this may seem a very simple topic, but the technical concept plays a major role in the whole project, and the technical details as well as the rights and obligations, whether under an agreement between Member States or a Commission

Implementing Decision 2019/1765, will determine the assessment. This illustrates the complexity of cross-border services.

For the sake of clarity, it is necessary to keep the information systems of the Member States, where the data are initially collected, separately, and distinguish between the contact points and the channel between them for sharing data from one country to another, and the information system of another country through which the data are finally processed (X-eHealth, 2021, page 38). In between, there is also technical processing, which translates the entered data into the language of the respective country (standardised processing) (X-eHealth, 2021, page 37). At the same time, the contradiction between the agreement and the Decision 2019/1765 has led to a confusion of roles. eHDSI is composed of core services and generic services. The core services are developed, deployed, and maintained by the European Commission; the generic services are developed, deployed, and maintained by the National Contact Points for eHealth, designated by each Member State (European Commission, 2019a, recital 7). It is mentioned in the agreement that the contracting parties (Member States) do not assume any liability for core services (European Commission, 2022d, Clause II.1.2). Despite that in the recital 20 of the Decision 2019/1765 we can find that the Member States, represented by the relevant National Authorities or other designated bodies, determine together the purpose and means of processing of personal data *through the* eHDSI and are therefore controllers (Art 2(1)(b), and art 7(1)). In the light of the above, it is inconsistent that if the objectives of the platform were agreed upon by the Member States, and the Commission is its processor, the Commission would be an autonomous controller in access management. The more proper role in this context would be that the Commission should rather be the processor authorised in the technical management of access (see for more in X-eHealth, 2021 point 4.1.4 and 4.1.7; see also art (7)(3) and recital 21 of European Commission, 2019a). The controller should be able to identify the persons to whom access to the data will be granted—the controller is responsible for the lawful use of the data and thus for the release of the data (GDPR, art 5(1)(a) and (2)). Thus, it may be concluded that the agreement and the explanations of the roles in implementing decision are confusing. This example illustrates the nuances and multiplicity of opinions involved in complex projects and probably makes joining them less attractive.

The most exciting part is probably the concept of personal data as such in the context of technological development (X-eHealth, 2021, point 4.1, especially point 4.1.6). Disputes with this content are also inherently challenging, as they require a similar understanding of technology and a constant update of privacy technologies.

As brought out already in Working Party 29's opinion 4/2007 on the concept of personal data, the scope of the data protection rules *should not be overstretched*. An undesirable result would be that of ending up applying data protection rules to situations which were not intended to be covered by those rules and for which they were not designed by the legislator (Article 29 data protection working party, 2007; see recitals 26 and 27 of the European Parliament and Council (1995)).

Where identification of the data subject *is not included* in the purpose of the processing, *the technical measures* to prevent identification *have a very important role to play. Putting in place* the appropriate state-of-the-art technical and organisational measure to protect the data against identification may make the difference to consider that the persons *are not identifiable*, taking account of all the means likely reasonably to be used by the controller or by any other person to identify the individuals. In this case, the implementation of those measures is not the consequence of a legal obligation (which only applies if the information is personal data in the first place), but rather a condition for the information precisely *not to be considered to be personal data* and its processing *not to be subject* to the Directive (Article 29 data protection working party, 2007, page 17).

In GDPR, the principles of data protection should apply to any information concerning an identified or identifiable natural person (GDPR art 4(1)). Personal data which have undergone pseudonymisation, which could be attributed to a natural person by the use of additional information, should be considered to be information on an identifiable natural person (European Parliament and Council, 2016a, recital 26). So, if it is possible to combine the data with additional data and if it is possible to "bring the ends together", we should consider this data as personal data.

The same question has also arisen in relation to cross-border data exchange and is also marked in X-eHealth informal paper. If one controller (Country X) encrypts data and it is transmitted via secure channel to another controller (Country Y), did the sender of the black box still process the personal data in the sense of GDPR? The channel as so-called TESTA system, provided by the European Commission, ensures the secure transport of the data in a more secure way than it would be possible through the public Internet (X-eHealth, 2021, point 4.1.6). Data entering the TESTA network are encrypted and the private keys will only be available to Member States' contact points (X-eHealth, 2021, point 4.1.5). It is also confirmed in art 7(2) of the Commission Implementing Decision 2019/1765 that the Commission *shall not have access to* patients' personal data processed through the eHDSI. The data must be encrypted by the Member State and that the decision explicitly states that the Commission does not process and cannot process the data.

In the EDPS informal comments on the Commission draft implementing decision repealing Commission Implementing Decision 2011/890/EU, the EDPS pointed out that the Commission is not allowed to "view" the data, and in other words, the Commission sees only a black box and data recipients as a secure channel provider (X-eHealth 2021, point 4.1.6; see page 38). It is mentioned in the x-eHealth information paper that it has been confirmed that the data must be encrypted by the Member State and that the decision explicitly states that the Commission does not process and cannot process the data. Encryption is a commonly known technique used *to protect the confidentiality of the transmitted information*, thus of the personal data involved. Implementation of encryption techniques is the process of encoding the information where personal data are included *in such a way* that *only authorised parties may access it* (X-eHealth, 2021, point 4.1.6, see page 39). But there is also a conclusion that it does not affect the fact that personal data, even though encrypted, are still personal data. That is why it opens a discussion about new technologies (see also Hamulák, 2018). If there is no key and the data is not technically visible, is it possible to say that the data processed really is "personal data" (X-eHealth, 2021, point 4.1.6)? Of course, it is possible to talk about personal data processing when it is clear who sent it and to whom (metadata), but the circle of questions arises precisely in the context of the aforementioned "black box" (X-eHealth, 2021, point 4.1.5, page 36). This means that, on the one hand, it is an obligation of the Member State to ensure that personal data are not processed (cannot be accessed or seen) without a clear purpose and legal basis and, on the other hand, it would be considered a breach of the obligations for the Member State if the Member State may be blamed of unlawful processing of personal data (X-eHealth, 2021, point 4.1.6, see page 38). However, the Commission is given the role of processor of *personal data*, including health data. As said in the informal paper, this gives rise to a substantive discussion whether new privacy technologies serve to ensure the protection of personal data, i.e., to keep it confidential (from being processed by third parties) (X-eHealth, 2021, point 4.1.6). It is especially significant if we look at the GDPR art 34(3)(a) which settles that *appropriate* technical and organisational protection measures are in particular those that *render the personal data unintelligible* to any person who is not authorised to access it, such as *encryption*.

Hence, the dilemma that remains to be shaped in the future is where to draw the line between the intended purpose of privacy technologies and the application of the GDPR. In essence, on the one hand, the obligation of data confidentiality must be complied with, i.e., personal data must not be disclosed, but on the other hand, it is still interpreted as processing of personal data. The two previous objectives are fundamentally contradictory, and the outcome of this discussion will certainly have a significant impact on privacy technology developers, who are motivated to create technologies that protect the confidentiality of data and prevent the processing of personal data without keys. As it is asked in x-eHealth informing paper: "Would interpretations change if the technology made it possible to ensure the processing of certain processes in such a way that data processing would not materially take place?" (X-eHealth, 2021, page 41). Hence, to minimise the processing of personal data (data minimisation is one of the core principles of GDPR according to GDPR art 5(1)c), technical solutions are the tools that may help us do this and achieve that goal (see more about Sharemind, Cybernetica AS, 2018).

3.3 EU Digital COVID Certificates: Another Example of Cross-border Health Data Processing

Digitization has seen its implementation accelerated by the recent COVID-19 pandemic. In practice, cooperation between Member States has improved with the recent implementation of COVID-19 certificates. On 30 January 2020, the Director-General of the WHO declared a public health emergency of international concern over the global outbreak of severe acute respiratory syndrome coronavirus 2 (SARS-CoV-2), which causes coronavirus disease 2019 (COVID-19). On 11 March 2020, the WHO made an assessment characterising COVID-19 as a pandemic. To limit the spread of SARS-CoV-2, the Member States adopted measures which have had an impact on the exercise by Union citizens of their right to move and reside freely within the territory of the Member States, such as entry restrictions or requirements for cross-border travellers to undergo quarantine or self-isolation or to be tested for SARS-CoV-2 infection (European Parliament and Council, 2021, recital 1–3).

The Regulation (EU) 2021/953 of the European Parliament and of the Council of 14 June 2021 on a framework for the issuance, verification, and acceptance of interoperable COVID-19 vaccination, test, and recovery certificates (EU Digital COVID Certificate) to facilitate free movement during the COVID-19 pandemic applies from first of July 2021. Good cooperation can be seen as a result of the necessity in a crisis situation which, in the wake of a global pandemic, forced countries to work towards a common goal. The use of certificates allowed people to travel between Member States. However, this is not an entirely new substantive obligation. For example, there have been countries where proof of vaccination or prophylaxis against, for example, yellow fever or malaria has always been required (World Health Organization, 2020a). Prior to the current digital evidence, the so-called yellow certificate "International certificate of vaccination" has been accepted on the basis of the evidence developed by the WHO, which is nowadays also updated and supplemented with a QR code (World Health Organization, 2020b). Therefore, it is not exactly a new obligation, but this kind of widespread use of health evidence has not been implemented in the EU in the past, and even less in a centrally organised way that ensures common principles, validated data, and service reliability.

Despite the above praise, it cannot be left unmentioned that, however, from a legislative point of view, it is questionable why the specific regulation trivialises the essence of the GDPR and repeats in this Regulation something what would follow from the GDPR anyway (e.g., articles 10(1)(2)—as art 5 principles in GDPR). As the EDPB and EDPS also recalled in their joint opinion, the principles relating to processing of personal data laid down in Article 5 of the GDPR may be continuously applied and integrated in any personal data processing operation (European Data Protection Board & the European Data Protection Supervisor, 2022a, point 3.1 (10) and 3.2(20)). In addition, the second paragraph of Article 1 makes it clear that the processing of information contained in this Regulation is in full compliance with the GDPR. The prohibition in art 10(8) that the processor may not transfer data to a third country is confusing in the context of the GDPR. First, even if a processor is used under art 28(3) of the GDPR, under art 28(3)(a), the processor should inform the controller anyway (unless it would be contrary to the public interest). The prohibition raises the theoretical question that, if the GDPR itself does not have such an imperative prohibition, for example for the processing of special categories of data in other contexts with data processor role, why is such a distinction justified by the three EU COVID certificates.

In a way, the implementation of the last sentence of art 10(3) can also be an obstacle in practice. Even though pursuant to national law, the competent authorities and cross-border passenger transport providers are required to apply certain public health measures during the COVID-19 pandemic—to check and confirm the vaccination status, test result, or recovery—then in reality, they are prohibited from retaining the processed personal data. It would seem to refer only to personal, i.e., physical control. The competent authorities of the country of destination or transit may use, for example, a developed QR code verification application. At the same time, with any technological device there is a certain informational footprint, and it is probably not possible to definitively state that personal data were not retained, even for a certain moment. Any data, including the fact of verification, is prohibited by the norm. The prohibition could specify, for example, that all the health data from the certificate are not retained, but the fact that the person held a valid certificate could be allowed for the purpose of meeting the aim of this Regulation. In addition, the same data may be unwantedly caught by an agency's camera (customs checkpoint, airport), which records data for an entirely different reason and which is clearly not immediately erased. Obviously, it would have been more appropriate to start from a minimum time limit for data processing (maintaining) rather than prohibiting retention as an absolute prohibition. Therefore, specific standards should consider the actual business processes and how the technology works—how a person moves around during the check and what is used to check the data. Otherwise, the norms may seem illusionary and cannot be fully met. EDPB and EDPS in their opinion only indicated that the necessary categories of personal data included in the QR code must be reassessed and different technical solutions improving data *minimisation* in different use cases may be needed. The EDPB and the EDPS therefore invited the Commission to assist the Member States in developing such technical specifications (The European Data Protection Board & the European Data Protection Supervisor, 2022a, point 3.2.2 (26)). To conclude, the principles of personal data processing must be always considered and taken into account—personal data processing should be limited to what is necessary in relation to the purposes for which they are processed (data minimisation principle).

3.4 Misunderstanding with Secondary Use meaning and New European Health Data Space

To bring new services and products to the market in the first place, it is necessary to proactively analyse data—to carry out various tests and research. This, in turn, may lead us to the question of whether the GDPR as a general framework allows this and whether or not there is a common understanding between Member States. The European Commission underlines in its Communication "A European strategy for data" (European Commission, 2020a) that it is precisely data-driven innovation that will bring enormous benefits to citizens. It even emphasises that the foundations for the competitiveness of the data-driven economy in the coming decades are being laid right now, so the EU should act now. At the same time, it emphasizes an equally European approach that strikes a balance between the flow and widespread use of data and high standards of privacy, security, safety, and ethics. Whether and how to ensure this as technology advances is a matter for discussion. In other words, the more obscure the rules and the more different they vary from country to country, the more difficult it is to carry out data analysis to generate new knowledge from the data.

By coupling information from registries, researchers can obtain new knowledge of great value with regard to widespread medical conditions such as cardiovascular disease, cancer, and depression. On the basis of registries, research results can be enhanced, as they draw on a larger population. Research results obtained *through registries provide solid, high-quality* knowledge which can provide the basis for the formulation and implementation of knowledge-based policy, improve the quality of life for a number of people, and improve the efficiency of social services. It is important to mention that, in the context of the GDPR, processing of personal data that would involve technological development should also be considered as research (European Parliament and Council, 2016a, recital 159). In order to facilitate scientific research, personal data can be processed for scientific research purposes, subject to appropriate conditions and safeguards set out in Union or Member State law (European Parliament and Council, 2016a, recital 157). Thus, the GDPR allows, even directs, to use data that is already collected from the register, to increase the benefit of society and innovation. Only one question remains—on what legal basis the purpose of such processing could be pursued?

Article 5(1)(b) of the GDPR settles that personal data shall be collected for specified, explicit, and legitimate purposes and not further processed in a manner that is incompatible with those purposes; *further processing for archiving purposes in the public interest, scientific or historical research purposes or statistical purposes shall,* in accordance with art 89(1), *not be considered to be incompatible with the initial purposes* ("purpose limitation"). Although the processing of sensitive or different categories of data is allowed under the GDRP in rather limited cases, the processing of such data is also supported as permissible for Article 89 purposes. Pursuant to art 9(2)(j) of the GDPR, it should be based on Union or Member State law which shall be proportionate to the aim pursued and provide for suitable and specific measures to safeguard the fundamental rights and the interests of the data subject. Thus, GDPR does not hinder the development of innovation; the resentment of researchers lies elsewhere.

In light of the above, a separate debate has arisen in the context of the concept of "secondary use," which is how most researchers understand data processing in the context of research. In essence, no such equivalence can be established. Firstly, the purpose of the initial data collection may already be for the purpose of conducting a study, and secondly, any subsequent processing may be for any other lawful purpose and is not limited to the purposes of GDPR art 89. For example, data may be processed for a purpose other than the original purpose, on any other legal basis (e.g., an individual gives consent to the processing of data for other purposes, or information is requested, e.g., in legal proceedings, for a processing purpose which is also different from the original purpose for which the data were collected, but which is itself also based on a separate legal basis). There is therefore no need to distinguish the concept of "secondary use" in substance. Data collection for statistical or analytical purpose may already be clearly foreseen at the time of initial collection of the data. However, if the data are needed by another third party for the purposes of the research, the third party must rely on Article 89 and, where applicable, on national law specifying safeguards, etc.

With the topic of secondary use of personal data, questions are arising regarding data subjects' consents (Peloquin et al., 2020). However, national jurisdictions differ in this respect. While in Estonia national law establishes a minimum framework for the processing of data without the consent of a person (Riigikogu, 2019, § 6), then this may not necessarily be the case in other countries (see obstacles to processing under data subject consent, Peloquin et al., 2020). However, in such a case, Europe will not be able to meet the objectives set out in "A European strategy for data", which is also supported by the GDPR itself (see recital 157, and art 5(1)(b), 9(2)(j) and 89). Of course, it must be considered that exceptions will remain in certain areas anyway, such as clinical trials of medications, which require the consent of the individual for research (European Parliament and Council, 2014b, and see also EDPB Guidance on Clinical Trials—consent to participate in the research *vs* legal basis for data processing for data controllers (European Data Protection Board, 2019). Despite various guidelines and GDPR and sectoral specific acts—which must also be in line with the GDPR—it is important to underline that GDPR does not in itself restrict research or the use of data. Rather, innovation is hampered by the different legal frameworks in different countries and a lack of understanding them. For example, in a situation where each country has different requirements for safeguards (GDPR art 89), the resentment of researchers is also understandable, for example, when merging personal data from different Member States can end up creating a huge mess. For example, when some of the personal data is consent-based and can be withdrawn at any time, while other personal data were provided on a different legal basis other than consent, in that case, the main objective may not be achieved when the consent is withdrawn. In addition, the person's consent is also inappropriate legal basis for personal data processing if it is not known in advance from whom to ask for it, i.e., the very act of identifying the target group would also require an analysis.

This is probably one of the reasons why a new proposal for a regulation has been initiated by the European Commission. The European health data space will support research on treatments, medicines, medical devices, and outcomes; also it will clarify the safety and liability of AI in health (European Commission, 2020b). When developing new regulation, it would certainly be worthwhile to pay attention to what safeguards can be put in place if, for example, in view of the development of technologies, it is not possible to predefine a specific research question (hypothesis). AI starts to look for links in the data on its own and finds interesting features that a person might not ever even hypothesise of. The specificity of genetic data and the safeguards for related research should definitely be considered further. Compared to other health data, genetic data has several features—for example the results obtained may also draw conclusions about a person's family members and descendants (Council of Europe, 1981, page 13, point 80). This new initiative would imply the possibility to process data in a secure environment without the data controller having to release the data for data reconciliation. This would create an infrastructure where data could be combined, but directly identifiable personal data would not be shared. Such measures would also increase people's confidence in research and data processing.

3.5 *Personal Data Protection and Financial Sector*

Digital and legal development in the European Union has been significantly visible in the FinTech, financial, and banking sector in recent years. One of the examples reflecting it is the Payment Services Directive 2 (PSD2) (European Parliament and Council, 2015) which, in addition to its open banking subject matter and all the other objectives set, *inter alia* supports, and helps to build stronger data protection framework for EU citizens with its confidentiality and strong authentication requirements. The core subject of PSD2 is that it allows payment service providers to obtain access to payment accounts of data subjects for the purposes of providing the data subject/consumer payment initiation services and account information services.

Generally, PSD2 requirements of strong customer authentication across the EU will help reduce the risk of fraud for online payments and online banking and protect the confidentiality of the user's financial data, including personal data (European Commission, 2019b). As the European Commission describes, the updated rules will help to facilitate innovation, competition, and efficiency in the EU online payments market (European Commission, 2019b). In this context, in the background of PSD2, it is worth noting that the GDPR also sets confidentiality requirement as one of the main data processing principles (GDPR art 5(1)f) thus creating a substantive bridge with PSD2. The last sentence of recital 39 in GDPR further describes this personal data processing principle and explains that personal data should be processed in a manner that ensures appropriate security and confidentiality of the personal data, including for preventing unauthorised access to or use of personal data and the equipment used for the processing. With the PSD2 and GDPR requirements both being applicable, EU citizens—whether either being in the role of the consumer or personal data subject—may be confident that they can benefit from enhanced level of safety and confidentiality while carrying out electronic payments with their personal data.

GDPR and PSD2 are both similarly forward-looking by supporting innovation and setting uniform standards for digital development. While GDPR is described as technologically neutral (in GDPR recital 15) meaning its rules for personal data processing shall apply to whichever technology that allows personal data processing, a somewhat similar comprehensive approach has been taken towards the target stakeholders group by the European Commission when describing that the PSD2 sets the stage for the future with the new rules being equally applicable to traditional banks and to innovative payment services and new providers, such as FinTechs (European Commission, 2019b). As a result, it can be concluded from the above that developments in the digital field and legal developments are comprehensive and do not significantly promote exemptions. In addition, the legislator has aimed at making it difficult, if not impossible, to circumvent the rules for both broader legal areas covered by PSD2 and GDPR. This should make further operating for stakeholders in the financial area predictable and provide greater confidence in competition because same rules apply for all service providers no matter their service provider form.

What is also spectacular about PSD2 is that the art 94 is entirely dedicated to data protection and in its Recital 89 the directive explicitly points out that "data protection by design and data protection by default should be embedded in all data processing systems developed and used within the framework of this Directive". This means that the principle of "data protection by design and data protection by default" was formulated in a recital of PSD2 even before it was formulated in the GDPR as an art 25. However, when comparing the wordings of PSD2 recital 89 and GDPR art 25, it can of course be concluded that the wording of the GDPR is much more comprehensive and specific about the principle mentioned above.

Although both PSD2 and GDPR meant to innovate and refine the area, there have also been some hesitation about the correct and proper implementation of the legislation. While providing open banking service, apparently no service provider wants to violate data protection rules and, thus, be a subject to administrative fines. As a result, to help overcome challenges and reduce misunderstandings, EDPB has helpfully prepared a guide (European Data Protection Board, 2020c) to explain the interplay of PSD2 and GDPR.

It is certain by now that both GDPR and PSD2 have reshaped the EU's financial sector. As of 2022, it can be stated that PSD2 has, for instance, stimulated innovation in the electronic identity and customer authentication area through the introduction of strong customer authentication (European Commission, 2020c). However, the European Commission has worded in its communication on a Retail Payments Strategy for the EU that the picture across the EU involves a plethora of different authentication solutions at national level with limited cross-border interoperability which may hinder further innovation and the development of new payment services (European Commission, 2020c). Hence, developing common authentication standards with broader possibilities of cross-border interoperability options may be one of the challenges for EU up next.

3.6 Artificial Intelligence (AI) and Personal Data Protection

In its communication to the European Parliament and the Council, the European Commission has described data protection rules as a pillar of citizens' empowerment and the EU's approach to digital transition (European Commission, 2020d). In the twenty-first century, it is difficult to imagine developments around digitization without proper personal data protection rules in place. The famous quote "data is the new oil" has been heard strongly in the background of recent legal personal data protection initiatives launched, including GDPR. The protection of personal data becomes more and more important when the topic of artificial intelligence and automatic decision-making by using algorithms becomes under discussion.

In April 2021, European Commission proposed the first ever legal framework on artificial intelligence (AI)—AI Act Proposal (European Commission, 2021d), which addresses the risks of AI and aims at strengthening the EU's ability to compete globally. This new Proposal is expected to further regulate the area and give more detailed specification on how the area will be regulated from the legal perspective in the future (see also Ramiro Troitiño et al., 2020). However, it is expected that GDPR will remain to regulate the personal data protection area as *lex generalis* and more specification and details are expected to be provided for the AI use only. This standpoint is also supported by the proposed new regulation on AI which explicitly mentions that the "proposal is without prejudice and complements the GDPR (European Parliament and Council, 2016a) and the Law Enforcement Directive (Directive (European Parliament and Council, 2016b)) with a set of harmonised rules applicable to the design, development and use of certain high-risk AI systems and restrictions on certain uses of remote biometric identification systems" (European Commission, 2021d, point 1.2). Nevertheless, it is noteworthy to mention from the innovation aspect that this new proposed regulation includes a specific chapter dedicated to innovation "Measures in support of innovation" that would hopefully contribute to the promotion of innovation even more. From a data protection point of view, it is gratifying that these articles belonging to the proposed innovation chapter already include and cover data protection supervision norms at a very early stage of creation of this regulation.

Although GDPR does not *expressis verbis* mention "artificial intelligence", it mentions "automated decision-making" which by its very nature can be linked with AI. The use of AI provokes debates and discussions from legal and ethical aspects. All relevant ethical and legal issues related to AI need to be addressed towards the progress and advancement of digitization. As is regulated in GDPR, automated decision-making must be clearly brought to data subject's attention with the right to information (GDPR articles 13 and 14) and right of access (GDPR art 15). The last sentence of GDPR recital 71 formulates that automated decision-making and profiling based on special categories of personal data should be allowed only under specific conditions. According to GDPR art 22(1) data subjects have the right not to be subject to a decision based solely on automated processing, including profiling, which produces legal effects concerning him or her or similarly significantly affects him or her. Hence, according to GDPR art 22, data subjects may lodge complaints with the data controller and request meaningful human intervention. For instance, this is exactly what the NGO noyb.eu did when they filed a GDPR complaint against Airbnb with the Data Protection Authority of Rheinland-Pfalz at the end of the year 2021 (Noyb, 2021). It has also been highlighted that administrations must be mindful with using AI for policymaking, as there is the tendency to believe that automated decisions fully based on data are automatically better than human decisions which is not necessarily true (Van Noordt & Misuraca, 2022). In parallel, it cannot be ruled out that similar tendencies could not be possible with regard to decisions on personal data where AI is involved. In view of the above, it can be concluded that the consequences of different activities must be taken with particular care and consideration when making decisions in relation to personal data with the help of AI.

Regarding AI and personal data protection challenges, it is important to solve the question of liability and clarify the question under what conditions may parties be responsible for breaches of data protection rules (Troitiño et al., 2020). It is notable that in the beginning of 2022, the European Data Protection Board (EDPB) has pointed out that while, under the GDPR, only controllers and processors would be liable, e.g., in case of a personal data breach, it is essential to consider the role and potential liability of providers of AI systems developed and made available in order to secure personal data processing (Jelinek, 2022). If this interesting theoretical proposal by EDPB would one day become reality or would be adopted as an applicable piece of EU legislation, it would mean extending the boundaries of data protection breaches from a liability perspective.

3.7 *Data Act Proposal*

One of the latest developments from a legal viewpoint with regard to innovation and data is the Data Act Proposal which includes measures for a fair and innovative data economy. As the European Commission has described, the Data Act will ensure fairness in the digital environment, stimulate a competitive data market, open opportunities for data-driven innovation, and make data more accessible for all (European Commission, 2022e).

The EDPB and EDPS note that Data Act Proposal would apply to a wide range of products and services, including the connected objects ("Internet of Things"), medical or health devices, and virtual assistants. In its joint opinion 2/2022 (European Data Protection Board & the European Data Protection Supervisor, 2022b), the EDPB and EDPS have shared their concerns and draw attention to several overarching concerns on the proposal. While the EDPB and EDPB welcome the efforts made to ensure that the Proposal does not affect the current data protection framework, the EDPB and the EDPS consider that additional safeguards are necessary to avoid lowering the protection of the fundamental rights to privacy and to the protection of personal data in practice (European Data Protection Board & the European Data Protection Supervisor, 2022b, p 2). What is more, the EDPB and the EDPS mention that they welcome the approach in Proposal's Recital 7, where it is explicitly mentioned that the Proposal complements and is without prejudice to Union law on data protection and privacy, in particular GDPR and ePrivacy Directive (European Data Protection Board and the European Data Protection Supervisor, 2022b, p 7). Hence, we will further see the developments in the area of digital environment, but it also looks as if the stakeholders consider that this new initiative is not designed to further affect the previously agreed basic data protection approaches.

4 Conclusions

This chapter focused on the recent data protection highlights, developments, and challenges in the EU. The emphasis of this chapter was on topics related to eHealth and financial issues due to the background of authors.

The EU has seen many developments during twenty-first century with regard to digital development, the use of AI, and personal data protection. As is brought to the reader's attention also in this chapter, there have been new fundamental reforms (such as GDPR) and more sector-specific developments (such as cross-border exchange of health data and its safeguard measures and developments in the financial sector such as PSD2 that have gone hand in hand with addressing data protection issues). Some developments have been due to an urgent need, such as recent implementation of COVID-19 certificates where digitization saw its implementation accelerated by the pandemic. Some developments related to personal data protection directly or indirectly are still unfolding (such as the Data Act) which means that "mid-term" summaries can also be prepared in the upcoming years.

As discussed in this chapter, eHealth and cross-border data exchange have brought numerable benefits, but the dilemma with the processing roles remains. In essence, on the one hand, the obligation of data confidentiality must be complied with, i.e., personal data must not be disclosed to unauthorized third parties, but on the other hand, the interpretation which data is counted as personal data can be problematic. The outcome of this discussion will certainly have a significant impact on privacy technology developers who are the creators of innovative technologies that protect the confidentiality of data.

Digitization has seen its implementation accelerated by the recent COVID-19 pandemic. In practice, cooperation between Member States has improved with the recent implementation of COVID-19 certificates. Good cooperation can be seen as a result of the necessity in a crisis situation which, in the wake of a global pandemic, forced countries to work towards a common goal. However, it cannot be left unmentioned that from a legislative point of view, it is questionable why the specific regulation trivialises the essence of the GDPR and repeats something what would already follow from the GDPR anyway.

Art 5 (1)(b) of the GDPR stipulates that personal data shall be collected for specified, explicit, and legitimate purposes and not further processed in a manner that is incompatible with those purposes: further processing. In the light of the above, a separate debate has arisen in the context of the concept of "secondary use". However, national jurisdictions differ in this respect. If in Estonia national law establishes a minimum framework for the processing of data without the consent of the person, then in several other countries it can only be based on the person's consent. Thus, it can be argued that there are different approaches to this issue in different regions. It may be added that innovation may be hampered by the different legal frameworks in different Member States. For example, in a situation where each country has different requirements for safeguards (allowed by GDPR art 89), the resentment of researchers may be understandable.

Since the main cornerstone of data protection in the EU is GDPR, it is evident, and described also in this chapter, that the different legal acts are closely interlinked and based on common basic principles regarding personal data protection. One of the examples brought in this chapter concerned PSD2 requirements of strong customer authentication which are in place to help to reduce the risk of fraud for online payments and online banking and protect the confidentiality of the user's financial data, including personal data. Since confidentiality principle is also stipulated in GDPR, it is easy to see the similarity of these two legal acts.

With its strategies (such as EU Digital Compass) set for the upcoming years and decades, EU has taken its standpoint to further promote, develop, and strengthen data protection and innovation all over the EU. The authors of this chapter believe that we will soon be able to look at these developments with interest. Additionally, as also previously mentioned in this chapter, the authors of this chapter believe that data protection issues will remain in focus in the years to come, as it may be almost impossible if not impossible at all to achieve all targets and goals the EU has set for the future without a high standard of data protection.

References

Article 29 data protection working party. (2007). *Opinion 4/2007 on the concept of personal data*. Adopted on 20th June 2007. Available from https://ec.europa.eu/justice/article-29/documentation/opinion-recommendation/index_en.htm (22.05.2022).

Article 29 data protection working party. (2017). *Guidelines on the application and setting of administrative fines for the purposes of the Regulation 2016/679*. Adopted on 3rd October 2017. Available from https://ec.europa.eu/newsroom/article29/items/611237/en (22.05.2022).

Council of Europe. (1981). *Explanatory report to the convention for the protection of individuals with regard to automatic processing of personal data*. Available from https://rm.coe.int/16800ca434 (22.05.2022).

Cybernetica AS. (2018). *Anonymisation 2.0: Sharemind as a tool for de-identifying personal data – Part 2: Sharemind and anonymisation*. Available from https://sharemind.cyber.ee/ (22.05.2022).

EAS. (2022). *e-Estonia. e-Identity*. Available from https://e-estonia.com/solutions/e-identity/id-card/ (22.05.2022).

European Commission. (2015). *Communication from the commission to the European Parliament, The Council, The European Economic and Social Committee and the Committee of the Regions a Digital Single Market Strategy for Europe*. Available from https://eur-lex.europa.eu/legal-content/EN/TXT/?uri=celex%3A52015DC0192 (22.05.2022).

European Commission. (2019a). *Implementing Decision 2019/1765 of 22 October 2019 providing the rules for the establishment, the management and the functioning of the network of national authorities responsible for eHealth, and repealing Implementing Decision 2011/890/EU (notified under document C(2019) 7460) (Text with EEA relevance)*. Available from https://eur-lex.europa.eu/legal-content/en/TXT/?uri=CELEX%3A32019D1765 (22.05.2022).

European Commission. (2019b). *Frequently asked questions: Making electronic payments and online banking safer and easier for consumers*. Available from https://ec.europa.eu/commission/presscorner/detail/en/qanda_19_5555 (22.05.2022).

European Commission. (2020a). *Communication from the commission to the European Parliament, the Council, the European Economic and Social Committee and the Committee of the Regions A*

European strategy for data. Available from https://eur-lex.europa.eu/legal-content/EN/TXT/?uri=CELEX%3A52020DC0066 (22.05.2022).

European Commission. (2020b). *Digital health data and services – the European health data space*. Available from https://ec.europa.eu/info/law/better-regulation/have-your-say/initiatives/12663-Digital-health-data-and-services-the-European-health-data-space_en (22.05.2022).

European Commission. (2020c). COM (2020) 592 final. *Communication from the commission to the European Parliament, the Council, the European Economic and Social Committee and the Committee of the Regions on a Retail Payments Strategy for the EU*. Available from https://eur-lex.europa.eu/legal-content/EN/TXT/?uri=CELEX%3A52020DC0592 (22.05.2022).

European Commission. (2020d). *Communication from the commission to the European Parliament and the Council*. Brussels, 24.6.2020COM(2020) 264 final. Available from https://eur-lex.europa.eu/legal-content/EN/TXT/PDF/?uri=CELEX:52020DC0264 (22.05.2022).

European Commission. (2021a). *Eurobarometer: Digital rights and principles*. Available from https://europa.eu/eurobarometer/surveys/detail/2270 (22.05.2022).

European Commission. (2021b). *Eurobarometer: Europeans show support for digital principles*. Available from https://ec.europa.eu/commission/presscorner/detail/en/ip_21_6462 (22.05.2022).

European Commission. (2021c). *Proposal for a Regulation of the European Parliament and of the Council amending Regulation (EU) No 910/2014 as regards establishing a framework for a European Digital Identity*. Available from https://eur-lex.europa.eu/legalcontent/EN/TXT/?uri=CELEX%3A52021PC0281 (22.05.2022).

European Commission. (2021d). *Proposal for a Regulation of the European Parliament and of the Council Laying Down Harmonised Rules on Artificial Intelligence (Artificial Intelligence Act) and Amending Certain Union Legislative Acts*. Available from https://eurlex.europa.eu/legal-content/EN/TXT/?uri=CELEX%3A52021PC0206 (22.05.2022).

European Commission. (2022a). *Europe's digital decade: Digital targets for 2030*. Available from https://ec.europa.eu/info/strategy/priorities-2019-2024/europe-fit-digital-age/europes-digital-decade-digital-targets-2030_en (22.05.2022).

European Commission. (2022b). *European digital identity*. Available from https://ec.europa.eu/info/strategy/priorities-2019-2024/europe-fit-digital-age/european-digital-identity_en#why-is-it-needed (22.05.2022).

European Commission. (2022c). *Personalised medicine. Background, conference reports, publications and links related to personalised medicine*. Available from https://ec.europa.eu/info/research-and-innovation/research-area/health-research-and-innovation/personalised-medicine_en#digital-health-and-care-ehealth (22.05.2022).

European Commission. (2022d). *Agreement between National Authorities or National Organisations responsible for National Contact Points for eHealth on the Criteria required for the participation in Cross-Border eHealth Information Services*. Available from https://ec.europa.eu/health/system/files/2017-07/ev_20170509_co06_en_0.pdf (22.05.2022).

European Commission. (2022e). *Data act: Commission proposes measures for a fair and innovative data economy*. Available from https://ec.europa.eu/commission/presscorner/detail/en/ip_22_1113 (22.05.2022).

European Council. (2021). *Confidentiality of electronic communications: Council agrees its position on ePrivacy rules*. Available from https://www.consilium.europa.eu/en/press/press-releases/2021/02/10/confidentiality-of-electronic-communications-council-agrees-its-position-on-eprivacy-rules/?utm_source=dsms-auto&utm_medium=email&utm_campaign=Confidentiality+of+electronic+communications:+Council+agrees+its+position+on+ePrivacy+rules (22.05.2022).

European Data Protection Board. (2019). *Opinion 3/2019 concerning the Questions and Answers on the interplay between the Clinical Trials Regulation (CTR) and the General Data Protection regulation (GDPR) (art. 70.1.b))*. Adopted on 23 January 2019. Available from https://edpb.europa.eu/sites/default/files/files/file1/edpb_opinionctrq_a_final_en.pdf (22.05.2022).

European Data Protection Board. (2020a). *Guidelines 03/2020 on the processing of data concerning health for the purpose of scientific research in the context of the COVID-19 outbreak*. Available from https://edpb.europa.eu/our-work-tools/our-documents/guidelines/guidelines-032020-processing-data-concerning-health-purpose_en (22.05.2022).

European Data Protection Board. (2020b). *Guidelines 05/2020 on consent under Regulation 2016/679. Version 1.1*. Available from https://edpb.europa.eu/our-work-tools/our-documents/guidelines/guidelines-052020-consent-under-regulation-2016679_en (22.05.2022).

European Data Protection Board. (2020c). *Guidelines 06/2020 on the interplay of the Second Payment Services Directive and the GDPR. Version 2.0*. Available from https://edpb.europa.eu/our-work-tools/documents/public-consultations/2020/guidelines-062020-interplay-second-payment_en (22.05.2022).

European Data Protection Board and the European Data Protection Supervisor. (2022a). *EDPB-EDPS Joint Opinion 1/2022 on the Proposal for a Regulation of the European Parliament and of the Council amending Regulation (EU) 2021/953 on a framework for the issuance, verification and acceptance of interoperable COVID-19 vaccination, test and recovery certificates (EU Digital COVID Certificate) to facilitate free movement during the COVID19 pandemic*. Available from https://edps.europa.eu/data-protection/our-work/our-work-by-type/edps-edpb-joint-opinions_en (22.05.2022).

European Data Protection Board and the European Data Protection Supervisor. (2022b). *EDPB-EDPS Joint Opinion 2/2022 on the Proposal of the European Parliament and of the Council on harmonised rules on fair access to and use of data (Data Act)*. Available from https://edpb.europa.eu/our-work-tools/our-documents/edpbedps-joint-opinion/edpb-edps-joint-opinion-22022-proposal-european_en (22.05.2022).

European Parliament and Council. (1995). *Directive 95/46/EC of the European Parliament and of the Council of 24 October 1995 on the protection of individuals with regard to the processing of personal data and on the free movement of such data*. Available from https://eur-lex.europa.eu/legal-content/EN/TXT/?uri=celex%3A31995L0046 (22.05.2022).

European Parliament and Council. (2011). *Directive 2011/24/EU of the European Parliament and of the Council of 9 March 2011 on the application of patients' rights in cross-border healthcare*. Available from https://eur-lex.europa.eu/legal-content/EN/TXT/?uri=CELEX%3A02011L0024-20140101 (22.05.2022).

European Parliament and Council. (2014a). *Regulation (EU) No 910/2014 of the European Parliament and of the Council of 23 July 2014 on electronic identification and trust services for electronic transactions in the internal market and repealing Directive 1999/93/EC*. Available from https://eur-lex.europa.eu/legal-content/EN/TXT/?uri=uriserv%3AOJ.L_.2014.257.01.0073.01.ENG (22.05.2022).

European Parliament and Council. (2014b). *Regulation (EU) No 536/2014 of the European parliament and of the council of 16 April 2014 on clinical trials on medicinal products for human use, and repealing Directive 2001/20/EC (Text with EEA relevance)*. Available from https://eur-lex.europa.eu/legal-content/EN/TXT/?uri=celex%3A32014R0536 (22.05.2022).

European Parliament and Council. (2015). *Directive (EU) 2015/2366 of the European Parliament and of the Council of 25 November 2015 on payment services in the internal market, amending Directives 2002/65/EC, 2009/110/EC and 2013/36/EU and Regulation (EU) No 1093/2010, and repealing Directive 2007/64/EC (Text with EEA relevance)*. Available from https://eur-lex.europa.eu/legal-content/EN/TXT/?uri=celex%3A32015L2366 (22.05.2022).

European Parliament and Council. (2016a). *Regulation (EU) 2016/679 of the European parliament and of the council of 27 April 2016 on the protection of natural persons with regard to the processing of personal data and on the free movement of such data, and repealing Directive 95/46/EC (General Data Protection Regulation)*. Available from https://eur-lex.europa.eu/legal-content/EN/TXT/PDF/?uri=CELEX:32016R0679 (22.05.2022).

European Parliament and Council. (2016b). Directive (EU) 2016/680 of the European parliament and of the council of 27 April 2016 on the protection of natural persons with regard to the processing of personal data by competent authorities for the purposes of the prevention, investigation, detection or prosecution of criminal offences or the execution of criminal penalties, and on the free movement of such data, and repealing Council Framework Decision 2008/977/JHA. Available from https://eur-lex.europa.eu/legal-content/EN/TXT/?uri=celex%3A32016L0680 (22.05.2022).

European Parliament and Council. (2021). *Regulation (EU) 2021/953 of the European Parliament and of the Council of 14 June 2021 on a framework for the issuance, verification and acceptance of interoperable COVID-19 vaccination, test and recovery certificates (EU Digital COVID Certificate) to facilitate free movement during the COVID-19 pandemic (Text with EEA relevance).* Available from https://eur-lex.europa.eu/legal-content/EN/TXT/?uri=CELEX%3A32021R0953 (22.05.2022).

European Union Agency for Fundamental Rights and Council of Europe. (2018). *FRA Handbook on European data protection law.* 2018 edition. Available from https://fra.europa.eu/sites/default/files/fra_uploads/fra-coe-edps-2018-handbook-data-protection_en.pdf (22.05.2022).

Hamulák, O. (2018). La carta de los derechos fundamentales de la union europea y los derechos sociales. *Estudios constitucionales, 16*(1), 167–186.

Information System Authority. (2022). *Estonian ID.* Available from https://www.id.ee/en/article/digital-documents-id-card-digital-id-residence-permit-card-and-e-resident-digi-id/ (22.05.2022).

Jelinek. (2022). *A. Jelinek's letter on behalf of European Data Protection Board to Commissioner D. Reynders.* Brussels, 22 February 2022. Available from https://edpb.europa.eu/system/files/2022-02/edpb_letter_ai_liability_out2022-0009_1.pdf (22.05.2022).

Noyb. (2021). *Complaint against Airbnb with the Data Protection Authority of Rheinland-Pfalz.* Available from https://noyb.eu/sites/default/files/2021-12/AIRBNB%20complaint%20-%20ENG%20redacted.pdf (22.05.2022).

Peloquin, D., et al. (2020). Disruptive and avoidable: GDPR challenges to secondary research uses of data. *European Journal of Human Genetics, 28*, 697–705.

Ramiro Troitiño, D., Kerikmäe, T., Guardia, R. M. D. L., & Pérez Sánchez, G. Á. (2020). EU in twenty-first century, does crisis mean opportunity? In *The EU in the 21st century* (pp. 3–9). Springer.

Riigi Teataja. (2008). *Tervishoiuteenuste korraldamise seaduse ja teiste seonduvate seaduste muutmise seaduse eelnõu seletuskiri.* Available from https://www.riigiteataja.ee/oigusuudised/eelvaadeSeadusUudis/370#2 (22.05.2022).

Riigikogu. (2019). *Personal data protection act.* Riigi Teataja. Available from https://www.riigiteataja.ee/en/eli/523012019001/consolide (22.05.2022).

Riigikogu (in force 2022). *Health Services Organisation Act.* Riigi Teataja. Available from https://www.riigiteataja.ee/en/eli/ee/508012018001/consolide/current (22.05.2022).

Salumaa-Lepik, et al. (2021). Data protection in Estonia. In E. Kiesow Cortez (Ed.), *Data protection around the world. Privacy laws in action* (pp. 23–57). Springer.

Technopolis Group et al. (2019). *eHealth – Future Digital Health in the EU. Final Report Version 25/03/2019.* Available from https://www.espon.eu/sites/default/files/attachments/Final%20report.%202019%2003%2025_final%20version_0.pdf (22.05.2022).

Troitiño, D. R., Kerikmae, T., Barbosa, P. A. R., & Shumilo, O. S. (2020). El libro blanco sobre inteligencia artificial: Análisis y comentarios sobre mercado, valores y cooperación europea. In *Inteligencia artificial: De la discrepancia regional a las reglas universales: Integración de percepciones políticas, económicas y legales* (pp. 303–318). Aranzadi Thomson Reuters.

Van Noordt, & Misuraca. (2022). *Artificial intelligence for the public sector: results of landscaping the use of AI in government across the European Union.* Government Information Quarterly, 2022. Available from https://www.sciencedirect.com/science/article/pii/S0740624X22000478?via%3Dihub (22.05.2022).

World Health Organization. (2016). *E-health in practice*. Available from https://www.euro.who.int/en/countries/estonia/news/news/016/03/e-health-in-practice (22.05.2022).
World Health Organization. (2020a). *International travel and health – 01 July 2020*. Available from https://cdn.who.int/media/docs/default-source/documents/emergencies/travel-advice/yellow-fever-vaccination-requirements-country-list-2020-en264bd6ca-d536–4835-93be-ef83c10f4b00.pdf?sfvrsn=3430b1a2_1&download=true (22.05.2022).
World Health Organization. (2020b). *Smart vaccination certificate working group*. Available from https://www.who.int/groups/smart-vaccination-certificate-working-group (22.05.2022).
X-eHealth. (2021). *Information paper on the current challenges in legal aspects of cross-border exchange of personal data*. D4.2.1. Available from https://www.x-ehealth.eu/wp-content/uploads/2022/01/D4.2.1-%E2%80%93-Information-paper-on-the-current-challenges-in-legal-aspects-of-cross-border-exchange-of-personal-data.pdf (22.05.2022).

EU Competition Law Goals and the Digital Economy: Reflecting Estonia's Perspective

Evelin Pärn-Lee

Abstract New competition dynamics in the digital economy raise questions as to the normative scope of competition enforcement. The question '*Is this a competition problem*?' has become common in the face of new business strategies, investments into new technologies, new forms of interaction with consumers, the accumulation of data, and the use of big analytics. This chapter attempts to outline the goals and values set by European Competition law and their application in promoting innovation and the digital economy. The discussion then moves on to explore the tension between the multitude of goals and economic analysis. It also reflects the different approach to state intervention rules in other jurisdictions, such as the USA and some Asian countries.

1 Setting the Scene

Recently, a Swedish scientist reported about an experiment with artificial intelligence algorithm GPT-3; namely the latter was tasked with writing a 500-word academic work on GPT-3, along with scientific references and quotes.[1] In few (!) hours, the paper was ready. Authors of the experiment are currently seeking to get it published in a peer-reviewed magazine. Another rather controversial reporting originates from China where researchers claim they have developed artificial intelligence that can read the minds of Communist Party members. The aim apparently is to measure the loyalty of Communist Party members.

The competition issues of using artificial intelligence or algorithms are no longer purely theoretical but have also come into the practice of the European Commission (Ezrachi & Stucke, 2019; Levitt et al., 2017). Studies conducted by the European Commission on e-commerce show that the use of algorithms is widespread, with more than two-thirds of companies using automated systems to monitor competitors'

[1] See https://www.giantfreakinrobot.com/tech/artificial-intelligence-academic-paper-gpt-3.html

E. Pärn-Lee (✉)
Tallinn University of Technology, Tallinn, Estonia

D. Ramiro Troitiño et al. (eds.), *Digital Development of the European Union*,
https://doi.org/10.1007/978-3-031-27312-4_11

prices, some of which also implement software solutions that adjust their own prices according to competitors' prices. In 2016, the German and French competition authorities investigated competition law issues related to the use of algorithms[2] (Autorite de la concurrence, Bundeskartellamet, 2016). As a result, the main concern seems to be how to prevent technology that could potentially benefit society from being misused to violate competition rules.[3]

As said above, the question no longer is merely theoretical, and the digital economy is a reality, which affects almost any field of life regardless of how

[2]The use of algorithms affects competition law in at least the following ways:

- The implementation of automated price observation and correction systems involves the risk of price agreements (cartels). It is important how these systems are implemented by the operators.
  - Entrepreneurs may use the algorithms deliberately to make price agreements (cartels) or to set a prohibited resale price.
  - Similarly, operators can be considered as participants in concerted practices simply by the way in which the algorithms they use operate: there are several forms of concerted practice and some of them may, according to the European Commission, take place in automated systems.
- When creating pricing algorithms, software developers need to be careful and make sure that they are not a concerted action or cartel by their nature. One thing is the competition liability of the users of the algorithms, but it should be assessed whether the person who created such software can be held liable.
- Even if previous practice has been to treat contracts concluded by people from the point of view of competition law, if the agreements are made by automated systems, this does not mean (read: should not mean) that the contracts concluded by computers (artificial intelligence) are excluded from competition law.
- The business automated systems have already led to competition law procedures in which the European Commission has analysed whether the implementation of algorithms in a certain way could constitute an abuse of a dominant position. For example, 2018 was penalised by Google for a fine of EUR 4.34 billion because it applied a so-called interconnection of products (*illegal tying*). In 2017, the Commission fined Google 2.42 billion euros because Google's search engine preferred its own price comparison service (so-called *Google Shopping Case*). Since 2016, the procedure under which the Commission is investigating Google is that Google restricts or displays certain third-party competitors' search ads (so-called *Google AdSense Case*).
- The development and use of technology has also raised questions about merger control; i.e. how to ensure that mergers that do not require the prior consent of the Competition Authority under the rules in force but that could potentially affect the product markets (reducing consumer choice or inhibiting innovation) are in the Authority's field of vision.

[3]As already pointed out, the European Commission is quite unanimous in saying that there is no need to change the EU's primary competition rules. Articles 101 and 102 TFEU are, in substance, abstract and flexible and provide implementers sufficient flexibility to apply the rules accordingly. However, according to the Commission, it is necessary to constantly review and, if necessary, supplement or amend the so-called EU secondary law, which is the regulations and directives. For example, when Facebook acquired 19 billion for WhatsApp, which had no profit and a low sales turnover, the big question was whether the merger should be preceded by a mandatory merger control known to be based on the parties' turnover.

advanced technologically one or the other state in fact is. Our way of living, making business, learning, and so on has changed forever (Kerikmäe & Pärn-Lee, 2018).

The EU competition rules stipulated in the Treaty on the Functioning of the European Union have perhaps had the greatest and most tangible effects on European companies as well as the European economy (Troitiño et al., 2020). It is remarkable that the primary rules set forth in Articles 101 and 102 but also 107–109 of the Treaty of Rome have stayed the same and have not been amended. The European Commission has repeatedly considered that there is no need to change EU competition rules. For example, in a speech given on 31 March 2022 Margrethe Vestager, the Executive Vice President of the European Commission and the Commissioner for Competition, noted that *EU competition rules have an in-built flexibility to address exceptional situations such as the Covid pandemic or the Ukraine war. And we've used this flexibility to the full, to support the European economy whilst ensuring that competition is not distorted within the Single Market. Because one of the big lessons of the past two years has been that Europe's resilience relies on a strong Single Market with competitive, innovative and agile companies.* (Vestager, 2022).

In 2012 the Commission undertook an ambitious reform to modernise EU state aid rules. In the wider context of fostering growth, the reform aimed at providing well-targeted state intervention tools and mechanism to deal with market failures and other EU objectives (European Commission, 2012). On 5 July 2022 the Commission presented the new European Innovation Agenda, which further stresses the importance of innovation in driving the EU's competitiveness and ensuring health and well-being of EU's citizens (European Commission, 2022). The Agenda also refers to the Commission launched review of EU state aid rules, amongst others the framework for research and development and innovation that is due in the last quarter of 2022, but also to climate change and environment protection as well as deep tech innovation coupled with close cooperation between the private and public sector. It seems as if we are truly standing at a crossroad. On the one hand, the EU's economy has surprisingly well recovered from the last global financial crisis and recent COVID-19 pandemic; one might say even that we have greatly benefited from these events as the financial crisis led to much stronger control of financial institutions and during the pandemic digital economy really started to flourish[4] (Jinzhu Zhang, 2022). On the other hand, we are in the process of achieving carbon neutrality, and at the same time we have been hit by a global energy crisis, which has at least potentially some impact on the competition and economy. And yet EU state aid rules in principle ban any state interventions. Is it possible to foster innovation without the true cooperation between the private and public sector, I

[4] According to the study conducted by a group of researchers on the COVID-19 impact on the digital economy revealed that digital industries in Armenia, Israel, Latvia, and Estonia have shown great growth potential during the epidemic. On the other hand, adverse impacts were brought to the digital industries in Ukraine, Egypt, Turkey, and the Philippines.

mean a cooperation where the risks of failure are taken by the public sector or at least shared?

With all this in mind, it is necessary to remind ourselves of the goals of EU competition law, especially state aid-related objectives, and see if they serve us also in the conditions of the digital economy.

2 What Is Competition?

According to Article 2 of the Treaty of Rome, the ultimate task of the European Community is *the economic welfare within the single market.* Article 3(f) of the Treaty of Rome specifies this from the competition point of view, noting the goal of *establishment of a system ensuring that competition in the common market is not distorted.* What is EU *common market* is clear; however, what is *EU's competition* seems less understandable. In practice, we proceed from the notion that *monopoly is bad* and *competition is good* because rivalry amongst vendors makes the producers to produce the best possible products whilst keeping prices and costs law.

Economists and competition lawyers seem to agree that there is no single definition of *competition*, nor is there a clear understanding on how to achieve *perfect competition*[5] as most real-world markets operate as *imperfect competition.*[6] If so, what is it that we protect under EU competition law (Zimmer, 2021)? Is it the perfect competition as a standard by which the effectiveness and efficiency of a real-world market operating in imperfect competition can be measured, or can we assume that the EU competition policy takes into consideration that hardly any market operates as perfect competition?

Competition can also mean different things in different jurisdictions, as well as can the objectives pursued with the competition policy differ. For example, Estonian competition law is not defining the term competition nor what competition policy the state is pursuing. Estonian Competition Act merely states that the scope of application of the law is *the safeguarding of competition in the interest of free enterprise upon the extraction of natural resources, manufacture of goods, provision of services and sale and purchase of products and services and the preclusion and elimination of the prevention, limitation or restriction of competition in other economic activities, as well as the processing of and liability for claims arising*

[5] Pure or perfect competition refers to a market structure that is entirely controlled by market forces and fulfils five key criteria: (i) the market is characterised by a large number of sellers and buyers and thus none can influence the price and all are price takers; (ii) all buyers sell identical products, meaning the products are homogenous; (iii) the information of the vendors as well as customers is perfect, meaning the purchases and the relevant terms thereof are known by everyone; (iv) vendors can freely enter and exit the market (no entry barriers exists); and (v) mobility of factors of production (including capital and labour resources are mobile).

[6] Imperfect competition exists whenever a market is not complying with the doctrine of neoclassical pure or perfect competition.

from damages.[7] According to the Estonian Competition Authority their activity is directed at ensuring fair competition and safeguard the functioning of market.[8] Latvian Competition Council declares on its website that their *main aim is to ensure possibility to every market participant to perform his economic activities in free and fair competition environment as well as to promote competition development in all sectors of national economy for the benefit of all society.*[9] *Germany Bundeskartellamt, on the other hand, has declared that its task is to protect competition in Germany and that the protection of competition is a key regulatory policy objective in a market economy.*[10]

Chilean Competition Tribunal declared in 2007 that their only objective of competition policy is to promote and protect competition; however, it is difficult to define legally what is *free competition* or why competition needs to be protected[11] (The Unilateral Conduct Working Group, 2007).

Policymakers often say that competition will lead to lower prices, higher quality, and bigger selection of services or products, as well as to innovation. But is it always so? Even though rivalry between businesses can have such effects, one cannot define competition based on such effects, as competition can as well lead to less innovation and less choice of products when the product homogeneity increases. As a matter of fact, a monopolistic market may be able to offer innovation as well as bigger selection of products and services, if the vendor is provided legal possibilities to exclude competitors from the market as it enables the monopolist to charge higher prices and use the profit in further innovation. On the other hand, such winner-take-all solution can also lead to a reduction in the number of competitors and less price or quality (or further innovation) related competition. Based on the above, do we need to see competition as rivalry, the more competitors the better, or rather as market performance, where companies price their products and services based on marginal costs, which are minimal because of internal efficiency. In the latter case, the market can be monopolistic and perfectly competitive, provided anyone can any time enter the market.

This raises the question whether we need to re-evaluate the mainstream concept of competition which compares the market with the standard of perfect competition

[7] Paragraph 1 Section 1 of the Estonian Competition Act, accessible here https://www.riigiteataja.ee/akt/130122021012

[8] https://www.konkurentsiamet.ee/en/competition-supervision-control-concentrations/competition-supervision/overview

[9] https://www.kp.gov.lv/en/about-competition-council-latvia

[10] https://www.bundeskartellamt.de/EN/AboutUs/Bundeskartellamt/bundeskartellamt_node.html

[11] It is interesting to note that in 2004 when the Chilean competition law was amended talks were held on whether *free competition* should be defined more narrowly as a right to participate in economic activities, a means of promoting economic efficiency, or a means of enhancing consumer welfare. In the end, it was decided by the legislators that the meaning of *free competition,* which means an effective competitive process, should be left to the Chilean Competition Tribunal's interpretation, on a case-by-case basis.

and introduces into policy measures and competition authority's toolbox concepts that enable to view markets and evaluate how they work on a case-by-case basis.

3 What Are the Main Objectives of EU Competition Law?

Academical discussion provides a multitude of objectives of EU competition law. Two primaries are market integration and consumer welfare. *Market integration* has been present from the start when first stated in Article 2 of the Treaty of Rome in 1957.[12] As a policy tool it foremost provides for harmonised application of EU competition within the EU. Further on, it is also relevant for complementing the four freedoms provided in Article 26 Sect. 2 of the Treaty of the Functioning of the European Union (Zimmer, 2021).

The policy goal directed to *consumer welfare*[13] was also already in the Treaties from the start[14] and has over time developed into the primary policy aspect. The European Commission has repeatedly stated that one of the main purposes of the EU competition policy is the promotion of consumer interest and that consumers benefit from the wealth generated by the European economy (Commission, 2002; Lowe, 2008).

Academical discussions also indicate *economic freedom, pluralism, and consumer choice* as competition rules objectives. Article 119 Sections 1 and 2 and Article 120 of the TFEU refer to *the principle of open market economy with free competition*. Even though the principle is set forth in the chapters of economic and monetary policy as well as economic policy of the TFEU rather than in the chapter regulating the rules on competition, some scholars think they apply as well in competition matters (Geradin et al., 2012). This policy in general maintains that all market players, including small and medium-sized players, must have the freedom to operate and should be protected from obstacles deriving from the private or public sector. It is believed that plurality of market participants increases social welfare, whereas economic concentration may enable to dominant company to influence

[12] According to Article 2 of the Treaty of Rome, the ultimate task of the European Community is *the economic welfare within the single market*. Article 3(f) of the Treaty of Rome specifies this from the competition point of view, noting the goal of *establishment of a system ensuring that competition in the common market is not distorted*.

[13] Consumer welfare is often considered together with economic efficiency, whereas efficiency is considered the primary goal of competition. In short, the competition authorities should refrain from enforcing competition rules in output-increasing conduct and incriminate output-reducing. The idea is that competition authorities should look at the efficiency of companies, which is the ratio of output to input used in production. In theory, distinction is also made between allocative and productive as well as dynamic efficiency, whereas allocative and production efficiencies are seen as static concepts whilst in dynamic efficiency the potential of the whole economy or of specific sector or business is considered.

[14] See Articles 2 and 3 of the Treaty establishing the European Community, as well as Article 3 of the Treaty of European Union.

democracy (Geradin et al., 2012). This is then again considered a justification for the claim that an undertaking having dominant position has a *special responsibility* which leads to the distinction that certain action that is not allowed to a firm in dominant position is considered lawful if executed by an undertaking not considered dominant. As a policy objective, the principle is seen by scholars and economists as controversial, because it encourages protection of competitors over competition (Ramiro, 2022). The European focus on protecting rivalry as a policy objective is in academical discussions one of the most referred differences between EU and US competition regimes (Smet, 2008).

Some commentators have said with reference to the preamble of the TFEU[15] that *fairness* is the main objective of EU competition law. The principle is based on the premise that conditions should be the same for everyone; in other words, competition enforcement should seek to *level the playing field*. In practice, the principle is seen being applied in cases where the European Commission has made companies having dominant position to share their intellectual property rights with competitors[16] or where the European Commission has sanctioned dominant companies who have manipulated official administrative procedures[17] (Lyons, 2009). Level-playing field is also used as a standard for state aid control. As stated above, the primary goal of EU competition has been from the founding treaties market integration and economic welfare, whereas it is more about protecting consumers than competitors. Obviously state aid regime is unique in the sense that it deals with states interfering with market functioning rather than with the conduct of market operators.[18] For most of the last few decades, the Commission has been clearly of the view that subsidies contribute little to lasting economic welfare; on the contrary, they lead to unfair competition between undertakings, to market distortions, and to an inefficient allocation of resources, putting additionally the achievements of internal market at risk as they also tend to create or increase barriers to trade. For the Commission the only benefit of state aid is the remedy of market imperfections, provided granting of aid takes place under precise conditions and strict Commission control. For example, SMEs, which are considered an important part of EU economy, access to capital markets has been considered limited; thus for the Commission an aid measure programme that aims at helping the SMEs to compete creates a level-playing field. The same concept is provided by the Commission regarding permitting state interference under regional aid, R&D, environmental programmes, and so on. This indicates that even though the EU state aid need to be interpreted and implemented in accordance with the aim and specific goals of the Treaties, they cannot be applied in vacuum, but in accordance with the relevant realities of economic, social, and

[15] Which states "Recognising that the removal of existing obstacles calls for concerted action in order to guarantee steady expansion, balanced trade and fair competition".

[16] Such as the famous European Microsoft case T-20/04.

[17] Such as the Bayer AG v Commission case (Adalat case) T-41/96.

[18] As do antitrust and merger rules set out in primary and secondary EU competition law, such as Articles 101 and 102, merger regulation (139/2004), and so on.

political environment within the EU but also beyond. For example, in 2014 Commissioner Joaquin Almunia said that the R&D&I Framework *will make it easier* for EU governments to invest more and more efficiently in R&D and innovation (Almunia, 2014). According to him, the rules will *help to unlock Europe's potential as Europe clearly is falling short of the Europe 2020 objective, ie spending 3% of GDP in R&D. The rules are expected to increase spending from their current level of 2% to 3%, that corresponds to the levels of US and Japan.*

Level playing field concentrates on achieving *ex ante* fairness, meaning that an aid measure is not distortive if it does not change the market position of competitors, which, as economists claim, is impossible to achieve and thus bears the risk that all aid is eventually considered *bad* aid, since it inevitably changes the market position (Friederiszick et al., 2006). Also, it is reported that the level playing field approach does not take sufficiently into account the other end of the spectre, i.e. consumer welfare, as a benefit received by the company as a result of subsidy may reduce its incentive to compete, thus reducing the innovation instead of increasing it. The above creates the question if the aim to level the playing field is the only possible and also feasible standard for EU state aid control and as such even above the aforementioned principles of internal market and economic welfare.

To the level-playing field concept also relates the recovery system that has been employed in the EU since 1973 when the Court confirmed first time that the Commission has the power to order recovery of unlawful and incompatible state aid (Friederiszick et al., 2006; Hofmann & Micheau, 2016). Today it is settled case law that the only logical consequence of illegal state is recovery thereof[19] together with default interest to restore the market situation that existed before the aid was granted. The fact that this is almost never possible is not relevant. In many cases, the subsidisation has lasted for years and recovery thereof with default interest may mean that the beneficiary goes into liquidation or even insolvency. When in other competition law areas, the ability of the beneficiary to pay the fine for antitrust law related wrongdoing is taken into account, for example the social aspect of people losing their work, then in state aid recovery no effect-based approach is applied. This is another example of why the level playing field may not serve as the best standard in state aid cases.

Finally, we should speak about *industrial policy* and R&D as EU competition policy goal. It cannot be denied that there is a connection between industrial and competition policy; however, it is more evident in state aid regulation where competition law has been applied as an industrial policy tool and less in other competition law areas (Zimmer, 2021). In state aid regime the policy refers to picking certain sectors and supporting these sectors with public funds. Such industrial policy is directed at picking the national winners and as such is highly problematic especially because of the EU state aid rules, which means that the support measures are invented at EU level, whereas the Member States' competence is

[19]Cases C-142/87 Tubemeuse [1990], para 66, C-305/89 Alfa Romeo [1991] para 41, C-404/00 Commission v Spain [2003], para 41.

essentially limited to enforcing EU-level decisions. This leaves only limited options for Member States to implement their industrial policies.

Commentators also note that industrial policy is not visible as a competition law objective because for a long time it was not clear what the EU industrial policy at the EU level actually was or whether one size fits all, and instead of an EU level approach each Member State in fact has its own industrial policy. In 2002, the Commission launched the debate over industrial policy,[20] focusing on achieving a competitive industry with priority for innovation and growth. Since then, the emphasis on innovation has been a driving force and essential factor in general competition policy but also cases dealt by the Commission. Even more, according to Article 173 Section 1 *The Union and the Member States shall ensure that the conditions necessary for the competitiveness of the Union's industry exist. For that purpose, in accordance with a system of open and competitive markets, their action shall be aimed at [. . .]—fostering better exploitation of the industrial potential of policies of innovation, research and technological development.* This indicates that the competition policy is nowadays closely related to EU industrial policy aiming competitiveness of the EU economy but also creation of employment and general growth. When competition policy aim is to make competition work and to protect it, then industrial policy is above all dealing with market failures.

It seems from the literature, Commission soft law, and the court practice that the initial goals of internal market and economic welfare are increasingly linked with other specific goals and objectives, such as efficiency, economic equity, social cohesion, employment, regional development, international competitiveness, and innovation as well as numerous other non-competitions related polices (Buttigieg, 2009). It is interesting to note in this context, however, that the US courts and government enforcement authorities, on the other hand, have reportedly moved away from multiple-goal approach to an approach that mainly focuses on consumer welfare, although not entirely coherent with the broader consumer well-being notion (Buttigieg, 2009).

---

[20] Commission (EC), Industrial Policy in an Enlarged Europa (Communication) COM (2002) 714 final, 11.12.2002; Commission (EC) Fostering structural change: an industrial policy for an enlarged Europe (Communication) COM (2004) 274 final, 20.04.2004; Commission (EC) Proactive Competition Policy for Competitive Europa (Communication) COM (2004) 293 final, 20.04.2004.

4 Are Incentive Effect and the Concept of Market Failure Prerequisites for State Intervention in Estonia?

In EU market failure does not seem to be a precondition for state intervention; however, Estonian lawmakers have put into hard law that state can interfere only when markets fail,[21] and aid granted must have an incentive effect.[22] Estonian Competition Act does not specify what is understood under the term *incentive effect* and how to furnish *market failure*.

On the European level, the concept of *incentive effect* has its origins in SAM initiative which aimed at facilitating sustainable, smart, and inclusive growth and resulted in adopting the so-called general block exemption (GBER),[23] which applies only to aid which has incentive effect. Aid is deemed to have an incentive effect if the beneficiary has submitted a written application for the aid to the Member State concerned before work on the project or activity starts along with information requested in Article 6 of GBER. In case of ad hoc aid[24] to large enterprises[25] aid is presumed to have incentive effect if the application for aid was made in writing in advance and the relevant Member State has verified prior to granting the aid that (i) in case of regional investment aid the project would not have been carried out without the aid, or (ii) in all other cases the aid contributed to material increase in the scope of project, material increase in the total amount spent on the project, or material increase in speed of completing the project.[26]

The concept of incentive effect is a clear prerequisite for compatibility under regional aid guidelines,[27] R&D&I framework,[28] guidelines for environmental protection,[29] and rescuing and restructuring guidelines.[30] Although the concept is

[21] Just recently, Estonian and Finnish governments agreed to the joint leasing of a floating terminal for liquid natural gas (LNG) in order to guarantee the supply of gas during the Russian energy crisis. In fact, an LNG terminal project was also on the table in 2014 and 2017. In 2014, it was rejected because Estonia granted the right to implement the project to Finland. In 2017, the matter was again on government table, and at that time the then prime minister Jüri Ratas stated that the LNG terminal can only be built based on market logic, which should probably be understood so that the state cannot build it itself nor support private sector construction, unless there is a market failure. When Russia threatened in the spring of 2022 to turn off the pipe gas the LNG terminal matter rose again along with the question if for the state to intervene there must be a market failure, and in which circumstances the public and private sector can cooperate to tackle the matter.

[22] According to paragraph 30 Subsection 2 of the Estonian Competition Act.

[23] Regulation (EC) no 800/2008 which was repealed and replaced in 2014 with the Regulation (EU) No 651/2014 applicable until 31.12.2020.

[24] Means according to Article 2 (17) of GBER any aid not granted on the basis of an aid scheme.

[25] As provided in Annex I of GBER.

[26] Article 6(3) of GBER.

[27] Section 26 (d) of Regional aid guidelines.

[28] Section 36 (d) of R&D&I Framework.

[29] Section 27 (b) of the Environmental guidelines.

[30] Section 38 (d) of Rescuing and Restructuring guidelines

considered clear and relatively simple to apply, wrongful application thereof both on the side of the beneficiaries and institutions granting aid has recently evoked a remarkable number of controversial recovery proceedings in Estonia. In one of such court case, the Supreme Court and the Court of Justice[31] reached even conflicting results. From the view of the Estonian domestic state aid policymakers, one cannot say for sure that the legal uncertainty and resulting recovery procedures were necessarily the result of the provision in the Estonian Competition law. However, this case clearly shows how important it is to know the policy and context behind the principle, as well as to understand the language used in relation to the legal concept established and applied by the Commission, as well as to take into consideration that law and interpretations as well as implementation thereof are not standing still and evolve constantly. And if institutions granting aid or national courts enforcing EU state aid regime need to apply the concept, they must take full account of the principles of equivalence and effectiveness.[32]

Problematic is also the requirement of *market failure* set forth in the Estonian national state aid law. There are currently 25 legal acts valid in Estonia that refer to the term; however, none of them define the term. Regulation no 38 enforced by the Minister of Enterprise and Information Technology on 13 July 2017 for supporting the development of local government housing considers a market failure to exist if the demand for rental premises exceeds the supply.[33]

The prerequisite of *market failure*[34] is not addressed in the EU state aid rules provided in TFEU or relevant secondary legislation such as GBER or de minimis regulation. Yet, it is seen as a cornerstone of compatibility analysis under Articles

[31] See for example Eesti Pagar AS vs Ettevõtluse Arendamise Sihtasutus and Majandus- ja Kommunikatsiooniministeerium C-349/17.

[32] Case C-368/04, *Transalpine Ölleitung in Österreich*, cited above footnote 8, paragraph 45; Joined Cases C-392/04 and C-422/04, *i-21 Germany*, [2006] ECR I-8559, paragraph 57; and Case 33/76, *Rewe*, [1976] ECR 1989, paragraph 5.

[33] Paragraph 1 (3) 1) of the Regulation no 38 dated 13.07.2017.

[34] Commission hard and soft law instruments are either silent or rather modest in defining the term. Only in its State Aid Action Plan from 2005, the Commission elaborated the concept thoroughly and according to this a "market failure" is consequently a situation where the market does not lead to an economically efficient outcome. Market failures have different origins, and notably:

– *Externalities*: externalities exist where actors do not take full account of the consequences of their actions on other actors in society. Market players may not have to pay for the full social cost of their actions (negative externalities) like in the case of pollution through industrial activity. Market players may also be unable to reap the full benefits of their actions (positive externalities) like in the fields of research and innovation.

– *Public goods:* public goods are goods which are beneficial for society but which are not normally provided by the market given that it is difficult or impossible to exclude anyone from using the goods (and hence making them pay for the goods). This can be the case of national defence and some type of public broadcasting.

– *Imperfect information*: imperfect information may lead to transaction costs, agency costs, moral hazard, or antiselection, which in turn lead to inefficient market outcomes. A well-known example of imperfect information can be found in the financial market, where start-up firms usually face problems in finding adequate funding.

107 (2) and (3) of TFEU and as such dealt with in the Commission soft law, such as for example R&D&I framework,[35] rescuing and restructuring guidelines,[36] guidelines for environmental protection,[37] or regional aid guidelines.[38]

In general, the starting point for the Commission is that competitive markets usually provide efficient results in terms of prices, output, and use of resources; thus, if markets do not deliver efficient outcome such market failure can be overcome with state intervention. The problem, however, seems to be that for the Commission the standard is the general EU objective of common interest, which not necessarily overlaps with the interest or industrial policy objective of a single Member State. Also, the market conditions are hardly homogeneous all over the EU. In addition, the market failure standard applied by the Commission seems to be different depending on the context and sector. For example, in R&D context business risk caused by imperfect and asymmetric information inherent to R&D projects is not considered a market failure in itself.[39] On the other hand, aid that can bring a material improvement that the market cannot deliver itself, for example by remedying a market failure or addressing an equity or cohesion concern, can be considered compatible by the Commission.[40] Thus, for compatibility, the presence of market failure is not a prerequisite but an alternative ground. The same is provided in the regional aid guideline[41] and rescuing and restructuring guidelines.[42] In the guidelines for environmental protection the Commission points out that a *mere existence of market failures is not sufficient to justify state intervention,*[43] whereas a remedy of a well-defined market failure is a clear prerequisite[44] for compatibility for aid.

Based on the above and even though the concept on market failure is a cornerstone of compatibility analysis under Articles 107 (2) and (3) of TFEU executed by the Commission, there is in fact no legal reasons why Estonian state aid policy should enable state to intervene only if market failure exists.

– *Coordination problems*: markets may also not function efficiently when there is a coordination problem between market actors. Coordination problems may exist for example in the field of standards setting, in transport infrastructures, or in the area of innovation.

– *Market power:* Another reason why the market may not lead to an efficient outcome is the existence of market power, for instance in a situation of monopoly.

[35] Framework for state aid for research and development and innovation. Official Journal C 198. 27.96.2014.

[36] Guidelines on state aid for rescuing and restructuring non-financial undertakings in difficulty. Official Journal C 249. 31.07.2014.

[37] Guidelines on state aid for environmental protection and energy 2014–2020. Official Journal C 200. 28.06.2014.

[38] Guidelines on regional State aid for 2014–2020. Official Journal C 209. 23.07.2013.

[39] Section 49 of R&D&I Framework.

[40] Section 36 (b) of R&D&I Framework.

[41] Section 26 (b) of Regional aid Guideline.

[42] Section 38 (b) of Rescuing and Restructuring guidelines.

[43] Section 36 of the Environmental guidelines.

[44] Section 27 (b) of the Environmental guidelines.

5 Is EU State Aid[45] Regime Affecting Estonian Digital Economy?

Estonian economy is almost fully digital; it has evolved over the last 30 years from planned economy into e-Estonia—one of the most wired and technologically developed countries in the world. There are little things you need to do in person, almost everything you can sort out electronically. Estonia is also the first country in the world that has introduced the e-residency—*a government-issued digital identity and status that provides access to Estonia's transparent business environment: a new digital nation for the world.* The COVID-19 pandemic only expanded the areas where things can be done electronically. You do not even need to go to court in person, if the judge agrees to an electronically held court session (Hamulák, 2018). Remote e-notary enables us to make notarial transaction from any point of the world, provided you have Internet and a possibility to digitally identify yourself. Only marriage and divorce confirmation cannot be executed by means of remote e-notary.

Thus, it is not possible to distinguish between ordinary economy and digital economy when it concerns Estonia. Also, Estonian market is considered one of the most open economies of the EU, with strong dependency of export as the export of Estonian goods and services exceeds 90% of GDP. As already provided above, subsidies do not necessarily mean general harm to the economic welfare or at least they are considered by economic theory not so harmful and inefficient compared to other trade measures (Hoekman & Kostecki, 2006), Even though economists agree that subsidy regime likely distorts resource allocation and harms competition, both from the perspective of unsubsidised competitors and the taxpayers, subsidies seldom harm consumers directly. On the contrary, the effects of subsidies on consumers welfare, at least good subsidies, are usually positive. For example, a subsidy in public transportation due to which the consumers benefit by enjoying lower prices is considered a good subsidy. Export subsidy, however, is reckoned as a bad subsidy, since it not only harms the international trade but also the domestic consumers, because export subsidy tends to increase the local price level.[46] As a

[45] Subsidies or state aid are no modern times feature and were common for military and political purposes already in the Roman Empire. Subsidies as an economy intervention, however, are reported to have started with mercantilism, whereas the common policies advanced involved protection of domestic industry, promoting exports and gaining positive balance of trade. In the nineteenth century, state was seen as an agent for industrialisation and public resources were employed for industrialisation and large infrastructure projects, whereas the twentieth century introduced protectionism along with subsidies as a measure for income redistribution and balancing externalities. After the Second World War world leaders started to look for options to remove obstacles from international trade leading eventually to the establishment of the World Trade Organization (WTO) together with a truly supranational subsidies control and ending the era where subsidies and other interference measures of economy were controlled at domestic level, if at all.

[46] But even export subsidy, although *per se* prohibited under WTO and EU rules, is not entirely bad from the economical point of view, as it is reported that foreign consumers will benefit from the subsidies to national undertakings.

consequence, the Estonian Enterprise Foundation,[47] which being the largest provider of business subsidies in Estonia, was some time ago forced to suspend and even terminate implementation of certain support schemes involving export activities, because there were no instructions of any kind, and it was unclear when an aid is export aid and when it is not. Also, is any aid to a company who exports its goods or services immediately export aid? There was not and still is no clear instruction by the European Commission on how to understand and apply the ban on export aid. This shows clearly that EU state aid rules cannot be fully understood without comprehending the interaction between the politics behind the rules and the economical motivations that lead Member States to support their national economic operators. Neither should one neglect the origins and sources of the rules, nor the global context.

On the one hand state intervention may distort competition, whilst strong competition is important for market-driven stimulations of innovation. For example, if a company is granted subsidies for innovation it may lose stimulus for being effective because it no longer needs to finance the innovation activities from its own budget. Receiving aid will then also strengthen the position of the aid receiver on the market and weaken the market position of its competitors, as it will cause the return of their investment into innovation usually to decrease. If the decrease is significant, the state intervention may result in reduction of innovation budgets, the direct result of which is that fewer new products and services are created. But this holds true mostly for EU internal affairs. On the other hand, we also need to take into consideration that Estonian but in fact also the companies of other Member States compete globally, as the digital economy has indeed abolished most of the trade restrictions and enables companies to operate all around the globe. If EU state aid rules limit subsidies to companies engaged in developing and commercialising new technologies, do similar rules apply to the companies outside the EU, e.g. in the USA and China? According to public sources The USA has not such a sophisticated and centralised legal regime as the EU (Wood). This does not mean that state intervention does not exist in the US economy; on the contrary state and local governments are rather eager to exchange state aid, which can be *inter alia* in the form of a tax credit for R&D, capital investment, and job creation, for promises from businesses either on state, federal, or local level. Mariana Mazzucato has repeatedly reported in her different publications on US government incentives to businesses (Mazzucato, 2021). She also has noted that in solving global issues we need to be innovative and apply *collaborative, mission-oriented thinking while also bringing a stakeholder view of public private partnerships which means not only taking risks together but also sharing the rewards* (Mazzucato, 2021). She exemplifies by presenting the recent industrial policy history of the USA by showing how the state has indeed been extremely proactive and entrepreneurial in developing and commercialising new technologies,

[47] Today Enterprise and Innovation Foundation, https://eas.ee/eas/

especially through such establishments as DARPA[48] and SBIR programmes.[49] It is also noteworthy that according to her, Apple—a company that we apprise for creating new products and markets—merely made use of publicly funded technologies. Apple's iPhone is smart because of features technologies such as the Internet, GPS, a touchscreen display, Siri, and so on, whereas all of these technologies were developed with public funds. Apple merely merged them into a product. She also strongly advocates that not only entrepreneurs but also a state can *create wealth* and, also, that it is naive to expect venture capital to step in at an early stage of innovation, as this is considered too risky by the private sector.[50] This may mean that for groundbreaking innovations within the EU and subsequent commercialisation thereof we need to consider public funding in much more relaxed terms and also accept the risks of complete failure as well as the fact that private funding enters only after the risks are tolerable.

Almost nothing is known about state intervention rules of China. According to public information China has a government fund amounting to 70 billion dollars yearly that is used for every area of economy, starting from national film industry to building ports and railway projects.[51] According to the European Commission[52] China has a Fair Competition Review System, which inter alia aims to prevent public policies from distorting and restricting competition whilst maintaining fair market competition and promoting a unified market. Also, the Commission entered with China into a Memorandum of Understanding to start a dialogue on state aid control,[48] which because China is the world's third largest economy and EU's second trading partner is an essential step forward for the EU in seeking fair competition on global level.

State aid may incentivise innovation and advance of the digital economy that would not otherwise take place. The Estonian Ministry of Economic Affairs and Communications initiated in 2018 a program called Accelerate Estonia (aE!) with the aim to turn wicked problems into economic development. Working closely with the tech ecosystem, the Estonian government explores areas that are ripe for disruption and build models that create that disruption. The programme experiments currently for example with community energy model,[53] taxonomy of sustainable

[48] The best-known inventions of DARPA are the Internet, GPS, and Siri.

[49] NASA's Small Business Innovation Research (SBIR), but also Small Business Technology Transfer (STTR) programs, provides early-stage funding for R&D, whereas they take zero equity meaning that the companies remain in control of their intellectual property.

[50] Usually, the private sector in the form of venture capital steps in when commercialisation becomes visible. This may happen 15–20 years after the start of the project.

[51] https://www.bloomberg.com/news/articles/2022-03-14/china-sets-63-billion-to-pay-subsidies-owed-to-renewables-firms

[52] See the Press release of the Commission from 2.6.2017, available here http://europa.eu/rapid/press-release_IP-17-1520_en.htm

[53] That aims to create an innovative community energy model for local communities and municipalities to prepare renewable energy projects and start producing locally renewable energy for local communities.

financing,[54] and so on. The aE! programme is launched under the auspices of the public procurement directives' R&D exemption and tries to adhere to EU state aid rules. This, however, may not always be possible or reasonable, as it first requires a market failure to exist[55] and, second, the state intervention must have an incentive effect. But can the mainstream understanding of market failure be present on the market that does not yet exist? Also, can one expect incentive effect to be there if no marketplace is there?

The US Apollo 11 mission took man for the first time to the moon, which was in 1969. Since 1972 no man has been on the moon. What is stopping it? Commentators address that NASA's moon project was built on the mission that man must go to the moon. All procedures, tools, and rules were built on this; amongst others NASA was released of public procurement rules and was entitled to conclude a contract with any person, a move that NASA representatives recall allowed them to seek the best possible solution without being hampered by bureaucratic procurement rules and forced to settle for a bid that met procurement rules but may not have offered the best innovation.

6 Conclusion

Technological advance and digital economy have not only given rise to new business models, new products, and markets, but here and there you can hear how economists and public policy researchers claim that digital economy is in fact disrupting old orthodox economic models. In a paper from 2021, Samuel Bowles and Wendy Carlin report a study they made involving almost 10,000 students of economic studies from 18 countries between 2016 and 2020 (Bowles & Carlin, 2021). The students were asked to name the most pressing problems today's

[54]The team aims to create a solution, which would make it easier for financial institutions and large companies to produce the required analyses and reports and at the same time would also be available and comprehensible to small businesses. The solution directly contributes to the necessary changes in the field of economy and climate set out in the "Estonia 2035" strategy. More specifically to the increase of cooperation between entrepreneurs and the public sector through the development of joint innovation platforms and data economy.

[55]Mainstream or orthodox economics presume that economy (market forces) is naturally stable and market forces (Adam Smith's invisible hand) move economy to equilibrium where demand equals supply. In case the invisible hand does not operate optimally, e.g. in case of externalities (such as environmental pollution), public goods (a good that benefits the whole society—police, education, healthcare), asymmetries of information (imperfect information on one side of the transaction), natural monopoly, and so on, intervention by state is necessary into the marketplace. Unorthodox economics, such as Hyman P. Minsky and his followers, however, claimed that the internal dynamics of modern economy are not equilibrium-seeking, the system is not stable, and in fact there is no invisible hand working that way. Moreover, stability of the market, if by any miracle that is to happen, is destabilising because of policymaking and business opportunities. Minsky also argued that stabile marketplace is difficult for businesses to find profitable opportunities as market tends to get saturated.

economists should be dealing with. The students named *inequality and climate change* as top issues of concern. The issue of inequality needs to be addressed both within the EU and globally. Certain EU Member States seem to have managed for years to evade the EU state aid rules. For example, Germany has Kreditanstalt für Wiederaufbau (Credit Institute for Reconstruction) or in short KfW, which is a creation of post-war German corporatism, established in 1948 as part of the US reconstruction plan for the Federal Republic of Germany to disburse Marshall Plan money. Created for a completely different political era, KfW preceded the 1957 Treaty of Rome and the EU. Some commentators have argued that *If the KfW was established today in its existing format, it too would face difficulties with the EU's state aid rules*. Only in 2002 KfW was restructured because of the pressure by the European Commission, which rules that KfW could retain German state loan guarantees provided it separates its commercial lending operation and creates a separate legal entity without state support. This was done to prevent KfW's commercial operation from having an unfair advantage over its competitors in the European single market. The above is not to claim that the similar inequality still exists within the EU Member States, but rather wants to draw attention to it and to the need to make sure that EU competition policy rules are implemented in a similar way in all Member States. *Global inequality* is also unfortunately not a new problem, is inherently wicked, and seems intractable. Already a decade ago Erik. S. Reinert provided in his book *How Rich Countries Got Rich . . . and Why Poor Countries Stay Poor* an historical overview of the development of rich countries through a combination of government intervention, protectionism, and strategic investment rather than through free trade. He challenges economic orthodoxy and questions if self-regulating markets are indeed the best possible answer to worldwide welfare. Even more, he claims that leaders of rich countries when guiding poor countries on the economical choices seem to have forgotten that their own economies have greatly benefitted from protectionism and the luxury of their free trade is the result of that (Reinert, 2008).

If we want to tackle the challenges of global inequality as well as climate change[56] we may need to rethink not only economy but also certain competition and state aid policies that impact the economy. It is certainly good achievement that the primary EU competition rules have proven to be sufficiently flexible and have enabled them to be applied in the changed economic situation without the need of changing them. Nevertheless, first, it is evident that there is no unified understanding of what competition is,[57] or a market failure or even the primary EU competition law policy goal. Second, studies have shown that EU single market and state aid rules tightly delimit and hamper digital economy and innovation and in order to promote digital economy and innovation-related thereto we ought to apply a mission-based

[56] *Climate change* has also been in the political agendas or quite some time. We all realise it will dramatically affect humans, animal, and other natural life on earth, yet no solution seems available.

[57] If the notion of *competition* is central when developing and applying economic theories behind it, maybe we ought to understand what it is?

approach and principles, rather than a highly detailed and bureaucratic list of dos and don’ts as we have them now.

Everything changes, including the underlying policies. It is not believable that the digital economy, which did not exist and could not be foreseen some 60 years ago, has not led to the need to change certain principles. EU competition law and state aid rules established more than 60 years ago simply do not reflect the changed economy, nor the fact that company’s rivalry is not only taking place in the EU but globally. We need to address that whether we want to continue to have a say in the global economy and the global deep tech race.

References

Almunia, J. (2014, May 21). *europa.eu*. Retrieved from europa.eu: http://europa.eu/rapid/press-release_SPEECH-14-402_en.htm

Autorite de la concurrence, Bundeskartellamet. (2016, May 10). *bundeskartellamt.de*. Retrieved from bundeskartellamt.de: https://www.bundeskartellamt.de/SharedDocs/Publikation/DE/Berichte/Big%20Data%20Papier.pdf;jsessionid=E5909CBE16BB3C1222F72F608ABD25CF.1_cid371?__blob=publicationFile&v=2

Bowles, S., & Carlin, W. (2021, March). *imf.org*. Retrieved from imf.org: https://www.imf.org/external/pubs/ft/fandd/2021/03/pdf/rethinking-economics-by-samuel-bowles-and-wendy-carlin.pdf

Buttigieg, E. (2009). *Competition Law: Safeguarding the Consumer Interest. A Comparative Analysis of U.S. Antitrust Law and EC Competition Law (International Competition Law Series, 40)*. Kluwer Law International.

Commission, E. (2002). europa.eu. *See XXIInd Report on Competition Policy (2002)*. Retrieved from europa.eu: https://ec.europa.eu/competition/publications/annual_report/2002/en.pdf

European Commission. (2012, May 8). *europa.eu*. Retrieved from europa.eu: https://ec.europa.eu/commission/presscorner/detail/nl/IP_12_458

European Commission. (2022, July 5). *eur-lex.europa.eu*. Retrieved from eur-lex.europa.eu: https://eur-lex.europa.eu/legal-content/EN/TXT/?uri=CELEX%3A52022DC0332

Ezrachi, A., & Stucke, M. E. (2019). *Virtual competition: The promise and perils of the algorithm-driven economy*. Harvard University Press. Reprint edition.

Friederiszick, H., Röller, L.-H., & Verouden, V. (2006). *European state aid control: An economic framework*.

Geradin, D., Layne-Farrar, D. A., & Petit, N. (2012). *EU Competition Law and Economics*. Oxford University Press.

Hamulák, O. (2018). La carta de los derechos fundamentales de la union europea y los derechos sociales. *Estudios constitucionales, 16*(1), 167–186.

Hoekman, B. M., & Kostecki, M. M. (2006). *The political economy of the world trading system: The WTO and beyond* (3rd ed.). Oxford University Press.

Hofmann, H. C., & Micheau, C. (2016). *State Aid Law of the European Union*. Oxford University Press.

Jinzhu Zhang, W. Z. (2022). *The Impact of Digital Economy on the Economic Growth and the Development Strategies in the post-COVID-19 Era: Evidence From Countries Along the “Belt and Road”*. Retrieved from frontiersin.org: https://www.frontiersin.org/articles/10.3389/fpubh.2022.856142/full

Kerikmäe, T., & Pärn-Lee, E. (2018). *Digitalisation and automatisationas challenges for changingthe European Union* (Vol. Chapter sixteen). Cambridge Scholars Publishing.

Retrieved from obchod.wolterskluwer.cz/cz/the-european-union-what-is-next-a-legal-analysis-and-the-political-visions-on-the-future-of-the-union.p4794.html.

Levitt, M., Schöning, F., Coulter, A., Stefano, G. D., & Citron, P. (2017, March 21). *competitionlawblog.kluwercompetitionlaw.com*. Retrieved from competitionlawblog.kluwercompetitionlaw.com: http://competitionlawblog.kluwercompetitionlaw.com/2017/03/21/eu-antitrust-enforcement-2-0-european-commission-raises-concerns-about-algorithms-and-encourages-individual-whistleblowers/

Lowe, P. (2008). The design of competition policy institutions for the 21st century — the experience of the European Commission and DG Competition. *Competition Policy Newsletter*.

Lyons, B. (2009). *Cases in European competition policy. The economic analysis*. Cambridge University Press.

Mazzucato, M. (2021). *Mission economy: A Moonshot guide to changing capitalism*. Harper Business.

Ramiro Troitiño, D. (2022). The essence of Europe: Understanding Europe through its designers. In *The European Union and its political leaders* (pp. 1–4). Springer.

Reinert, E. (2008). *How rich countries got rich ... and why poor countries stay poor*. Public Affairs.

Smet, D. D. (2008). The diametrically opposed principles of US and EU antitrust policy. *European Competition Law Review*.

The Unilateral Conduct Working Group. (2007). *Report on the objectives of unilateral conduct laws, assessment of dominance/substantial market power, and state-created monopolies*. Retrieved from https://centrocedec.files.wordpress.com/2015/07/repo

Troitiño, D. R., Kerikmäe, T., De la Guardia, R. M., & Sánchez, G. Á. P. (Eds.). (2020). *The EU in the 21st century: Challenges and opportunities for the European integration process*.

Vestager, M. (2022, February 23). *europa.eu*. Retrieved from europa.eu: https://ec.europa.eu/commission/presscorner/detail/en/SPEECH_22_2203

Zimmer, D. (2021). *The goals of competition law*. Edward Elgar.

Evelin Pärn-Lee is the General Director of the Estonian Competition Board (www.konkurentsiamet). She is also junior researcher and a PhD student at Tallinn University of Technology (TalTech). She has been lecturing EU competition law and policy at TalTech and Tartu University over 10 years. Her research is related to the interface of EU competition law, intellectual property law, innovation, and ICT in general, as the private and public sector endeavours to innovate create different challenges for competition regimes. In her doctoral thesis she studies if the EU State aid framework and implementation thereof, when fostering innovation, is adequately considering the ultimate aim of the EU's competition law and policy, which is protecting the competition in order to maintain a level playing field in the Digital Single Market. She has authored several publications in journals, but also as book chapters and books (https://www.etis.ee/CV/Evelin_P%C3%A4rn-Lee/eng?lang=ENG). Apart from academic work, she has over 25 years of international consultancy experience as an attorney. She has also acted as a legal expert to European Commission and Estonian public authorities with regard to innovation, research, and technology projects but also artificial intelligence studies. Time to time she also provides public lectures or training on topics of state aid, innovation, procurement, etc. Her email is evelinparnlee@gmail.com

Digitally Sovereign Individuals: The Right to Disconnect as a New Challenge for European Legislation in the Context of Building the EU Digital Market

Lucia Mokrá

Abstract Digital tools have increased efficiency and flexibility for employers and employees but have also created a constantly on-call culture, with employees being easily reachable anytime and anywhere, including outside working hours. Technology has made teleworking possible, while the COVID-19 pandemic and the lockdowns have made it widespread. On the one side the online and off-side working has sustained the employment rate even during the pandemic, but it has also blurred the distinction between work and private life. Between 2020 and 2021 during the lockdown in EU member states, 37% of workers had started working from home. According to Eurobarometer, 27% of people who work from home worked outside working hours. The European Parliament had called on the Commission to come up with a law containing the right to disconnect, as it is not part of the EU law, although the right to work and related free movement of workers is the essential part of the EU internal market. The EP Resolution adopted on 21 January 2021 defined the right to disconnect as "the right for workers to switch off their digital tools including means of communication for work purposes outside their working time without facing consequences for not replying to e-mails, phone calls or text messages". It is defined as "a fundamental right which is an inseparable part of the new working patterns in the new digital era" and simultaneously also as "an important social policy instrument at Union level to ensure protection of the rights of all workers".

1 Introduction

Digital technologies contribute to the transformation of society and business, in combination with the development of artificial intelligence. They are definitely running factors of the technical revolution, so-called 4.0. The advantages of digitalisation in industry and also in the labour market became strategic on the

L. Mokrá (✉)
Faculty of Social and Economic Sciences, Institute of European Studies and International Relations, Bratislava, Slovakia
e-mail: lucia.mokra@uniba.sk

D. Ramiro Troitiño et al. (eds.), *Digital Development of the European Union*,
https://doi.org/10.1007/978-3-031-27312-4_12

level of the European Union as well. The European Commission with the ambition of making the EU's single market fit for the digital age presented the digital single market strategy already in 2015 (European Commission, 2015), with the aim of moving all EU member states' national digital markets into a single one.

The digital single market strategy contains 10 priorities, which are divided into three fundamental pillars: improving access to digital goods and services, an environment where digital networks and services can prosper, and digital as a driver for growth. As such, it was mainly market oriented, focused on the development of the single market. The European Commission after elections in 2019 had presented its priorities and working programme, underlining the importance of the digital single market: "The EU's digital strategy aims to make this transformation work for people and businesses, while helping to achieve its target of a climate-neutral Europe by 2050" (European Commission, 2019). When the COVID-19 pandemic occurred, the circumstances for the implementation of the digital strategy shifted as well. The main goals sustained, but the approach responded to the necessity of anthropo-centre perspective. On 9 March 2021, the European Commission proposed Digital Compass for evolving digital transition in the next decade.

This initiative had been followed by the Commission's proposal of inter-institutional declaration on digital rights and principles for the digital decade of 26th January 2022. "The digital rights and principles outlined in the declaration will complement existing rights, such as those rooted in the Charter of Fundamental Rights of the EU, and data protection and privacy legislation (Kerikmäe et al., 2019). They will provide a reference framework for citizens on their digital rights, as well as guidance for EU member states and for companies when dealing with new technologies. They are intended to help everyone in the EU get the most out of digital transformation.

The proposed rights and principles are:

1. Putting people and their rights at the centre of digital transformation
2. Supporting solidarity and inclusion
3. Ensuring freedom of choice online
4. Fostering participation in the digital public space
5. Increasing safety, security, and empowerment of individuals
6. Promoting the sustainability of the digital future (European Commission, 2019)

The so-called human-centred digital agenda is also reflecting contemporary societal development.

Digital transformation may have a positive impact on the unexpected situation related to the COVID-19 pandemic. On the one side the online and off-side working has sustained the employment rate even during the pandemic, but it has also blurred the distinction between work and private life. Between 2020 and 2021 during the lockdown in EU member states, 37% of workers had started working from home. According to Eurobarometer, 27% of people who work from home worked outside working hours. The European Parliament had called on the Commission to come up with a law containing the right to disconnect, as it is not part of the EU law, although the right to work and related free movement of workers is the essential part of the EU internal market. The EP Resolution adopted on 21 January 2021 defined the right to disconnect as "the right for workers to switch off their digital tools including means of communication for work purposes outside their working time without facing consequences for not replying to e-mails, phone calls or text messages". It is defined as "a fundamental right which is an inseparable part of the new working patterns in the new digital era" and simultaneously also as "an important social policy instrument at Union level to ensure protection of the rights of all workers".

This chapter focused on the analysis of the implementation of the right to disconnect in the framework of the EU digital market development, considering benefits and risks of the coordinated approach to the existing legal regulation and case law of the Court of Justice related to the free movement of workers in the European Union. The chapter focuses specifically on the draft of European legislation and the need of its application in the member states of the European Union.

2 Digitalisation and Protection of Human Rights

In relation to the digitalisation of the single market, especially two fundamental rights are in the centre of the Commission's focus. These are guaranteed by the Charter of Fundamental Rights as well as by secondary legislation. There are especially the right to privacy and the right of personal data protection.

The right to privacy has the greatest practical significance in connection with the protection of personality since it serves its free development (Schorkopf, 2007). Although the right to privacy (or the right to private life) is explicitly protected by Article 7 of the Charter of Fundamental Rights, the Court of Justice of the European Union mostly focused its interpretation on particular aspects as the right to privacy of information of health status, inviolability of the home and office, and right to privacy in communication. The issue of privacy in general or in connection with digitalisation has not come to the deliberation of the Court yet. If it happens, due to specific conditions of single market, the Court may have limited discretion in reference to the case law of the European Court of Human Rights, which has rather sustained and developed case law in the area.

The different situation exists in the case of protection of personal data, guaranteed by Article 8 of the Charter of Fundamental Rights. It is a novelty in comparison to the Convention. It is generally interpreted as a protection of fundamental freedoms and rights of natural persons, in particular of their privacy, in connection with the processing of personal data. "Article 7 of the Convention is therefore considered to be complementary to this article" (Pikna, 2010: 84). Its importance is particularly reflected in the European area of freedom, security, and justice. It is a good example of the right closely connected to the running and acting of EU institutions.

The legal basis of the provision of Article 8 of the Charter of Fundamental Rights can be found in Article 16 of TFEU and 39 of TEU, Directive 95/46/EC on the adequate protection of personal data, and Regulation 45/2001 on the protection of individuals with regard to the processing of personal data by the Community institutions and bodies and on the free movement of such data, later replaced by Regulation (EU) 2018/1725 of the European Parliament and of the Council of 23 October 2018 on the protection of natural persons with regard to the processing of personal data by the Union institutions, bodies, offices, and agencies and on the free movement of such data, and repealing Regulation (EC) No 45/2001 and Decision No 1247/2002/EC.

Protection of personal data in the European Union was interpreted by the Court of Justice, especially in relation to its compatibility with:

(a) The right of citizens to move and reside freely within the territory of the member states
(b) The right to privacy
(c) The right to the protection of personal data
(d) The right to freedom of expression
(e) The right to good administration

when in the joined cases Digital Rights Ireland Ltd. and Kärntner Landesregierung (CJ EU, 2014), the Court ruled that whenever the personal data protection and public interest are challenging each other, the proportionality should be applied. The proportionality as guiding principle had been followed by the Commission in a number of legislative proposals related to the digitalisation of internal market (Blažo, 2022: 119).

Digitalisation and the development of technologies have made teleworking possible, while the COVID-19 pandemic and the lockdowns have made it widespread. This was exactly the situation when personal privacy (right to private life) and public interest (economic development of the country and sustainability of the public administration) came into conflict with each other. Digital tools have increased efficiency and flexibility for employers and employees, but also created a constantly on-call culture, with employees being easily reachable anytime and anywhere, including outside working hours (Troitiño et al., 2020). This remains an open question: how to guarantee the right of workers to rest between working shifts, balancing the private life with the ambition of employer for the economic profit?

3 Right to Disconnect: Balancing between Digitalisation in Public Interest and Rights of Workers

In relation to pandemic's influence on the employment, a number of employers transferred their working environment online. The combination of the pandemic restrictions for movement together with the fear of losing the job led to the situation where the boundaries between working hours and personal time have become blurred.

The working time is on the EU level regulated by the EU working time directive (European Parliament and Council, 2003). The directive applies to (a) minimum periods of daily rest, weekly rest and annual leave, breaks, and maximum weekly working time and (b) certain aspects of night work, shift work, and patterns of work. Under the text of the directive, it has guaranteed daily rest of 11 consecutive hours per 24-hour period as well as breaks; where the working day is longer than six hours, every worker is entitled to a rest break. Member states are entitled to set details of duration and terms of the breaks in the work, especially in the form of the collective agreements. If collective agreements are not possible to be achieved, national legislation has to grant the breaks within the working time.

The expectations that people who regularly telework are more than twice as likely to work more than the maximum working hours set down in the EU's working time directive (European Parliament and Council, 2003), than those who do not, had been already confirmed by Eurofound research. This research confirmed that "working from home has increased by almost 30%. Almost 30% of those working from home report working in their free time every day or several times a week, compared to less than 5% of office workers" (Eurofound, 2021).

All these circumstances and experience led the European Parliament to call for an EU law that grants workers the right to digitally disconnect from work without facing negative repercussions. In their legislative initiative of 21 January 2021 that passed with 472 votes in favour, 126 against, and 83 abstentions, MEPs call on the Commission to propose a law that enables those who work digitally to disconnect outside their working hours. It should also establish minimum requirements for remote working and clarify working conditions, hours, and rest periods (European Parliament, 2021a).

The European Parliament adopted Resolution (European Parliament, 2021b) building upon previous legal regulation connecting space of single market and free movement of workers, safety and health of workers at work, organisation of the working time, transparent and predictable working conditions, Charter of fundamental Rights, European Pillar of Social Rights while keeping in mind international obligations especially to UN agreements, European Social Charter revised, and ILO, underlining necessity to keep agreements with social partners on telework and digitalisation in mind.

The European Parliament's Resolution (hereinafter as Resolution) not only refers to necessity to regulate working time, resting time, and breaks but also corresponds to findings from different studies and reports of Eurofound, which confirms that 27% of respondents to its research were working from home in their free time to meet work demands.

Among the basic grounds justifying the need to regulate the right to disconnect as part of the right to work within the internal market, the parliamentary Resolution mentions that although telework has been instrumental in helping to safeguard some employment and business during the COVID-19 crisis, but reiterates that, because of the combination of long working hours and higher demands, it can also pose higher risks to workers with a negative impact on the quality of their working time and their work–life balance, as well as their physical and mental health. The Resolution stresses that the right to disconnect allows workers to refrain from engaging in work-related tasks, activities, and electronic communication, such as phone calls, e-mails, and other messages, outside their working time, including during rest periods, official and annual holidays maternity, paternity and parental leave, and other types of leave, without facing any adverse consequences, especially the loss of the job itself.

The provisions of the Resolution reiterate that respect for and predictability of working time are considered essential to ensure the health and safety of workers and their families in the Union and are aligned with previous legislation, particularly Directive 2003/88/EC concerning certain aspects of the organisation of working time, in particular as regards the right to paid annual leave, in Directive (EU) 2019/1152 on transparent and predictable working conditions, in Directive (EU) 2019/1158 on work–life balance for parents and carers and in Council Directive 89/391/EEC on the safety and health of workers as well as the international obligations in relation to the International Labour Organization (ILO), in particular the 1919 Hours of Work (Industry) Convention (No. 1), the 1930 Hours of Work (Commerce and Offices) Convention (No. 30), the 1981 Collective Bargaining Recommendation (No. 163), the 1981 Convention on Workers with Family Responsibilities (No. 156) and its accompanying Recommendation (No. 165), as well as the 2019 ILO Centenary Declaration on the Future of Work, and the Council of Europe's Revised European Social Charter of 3 May 1996, and in particular Article 2 (regarding the right to just working conditions, including to reasonable working hours and to rest periods), Article 3 (regarding the right to safe and healthy working conditions), Article 6 (regarding the right to collective bargaining), and Article 27 (regarding the protection of workers with family responsibilities.

4 Commission's Turn and Guide from Court of Justice of the European Union (?)

The European Parliament by its Resolution (European Parliament, 2021b) called the European Commission to draft the legislation in relation to the right to disconnect. The text of the Resolution is accompanied by the proposal of the Directive on the right to disconnect. Both the text of the Resolution and the proposal of the Directive refer to settled case law of the Court of Justice of the EU on the definition of worker, working time, and rest time. The innovation regards the right to disconnect is especially in the definition of worker, to whom this right should be granted. Although the concept of "worker" should be defined by national law as stated in the Point (a) of Article 3(1) and Article 3(2) of Directive 2008/104/EC, for the purpose of the right to disconnect, the Court of Justice of the EU ruled that "the legal characterisation under national law and the form of that relationship, as well as the nature of the legal relationship between those two persons [are not] decisive in that regard" (Court of Justice of the EU, 2016: 27). It means that the form, length, or type of labour contractual relations between employer and employee should not impact the effective application of the employee's right to disconnect and the protection of social rights and right to private life as granted by the Charter of Fundamental Rights and other provisions of EU law. For this purpose, it is the obligation of the EU member state to ensure that national legislation will respect the limits of working time and that national legislation will transpose properly obligation from EU law "to ensure employers set up an objective, reliable and accessible system enabling the duration of time worked each day by each worker to be measured" (Court of Justice of the EU, 2019: 29), to guarantee a balance between working hours and resting time and the related right to disconnect if employee is teleworking.

It is case law of the Court of Justice of the European Union which confirms that on-call time, during which a worker is required to be physically present at a place specified by the employer, is to be regarded as "wholly working time [...], irrespective of the fact that, during periods of on-call time, the person concerned is not continuously carrying on any professional activity" (Court of Justice of the EU, 2004: 93), and that standby time, which a worker is obliged to spend at home, while being available to the employer, is to be considered to be working time (Court of Justice of the EU, 2018: 66). Moreover, the CJEU has interpreted minimum rest periods as "rules of Community social law of particular importance from which every worker must benefit as a minimum requirement necessary to ensure protection of his safety and health" (Court of Justice of the EU, 2006: 38) The CJEU has confirmed that Directives 89/391/EEC and 2003/88/EC require employers to set up a system enabling the duration of time worked each day by each worker to be measured and that such a system be "objective, reliable and accessible" (Court of Justice of the EU, 2019: 60).

The Court of Justice of the European Union is by its extensive and innovative, as well as continuously settled, case law providing a helping hand to Commission to draft the Directive, which will reflect changing circumstances in the area of working conditions, social rights, and digitalisation. Although the three areas have different addressees: member states, workers, and the European Union, they are interlinked, and without the effective exercise of the right to disconnect, the progress and development of the digital market will be fundamentally constrained not by the physical but above all by the mental capacities of employees. The approach and justification provided by the Court of Justice is also addressed to social partners, who should ensure and coordinate common approach to the existing working conditions without being detrimental to social rights and mobility within the Union.

5 Conclusion

The right to disconnect is not explicitly regulated in Union law; however, its implication may be found in connection with existing fundamental rights granted by the Charter of Fundamental Rights as well as the digital rights as presented in the Digital Single Market strategy and Digital Compass. Although the first ambition of digitalisation of the EU single market focused on the protection of data, development of the business, and adaptation to changing economic environment in times of digital *revolution*, the COVID-19 pandemic since 2020 has shown the negative impact of digitalisation on workers, especially on their physical and mental health, private life, and personal protection. Therefore, the European Parliament took the initiative and urged the European Commission to propose legislation regarding right to disconnect. The Resolution of 21 January 2021 identified the right to disconnect as "the right for workers to switch off their digital tools including means of communication for work purposes outside their working time without facing consequences for not replying to e-mails, phone calls or text messages" (European Parliament, 2021b). As the definition of the worker is not decisive here, it should be applicable to both the public and private sector and in relation to all types of work, including atypical workers. The obligation on the employers to ensure the workers' right to disconnect has to become regular part of labour law both on European and national level.

The European Commission may for this purpose use two helping hands—of the European Parliament which had already annexed the proposal of Directive on the right to disconnect to the Resolution as well as the one of the Court of Justice of the European Union, which progressively and continuously provides a broader interpretation of social rights in the framework of EU secondary law and application practice at national level through preliminary rulings.

The balance between the workers' right to disconnect and the EU's ambition to build digital single market is the question of the incoming legislative proposal. However, these two objectives are not mutually exclusive, on the contrary. Their complementarity and effective application can help to propel the European Union in the global economic arena and at the same time confirm the EU's observation of human rights as a fundamental value of the EU under Article 2 TEU.

Such a combined approach can already be seen in the emphasis on the requirement of digital transformation in the EU member states respecting the human rights needs. It is also reflected in the requirement of the allocation of finances in the Recovery and Resilience plans for each member state. Each EU member state is required to dedicate and spend at least 20% of the Recovery and Resilience to the digital transition, respecting rules and principles stated both in digital single market strategy and in Digital Compass.

References

Blažo, O. (2022). *The Digital Markets acts: Between market regulation, competition rules and unfair trade practices rules* (Vol. 66(1), pp. 117–136). Strani Pravni Zivot.

Court of Justice of the EU. (2004). *Pfeiffer and others, C-397/01 to C-403/01*, ECLI:EU:C:2004:584.

Court of Justice of the EU. (2006). *Commission v United Kingdom, C-484/04*, ECLI:EU:C:2006:526.

Court of Justice of the EU. (2014). *C-293/12 Digital Rights Ireland Ltd and C-594/12 Kärntner Landesregierung* [2014] ECLI:EU:C:2014:238.

Court of Justice of the EU. (2016). *Betriebsrat der Ruhrlandklinik, C-216/15*, ECLI:EU:C:2016:883.

Court of Justice of the EU. (2018). *Matzak*, ECLI:EU:C:2018:82, C-518/15.
Court of Justice of the EU. (2019). *Federación de Servicios de Comisiones Obreras (CCOO)*, C-55/18, ECLI:EU:C:2019:402.
Eurofound. (2021). *Telework and ICT-based mobile work: Flexible working in the digital age, 2021*, online: [https://www.eurofound.europa.eu/sites/default/files/ef_publication/field_ef_document/ef19032en.pdf].
European Commission. (2015). *Communication from the Commission to the European Parliament, the Council, the European Economic and Social Committee and the Committee of Regions*. A Digital Single Market Strategy for Europe. Brussels, 2015, COM(2015)192 final, online: [https://eur-lex.europa.eu/legal-content/EN/TXT/?uri=celex%3A52015DC0192].
European Commission. (2019). *A Europe fit for the digital age*. Brussels, European Commission, 2019, online: [https://ec.europa.eu/info/strategy/priorities-2019-2024/europe-fit-digital-age_en].
European Parliament. (2003). *Directive 2003/88/EC concerning certain aspects of the organisation of working time*. European Parliament and Council, 2003, online: [http://data.europa.eu/eli/dir/2003/88/oj].
European Parliament. (2021a). *'Right to disconnect' should be an EU-wide fundamental right, MEPs say*. Brussels, European Parliament, 2021, online: [https://www.europarl.europa.eu/news/en/press-room/20210114IPR95618/right-to-disconnect-should-be-an-eu-wide-fundamental-right-meps-say].
European Parliament. (2021b). *Resolution of 21 January 2021 with recommendations to the Commission on the right to disconnect (2019/2181/NL)*, online: [https://www.europarl.europa.eu/doceo/document/TA-9-2021-0021_EN.html].
Kerikmäe, T., Troitiño, D. R., & Shumilo, O. (2019). An idol or an ideal? A case study of Estonian e-Governance: Public perceptions, myths and misbeliefs. *Acta Baltica Historiae et Philosophiae scientiarum, 7*(1), 71–80.
Pikna, B. (2010). *Evropský proctor slobody, bezpečnosti a práva (prizmatem Lisabonské smlouvy)*. Praha.
Schorkopf, F. (2007). Human dignity, fundamental rights of personality and communications. In D. Ehlers (Ed.), *European fundamental rights and freedoms*. De Gruyter Textbooks.
Troitiño, D. R., Kerikmäe, T., De la Guardia, R. M., & Sánchez, G. Á. P. (Eds.). (2020). *The EU in the 21st century: Challenges and opportunities for the European integration process*.

Lucia Mokrá is Associate Professor at Comenius University in Bratislava, Faculty of Social and Economic Sciences, Institute of European Studies and International Relations (Slovakia). Email: lucia.mokra@uniba.sk.

Nordic Roadmap Toward an EU-Wide and Seamless Cross-Border Cooperation on Judicial Matters

Alina Hruba and Maria Claudia Solarte Vasquez

Abstract This chapter reviews the digitization process of the courts in the five selected Nordic countries and the regulatory components that support the evolution of e-governance solutions in this field in the region. These states have maintained a paradigmatic democratic prosperity where the judicial activity and its procedural justice elements have been considered central to the proper implementation of the rule of law. The chapter is based on a case study that observes, from a historical perspective, the stages in the development of the Scandinavian legal culture giving shape to the specific features of a sub-family of law. More narrowly, it looks for e-justice benchmarking reform processes in five countries, using a comparative approach, and refers to some shortcomings that may be affecting the Nordic judicial cooperation. The chapter shows that both the reforms and shortcomings influence the extent to which the countries can advance in the direction marked by the European Union (EU) policies on cross-border cooperation in judicial matters.

1 Introduction

The transformation of the judiciary continues to rank high within the European Commission goals of achieving the digital transformation of the public services by 2030 while ensuring their democratic sustainability based on accessibility, excellence, usability, and security standards (European Commission, 2021; Nylund, 2021b). Advancing with the digitalization of justice continues to be one of the

A. Hruba
Yaroslav Mudryi National Law University, Kharkiv, Ukraine

University of Helsinki, Helsinki, Finland
e-mail: alina.hruba@helsinki.fi

M. C. S. Vasquez (✉)
Tallinn University of Technology, Tallinn, Estonia
e-mail: maria.solarte@taltech.ee

D. Ramiro Troitiño et al. (eds.), *Digital Development of the European Union*,
https://doi.org/10.1007/978-3-031-27312-4_13

most important priorities of Europe's Digital Decade Strategy (EDDS),[1] as it was established in 2019 in the Action Plan on e-justice 2019–2023: OJ C 96, 13.3.2019, pp. 9–32.[2] In consequence, it is one of the Commission's ambition to support the efforts to modernize the European justice systems and make them more efficient (European Commission, 2021, p. 12).

The European courts are at various stages of digitalization. Although the use of computers has become widespread, harnessing the potential of the powerful technologies available remains a challenge for some countries while highly elaborate approaches (both technologically and legislatively) are in place in others. There is no systematic correlation between the level of sophistication of information technologies and their efficiency, effectiveness, or impact. Several European jurisdictions have reported considerable operational difficulties following the testing or implementation of simple electronic case management systems. Clearly, the scope of digitally transforming the public sector, the judicature above all, is a sociotechnical process that goes far beyond digitizing, and it may be helped or delayed by the laws in place, the courts markedly (Velicogna et al., 2020; Council of Europe, 2018). Having said this, that the courts in Scandinavia are not at the forefront of the digital EU transformation was anticipated. Like everywhere else in Europe, the most common initiatives in those jurisdictions are the transfer of administrative and secondary court tasks to platforms, and the conduction of proceedings online. Nylund (2021b) finds that digitizing has not had a meaningful impact in the structural organization of the court system or in substantial aspects related to proceedings. Nonetheless, the common or individual decisions they have taken to upgrade the courts and the judiciary could ultimately lead to a fundamental and principled digital transformation, more apt for expanding judicial cooperation strategies in the long run. Not only drawing conclusions about impact, especially if to refer to the greater institutional context, is still premature, but crucial aspects related to digitalization such as procedural justice reforms cannot be assessed lightly. Hjort (2021) agrees, stating that the process at the national and supranational levels is paced, providing the means to gradually transition from what has been called "first-generation goals" such as determining the due processes regarding decision making, toward "second and third generation goals," like effective access to justice in minor lawsuits, and the use of information and communication technologies (ICTs) to facilitate and make adjudicatory methods more efficient, respectively (Velicogna et al., 2020; Hjort, 2021, pp. 78–79). Needless to say, social and institutional progress needs to be founded on technical capabilities such as enabling and fit to purpose technologies.

In light of the above, the experience of the five Nordic countries, namely Denmark, Sweden, Iceland, Norway, and Finland that have achieved a respectable level of digitalization, is seen as a good case study that could show some of the

[1] https://ec.europa.eu/info/strategy/priorities-2019-2024/europe-fit-digital-age/europes-digital-decade-digital-targets-2030_en

[2] https://e-justice.europa.eu/home.do?plang=en&action=home

practices defining the concept of digital courts in the EU as well as some patterns of judicial cooperation. The court systems of three of these countries that are EU Member States have demonstrated resilience during the COVID-19 pandemic, as the data of the EU Justice Scoreboard reports of 2021(COM(2021)389) and 2022 (COM(2022) 234) suggest (European Commission, 2021, 2022a, 2022b), and their regulatory frameworks are highly compatible. Unsurprisingly, they have been ahead in the *public sector* digitalization processes in Europe, for more than a decade (Wallström et al., 2009). In 2022, Finland ranked first, according to the European Commission's annual Digital Economy and Society Index (DESI), which tracks the digitalization progress of EU member states. Denmark and Sweden come second and fourth, respectively (European Commission, 2022a). In addition, they are leaders in closely related respects, as seen from their positioning in the World Justice Project Rule of Law Index,[3] to becoming among the best locations to build data centers for colocation and connectivity services.[4]

The Nordic legal systems are not often identified as a distinct legal tradition but considered as a particular subset of the Civil Law (Husa et al., 2007), also known as the Scandinavian subfamily of law. However, the Nordic denomination is seen as a useful categorization in itself, as explained by Lukianov (2015), and has permitted a closer analysis. Nylund (2021a) has observed the existence of "*a distinctly Nordic procedural or court culture*," that is, a set of ideas and values that in combination constitute the core of the legal culture and the systems of the countries in that area. The general principles and rules of administrative law are similar, although it still reflects historical and perhaps unsurmountable differences that further separate the Nordic countries into two groups: one including Denmark, Norway, and Iceland and the other formed by Finland and Sweden.

This study opts for a simplified functional comparative approach. It contrasts the reported advancements and the doctrine against the background that the EDDS and its deriving policies mark, when relevant to cross-border cooperation in judicial matters,[5] paying special attention to the advancements on e-justice attained during the COVID-19 pandemic and annotating some cooperation challenges arising among the Nordic countries. This combined assessment helps reveal the strengths and difficulties of digitization, which is viewed as a foundational, often underestimated, and resource-intensive stage to consider and monitor in the ongoing transformation of the public sector taking place globally.

[3] https://worldjusticeproject.org/sites/default/files/documents/WJP-INDEX-21.pdf

[4] https://techmonitor.ai/technology/data-centre/best-place-data-centre-denmark-google

[5] https://ec.europa.eu/info/law/cross-border-cases/judicial-cooperation_en

2 Historical Background and the Scandinavian Legal Culture of the Nordic Legal Systems

As mentioned, the European northernmost countries located in the Scandinavian peninsula and referred to as Nordic are Denmark, Sweden, Norway, and Finland. They have formed a distinguishable block along with Iceland, which displays extensive compatibilities on the fundamental values and principles enshrined in their legal systems, from times immemorial (Von Eyben, 1962). The Nordic law identity is deeply rooted and well established in the comparative law literature. It represents the Scandinavian legal subgroups of the civil law tradition in the legal landscape of Europe (Bernitz, 2000; Kischel, 2019).

The relationship between geography on the one hand, and governance on the other, has been crucial when it comes to the regulatory development of the Nordic countries, which have vast territories and small populations. In addition, they have been kingdoms that exercised a powerful influence over one another in different periods of time, more concretely, between the XI and XIV centuries. The Norwegian king ruled the Orkney, Shetland, Faroe Islands, Greenland, and Iceland, besides its own territory in the peninsula. The Swedish kingdom included Finland (from 1104 to 1809) and Norway for more than half a century until 1397 (Sunde, 2021, p. 55).

Over the centuries, the communities inhabiting the region have been interrelated. Based on the Kalmar Union of 1397, the Danish kingdom was an association of Sweden and Norway, with Finland and Iceland. Later on, this relationship took on a different character but lasted well until the nineteenth and twentieth centuries (Zinchenko, 2017, p. 59). Thus, Denmark, Norway, Iceland, and the Faroe Islands formed the West Scandinavian Kingdom, while Sweden and Finland formed the East Scandinavian Kingdom. In the seventeenth century Sweden also ruled the Baltic countries and territories extending far into the European continent, while Denmark would acquire small colonies in India, Africa, and the Caribbean (Sunde, 2021, p 55, 56).

Denmark had to cede Norway to Sweden in 1814, but with an independent constitution, which resulted in a new kind of relationship between the countries. Norway and Sweden formed a short-term alliance that was dissolved in 1905. Iceland became an independent state under the Danish-Icelandic Union Act of 1918 but recognized the Danish king as sovereign until World War II, when the country became a republic. Finland was part of the Swedish kingdom until it was conquered by Russia in 1809. During the Russian rule, it managed to have autonomy in domestic affairs. Finland finally gained independence in 1918 (Bernitz, 2007, p. 16). These few examples illustrate the development path of the Nordic countries, where the roots of a shared historical and legal heritage can be traced. They are clear instances of integration through power and collaboration arrangements that required "cross-border co-operation" and presumably affect it today. Notwithstanding, Bernitz (2007) showed that the similarities are limited, suggesting why the eastern part of Scandinavia covering Denmark, Norway, and Iceland may still separate itself from the eastern part where Finland and Sweden are located. These events also resemble the rise, evolution, and fall of the Roman empire and determined the origins and establishment of the civil law tradition and its current status.

The Scandinavian law has evolved independently from other European legal systems, because the Nordic kingdoms were remote and inhospitable. They remained to an extent in isolation, largely responding to their common social, economic, political, and cultural needs. In the past 100 years intensive formal and informal cooperation initiatives between the Nordic countries have been initiated, in an attempt to harmonize their legislation (Hasselbalch, 2017). These commonalities sustain a fairly identical stance about central regulatory matters, like the legal values and principles their systems protect, and the boundaries between the public and the private sector that have resulted in a clear demarcation of the questions that need to be resolved by the state. Some examples are their commitment to secularism, the conception of equity and fairness (Husa et al., 2007; Letto-Vanamo & Tamm, 2019), social democracy, state paternalism, contractarianism (Skogh & Stuart, 1982), and universal welfarism (Keller, 2009; Nousiainen, 2018). The Nordic countries are perhaps best known for being able to create a functioning welfare state based on Western liberal values (Letto-Vanamo & Tamm, 2019) but with a legal infrastructure that enshrines social rights and the protection of the vulnerable population. The Nordic welfare "style" classifies as universalist in that it is status oriented, promoting equality on its own terms: reaching the highest standards rather than promoting the satisfaction of "minimum" needs (Husa et al., 2007, p. 19). In the catalogue of global comparative law studies, the Scandinavian legal systems appear to be relatively uniform, at least enough for the literature to recognize the existence of a distinct legal culture (Husa et al., 2007, pp. 28, 29). The question that arises is whether regional integration processes, globalization, and the digital transformation of governance and regulatory models could eventually blur the boundaries between the subgroups within legal traditions.

The golden age of Nordic legal cooperation is associated with the 1950s and 1960s, even if some efforts made then, and the fundamental work done at that time, only came to fruition later. A number of highly technical legislative "products" were being conceived in important fields of law by legislative committees set up simultaneously in Denmark, Finland, Norway, and Sweden, working on a long-term basis. Bernitz (2000, 2007) lists outcomes related to obligations, contracts, torts, family law, maritime law, intellectual property, and trade law. The benefits of this coordination have translated into remarkable social and economic progress. A legal system is shaped in the way any other cultural manifestation is, telling the story of groups, communities, society, peoples, and civilizations. Being the laws part of the culture, and culture a cooperative effort, the fruitful exchange between the countries allowed legal doctrines and values to transcend borders. Nowadays, legal cultures and legal systems may have a global impact (Scott, 2014). Core examples are the notion of democracy, the rule of law, human rights and fundamental freedoms, and others, which have become strengthened with conventions and political support, as in the case of the EU General Data Protection Regulation (GDPR) (Regulation (EU) 2016/679, OJ 2016L 119/1), to mention a more recent and well-known instance emerging from the EU landscape.

The Nordic countries have their own models of legislative cooperation, which are constantly expanding into new areas. Moreover, they have agreed to adopt identical laws whose contents have been drafted despite the absence of an international instrument (Petrén, 1968; Letto-Vanamo & Tamm, 2019). At the same time, the expansion of the EU legislative competences has caused a reduction of the Nordic legislative integration activities. However, the countries in the region have sought to maintain their pre-existing institutional arrangements and orientation in the field of private international law (Karhu, 2010). The Nordic legal community of practice, in particular scholars and other experts, meets regularly at inter-Nordic congresses and seminars. This ensures cohesion and the preservation of the legal culture identities that are represented while allowing mutual influence for their further development. Sunde (2021) points out that Nordic judges with international experience confirm to have the same views and positions rather often and feel comfortable to exchange opinions informally. This may be a very simple observation but shows that there is sense of "Nordicness" and a climate of cooperation in general.

The Nordic law is characterized for its solid legislative and judicial tradition, the accessibility of the law in terms of the use of plain language in the drafting of legal texts and court decisions, oral legal proceedings, a limited number of professional lawyers, and a concrete and pragmatic legal science (Sunde, 2021, p. 53). There is no tendency toward conceptualism and the construction of large-scale theoretical systems. This is mainly because the Scandinavian consciousness has never embraced regulatory scientism or theorization and is accompanied by a more action-oriented legal thinking than what could be expected from the Germanic tradition (Husa et al., 2007). The Scandinavian law is inspired by an idea of simplicity, which is not a feature to be found in other legal sub-families. Presumably, one of the qualities of Nordic legal thinking is its realistic nature (Husa et al., 2007). The current focus is on supporting countries to build a society where the rule of law prevails, which means, in practice, to complete a high level of democratic development. In Denmark, such ideas are openly based on the belief that what we may call the Nordic approach to the rule of law is superior to others. What is striking in this rule of law and development-oriented thinking is the naivety about the attempt to use law as a tool for development. Husa (2015, p.19) argues that critical and well-known failures, problems, and inherent flaws in the thinking about law and development as a practice and as an academic field are conspicuously absent.

In the public sector, authorities face the challenge of having to improve the efficiency, impact, and quality of the services they deliver. It becomes an e-government project once the upgrading of equipment and ICTs is combined with organizational change, the improvement of public services, democratic processes, and public policy (Saarenpää, 2004). Pöysti (2010) classified e-government types in four generations, according to their scope. The first is where the use of electronic tools enables interactions with public authorities. The second focuses on the "platformization" of state services and the public administration. The third generation enhances the value of electronic services applying a user-centric perspective and expanding delivery channels. The last works on the trustability and quality of the technology-based services. To expand the scope and capacities of the public administration using technology is imperative, and although digitizing appears to be central to the first-generation models only, it is an obvious condition for the adequate functioning of the rest. This is why there is a relative advantage for states that were not early adopters of the ICTs and other technologies. As the saying goes, good things come to those who wait; they may be now deploying much better versions and preventing premature disruption risks (Sheppard, 2015).

The Nordic countries are also unique in their e-government approach. They have invested in technologies and introduced rules that allow, and in some cases impose, digital communication channels to the public. What is more, their concerns for organizational change strategies, user centricity, trustworthiness, and quality of the public services are proactive and pre-digital. Denmark has chosen a radical path, where the administration can, except in a few narrowly defined situations, send information to electronic messaging boxes that the users must have set up in advance. In numerous instances, applying for permits, services, etc. can only be done digitally (Mäenpää & Fenger, 2019). These countries are steadily advancing with the digital transformation of their public sector, and regarding e-governance, their enhanced e-justice proposal seems to stand out. Still, they have not been unaffected by the contingencies caused by the pandemic, and some decisions deviated from the planned system design pathway as they were forced upon the judiciary, especially those concerning procedural rules to guarantee basic access to justice in times of crisis.

3 The Scope of e-Justice Implementation Processes in the Nordic Countries

The previous section mapped the roots of the shared historical and legal Scandinavian heritage, leading to the development of the Nordic country's public sector capacities for modernization and cross-border integration and cooperation. Various resources are available to obtain more information regarding the features and peculiarities of the states' wider e-governance and digitalization models. For example, an overview of digital public administration and interoperability state that summarizes the progress and milestones of European countries in those areas is published yearly, by the National Interoperability Framework Observatory (NIFO).[6] The report is made on the state of play of each Member State and associated countries.[7] The latest addition are the factsheets and infographics version 2022.[8] This part summarizes the court digitization status in each country and some features

[6]NIFO - National Interoperability Framework Observatory I Joinup (europa.eu)

[7]https://joinup.ec.europa.eu/sites/default/files/inline-files/DPA_Factsheets_2021_Denmark_vFINAL.pdf, https://joinup.ec.europa.eu/sites/default/files/inline-files/DPA_Factsheets_2021_Sweden_vFinal.pdf, https://joinup.ec.europa.eu/sites/default/files/inline-files/DPA_Factsheets_2021_Iceland_vFinal.pdf, https://joinup.ec.europa.eu/sites/default/files/inline-files/DPA_Factsheets_2021_Norway_vFinal.pdf, https://joinup.ec.europa.eu/sites/default/files/inline-files/DPA_Factsheets_2021_Finland_vFinal.pdf

[8]https://joinup.ec.europa.eu/collection/nifo-national-interoperability-framework-observatory/digital-public-administration-factsheets-2022

of their e-justice proposals and comments on the readiness of the systems for EU-wide cooperation.

3.1 Denmark

Several elements of e-government and e-justice have been introduced since the mid-2000s, with the aim of adding transparency and accessibility to the administration. Nevertheless, the digital communication between the public and the courts remained long limited, due to a few residual regulatory obstacles. The Danish court proceedings are governed by the Danish Administration of Justice Act, which is a complex set of provisions that were originally adopted in 1916 (Jensen, 2000). A major revision took place in 2008, for the purposes of increasing the efficiency of the court activities and speeding up the handling of cases. In 2014, the court administration adopted a digitization strategy for the Danish courts, aiming to ensure that all civil proceedings can be conducted digitally (Juul-Sandberg, 2016). Paper vouchers and postage in civil proceedings disappeared in 2016. Since 2018, the judiciary began to adapt to the laws and policies requiring up-to-date and technology-based solutions more decisively. Also, the expectations of the public were considered. Internal procedures are supported by ICTs (such as the Civil System for emails and communications) and external (secure) electronic communications and document exchange between users and courts. Their handling has also been optimized. Today, a self-service digital portal can be used to file a lawsuit, pay court fees, obtain information, and guidance through a new text library and electronic communication (Hjort, 2021, p. 78). The portal Minretssag is where all the related documents (defense, pleadings, appeals, appendices, etc.), correspondence, and materials related to the cases are recorded.[9] Danish courts are compiling a free access database of judgments that are selected based on the principles of essentiality and public interest, including historic rulings from civil and criminal cases. The new case management system for the courts is similar to the online file management information system "E-File" that operates in Estonia[10] (Waage & Motzfeldt, 2022, p. 14). All data is now centralized and available electronically to enable the automatic processing of simple cases, thereby adding efficiency to the administration of the courts. Even if not proven to effectively contribute to ensuring access to justice, this "electronic case management" instance is a valuable upgrade that saves the state and the parties to a dispute, time, and money (Lupo & Bailey, 2014; Kramer, 2022; Loebl, 2019).

For context, it is worth mentioning that the Danish government made compulsory for all ministries to assess whether new legislation is ready for digitization (Plesner & Justesen, 2022). Thus, complex text in legislation (with exceptions), vague terms,

[9] www.minretssag.dk

[10] https://www.rik.ee/en/e-file

and unjustified procedural requirements are discouraged, not to hinder the optimization of the digital public administration.[11] Technology adoption by the Danish government has boosted the plain language approach to facilitate its codification and integrate case processing in state courts and technology: "*Unnecessary and complex legislation should be simplified and new legislation should be easily understandable and digitally processable. Future legislation must be drafted in a way that is both easily manageable and enables digital administration*" (Voss & Rego, 2019, p. 28). Deliberately or not, these are crucial factors that facilitate access to justice (Gras, 2021). According to the Danish Digitization Agency, *"digitization-ready legislation will underpin an easier everyday life for public employees and an efficient public sector that is capable of serving both citizens and corporations"* (Voss & Rego, 2019, p. 28). Denmark is ahead of other Nordic countries in the digitalization of the courts and has set the standard on digital ready legislation, which resonates abroad, but according to Waage and Motzfeldt (2022), it still has a lot to learn from the Baltic countries. This is especially true of further progress toward interoperability (Hanseth, 2014) in terms of the EU digitalization path, for strengthening judicial cooperation.

3.2 Sweden

The Director-General of the Swedish National Courts Administration (SNCA) declared that *"during the pandemic, the Swedish courts have continued their vital operations with great respect and care, both towards its employees and the affected parties"* (Rolén, 2021).[12] He explained that the SNCA itself has been adjusting its processes by using digital means and providing training, stressing the need to uphold the rule of law principles and standards on procedural justice. Presumably, one important aspect of the courts' ability to ensure continuity of operations was the use of video technology, which became widespread during this period, as it was also reported; by the end of 2020, the number of video calls included in the hearing process reached almost 4500 calls per week, which was more than double of that recorded in 2019. The SNCA has been able to adapt quickly and efficiently to the available technology platforms, and this is viewed as a big accomplishment in the digitization process of the Swedish courts, considering how traditional they are, for example, the power of attorney that had to be appended to the document initiating the case could not be submitted electronically before the electronic claim filing service was launched in 2021, and by then, there was no case handling portal, understood as a software-based solution to automate at least some processes from end to end in Sweden. Nevertheless, digitization steps were taken much earlier: since at least the early 1990s when e-mail communications with and from the courts

[11] International Recognition of Denmark's Work on Digital-Ready Legislation (digst.dk)

[12] The Court Administrator_Volume 9_Winter 2021.pdf (iaca.ws) p.16.

became a valid way of interaction with the judiciary. A program has been in place for almost a decade in local district courts, where oral evidence such as witness hearings are registered on video. In case of appeals, the upper courts use the recording sessions, and witnesses do not need to be called to the courtrooms again, as the 2019 Data Protection and Digitalization meeting report of the International Labour Organization states.[13]

In the field of criminal justice, in order to meet the challenges facing the judiciary, as well as to improve safety, and ultimately, reduce crime, the Swedish Ministry of Justice has been implementing a project that is especially important to handle cases more effectively. To this end, it is focusing on ICTs-based judicial collaboration.[14] Moreover, this project "*entails increased service to citizens and better data for knowledge, analysis and follow-up in the entire judicial chain*" (The Ministry of Justice, 2015, p. 29). Unlike before, cases can be tracked electronically, opening new possibilities for the introduction of knowledge or data-science-based enforcement. The digitization of information exchange in this context logically allows for a more systematic management and efficient use of resources in the judicial system that logically prepares at a manageable scale for the necessary cooperation at the EU-wide level.

The initiatives at the Ministry level do not necessarily seek to disrupt the administration of justice, or the way the judiciary runs, but to take advantage, gradually, of the technologies at its disposal, according to the purposes dictated by the law and policies. The pandemic imposed certain requirements, and the commitments of the country at the supranational level, others. In the international field, the ministry manages all matters connected to cross-border cooperation that are relevant to Sweden. The Division for Criminal Cases and International Judicial Cooperation is the designated central authority in charge.[15] There are no remarkable advancements in this area, but rather an obvious use of the equipment that is not only available but ubiquitous. Remote conferences by phone or using mediating devices are common, provided that the location of the parties can be ascertained. In cases where proceedings could require common resources, transfers and direct communication are possible, and effectively take place between the Nordic countries. Sweden, according to the most recent report about the digitalization of the courts, may be the slowest of the five moving in institutionalizing changes, but also the one that does it more responsibly, if to think of this objectively, as under strict observance of the laws and policies.

[13] https://www.ilo.org/wcmsp5/groups/public/%2D%2D-ed_dialogue/%2D%2D-dialogue/documents/meetingdocument/wcms_720830.pdf

[14] https://www.government.se/49ec0b/contentassets/9ebb0750780245aeb6d5c13c1ff5cf64/the-swedish-judicial-system.pdf

[15] International Judicial Co-operation - Government.se

3.3 Iceland

Hreinsson (2004) spoke almost twenty years ago about the country being concerned about technical interoperability issues in general, and harmonizing software, hardware, and procedures within the whole administration, inter alia, to facilitate electronic communication. An internet legal database was launched for the general public to access all laws, regulations or other administrative acts, international treaties, court decisions, and final rulings of the authorities that have general application or set a precedent, on the same portal. Along with other public services, all the information a citizen may want and the interactions they may require could be obtained through the Island.is. This is the point of entry to the state ecosystem, built on an "X-road" making every service completely interoperable with the rest, and providing a functional infrastructure that is maintained and backed up by the state and "populated" by Icelandic government agencies.

One of the first and most known digitization initiatives concretely related to the judicial administration reported by the court administration was the implementation of a new and centralized case management system in the courts in COVID-19 times.[16] It allows the electronic filing of documents to the court and transfers of case-related information across instances. At the same time, the judiciary carried out a comprehensive upgrade of the courts' computer systems, which was a prerequisite for extending the use of electronic ruling of cases by the court.

To address the implications of the COVID-19 pandemic, the law was amended to ensure that the courts could continue to perform their statutory functions seamlessly while providing legal certainty (Kristjánsson, 2020). Furthermore, the Director of the Icelandic Judicial Administration has explained that explicit permission was granted for the use of teleconferencing equipment during court proceedings, including hearings of statements of parties, suspects, and witnesses. In addition, it restated the validity of electronic communications between the parties and the courts. The Icelandic Judicial Administration has observed that from the beginning of the pandemic, the community has demonstrated a growing acceptance of digital and electronic solutions, which comes as a benefit for the courts, especially, given the advantages their day-to-day operations may experience in terms of time and cost reductions: *"electronic procedure in the process of legal proceedings requires a clear vision of the extent to which electronic solutions can replace more traditional procedures in the process of justice. A review must be conducted of legislation on the administration of justice and a consensus will need to be achieved on the final objective of this trend."*[17] Iceland has improved its competitiveness, public services, infrastructure, and security by innovating radically, in the same way Estonia did (Kalvet, 2012). Because of its government's focus on digital public services, and

[16] https://www.iaca.ws/assets/The_Court_Administrator/The%20Court%20Administrator_Volume%209_Winter%202021.pdf

[17] https://www.iaca.ws/assets/The_Court_Administrator/The%20Court%20Administrator_Volume%209_Winter%202021.pdf

making policy changes to become a global leader in digital technology, Iceland leads the digital ranking in Europe and is climbing fast, like Malta, Estonia, and Luxembourg, according to eGovernment Benchmark 2022 (European Commission, 2022b). The advanced interoperability and the cooperation practice with other Nordic countries suggest that the pathway to participating in the creation of a European judicial ecosystem is already paved for Iceland.

3.4 Norway

The use of electronic technology in the courts of law of Norway is institutionalized and regulated by statutes such as the Digital Court Act and under the procedural rules of the Dispute Act (Lov 17. juni 2005 nr. 90 *om mekling og rettergang i sivile tvister* [Act of 17 June 2005 no. 90, relating to mediation and procedure in civil disputes])[18] (Backer, 2007). This and other related statutory material, such as the Act on Disputes over the Use of Electronic Communications during Hearings, is available online in the State portal Lovdata.no, but only some sections are published in English.

The upgrade of the judiciary has been evolving for many years, even though it is considered that Norway has been somewhat delayed compared to other Nordic countries as a consequence of a lack of general political awareness and, thus, the necessary budget allocation for projects to be initiated. The slow pace of e-justice reform in the country was attributed some years ago to an overall satisfaction with the non-digital civil justice system as it was (Fredriksen & Strandberg, 2017, p. 80), At the time, the global World Justice Project's *Rule of Law Index* ranked Norway second in its general assessment, and in regards to civil justice specifically.[19] The situation has drastically changed in the country and the region where it is widely acknowledged that a faster adoption of digital technologies is necessary to improve the efficiency of the judicial system and prepare for interoperability and Europe-wide cooperation (Organisation for Economic Co-operation and Development, 2021). The digitization process of the judiciary has been running for some almost 20 years and carried out in several stages. The Norwegian courts of general jurisdiction will soon, in fact, complete their digitization process, meaning that the use of electronic equipment is principally implemented. From 2003, § 197a of the Court Act, for example, included detailed provisions regulating the use of electronic communications in cases pending before the courts of general jurisdiction. The Royal Decree No. 1258 of October 2016 (FOR-2016-10-28-1258) on electronic communication with the general court stipulated that a special portal had to be launched to host the communication with the public (for sharing information about cases, dispute papers, judgments in compiled form, and the submission of fees in both civil and criminal cases). Because all the pre-hearing formalities are now

[18] http://app.uio.no/ub/ujur/oversatte-lover/english.shtml

[19] https://worldjusticeproject.org/sites/default/files/documents/roli_2015_0.pdf

concluded through the portal, the use of the Actor (Aktørportalen) has become compulsory for the legal counsels.[20]

The Norwegian Judicial Administration set up the Actor portal, where the documents obtained by a court of general jurisdiction become accessible to other participants in the proceedings. In hearings, however, only some electronic resources are used, at the discretion of the judge. Norway is one of the few European states where hearings are often not recorded. According to § 13-7 of the Disputes Act, the testimony of parties and witnesses during the main hearing must be noted down electronically. The rules for recording evidence were published in a royal decree on 28 September 2018, but in § 13-7(2)(b), it states that the electronic recording option can be waived if the necessary recording facilities are not available in the premises. In practice, since only some courts have such equipment, depositions are not recorded digitally. Fredriksen and Strandberg (2017, p. 88) add that in Norway the prevailing view is that the "best" way to proceed in courtrooms is in the presence of the parties and witnesses.

The law allows litigants, witnesses, and experts to be cross-examined remotely, when a direct interview is not possible or it would be especially burdensome or expensive. Extramural interrogations and other proceedings should not take place where the testimony may be of particular significance or when it is deemed inappropriate for other reasons. Cross examinations and questioning in absentia can always be carried out if the cost or inconvenience of administering the proceeding before the adjudicating court is substantial in relation to the importance of the dispute for the parties, or for the testimony of experts who have filed a written submission with the court.

3.5 Finland

Recent improvement areas and country-wide projects in Finland have focused on increasing the use of technology and boosting the capacities of the judicial services. Like in most European countries, the priority has been to find solutions for digitization, standardization, and harmonization of the court procedures and methods (both within and between courts), as well as the reduction and equalization of case processing times that are the result of the different schemes of requirements. One of the supporting tools for the judges was the creation of a digital "judge's portal" compiling all the information needed to facilitate the work of the courts (Contini et al., 2017), but the use of digital technologies in the work of the courts dates back to the 1980s when Finlex, the Finnish Internet service on legal information, was introduced.[21] By the end of the decade all courts were equipped with personal computers with access to the mainframe systems implemented by the state (Kujanen,

[20] Aktørportalen | Norges domstoler|Aktørportalen | Norges domstoler

[21] https://www.finlex.fi/en/

2004). The law on the use of telecommunications in judicial matters was enacted in 1993. After that, documents could be delivered to the court by telefax, by e-mail, or by direct computer transfer of the data (Hamulák, 2018). The Act on Electronic Services and Communications in the Public Sector was adopted in 2003. It sought to improve the smooth and speedy provision of services and communications, information security in the judicial bodies, and the executive branch, by promoting further the use of electronic data transmission (European Labour Court, 2019). Public officials underwent regular training in preparation for the transition to the electronic data system that is in place now. The Material Bank System (AIPA) of the National Prosecution Authority and general courts promised to create an electronic system for the administration of justice, hosting a database related to court cases handled by prosecutors, district courts, courts of appeal, and the Supreme Court. AIPA was intended to enable the transition to a procedure whereby case initiation, case management, case decision making, and archiving should be digital (Contini et al., 2017, pp. 48, 49). By now, Oikeus.fi does so, making the digital ways of working for courts more agile and offering citizens improved services when interacting with the judiciary.[22]

From the approaches that Finland has been developing in the past few years to prepare for a proper digital transformation, two advanced digitization systems related to courts are worth mentioning: the case management system TUOMAS and the electronic transfer system SANTRA (Velicogna, 2011). Courts receive applications electronically through the latter or via e-mail and fax. Notably, digital filing is compulsory in civil cases. Additionally, claimants using SANTRA transmit data on all their applications to the court's general "mailbox." SANTRA then forwards the statements to the courts' individual mailboxes. The courts, in turn, update their TUOMAS systems based on what they received. Usually, summons are sent through the national postal services. The Finland Post operates an electronic version (EPS) that can be used by courts and tribunals. TUOMAS creates the documents or files needed to receive subpoenas, and thus, the sending of files to Finland Post is automated. TUOMAS keeps track of the deadlines given to defendants for recourses. If the deadline has expired, the system issues a judgment based on the data in the application and the summons. It stores and keeps track of all case documents, forming records and enriching the database for future references. As a result, judges can use the texts of the application and summons when writing the decision if they are stored in TUOMAS. In debt collection cases, for example, the plaintiff using SANTRA will also receive the judgment back into their data system. The courts can use the data to apply for enforcement. In addition, electronic evidence is generally a prerequisite for successful investigations and prosecutions (Center for the Study of Democracy, 2011; Riekkinen, 2016).[23] In 2018, the possibility to use video links for hearings was expanded to criminal cases. In consequence, defendants

[22] home page - Oikeus.fi

[23] https://www.dadinternational.org/images/PDF/4_Finland.pdf

are now able to follow the entire hearing remotely, and this kind of participation is considered equivalent to physical presence (Hjort, 2021, p. 76, 77).

The country's e-governance model is similar to the Estonian model where every citizen can create their account to interact directly with the public authorities. The user/citizen can view the status of his/her case, receive the related decisions and notifications electronically, submit, upload, and store documents that need to become accessible to the authorities, and manage contact information. This is a secure environment where people need to present personal identification numbers and a unique electronic ID with appropriate credentials and certificates (Thomas: Hoffmann & Solarte Vasquez, 2022). The institutionalization of such forms of interaction based on technologies and with user control contributes to the current collective efforts toward cross-border cooperation and advanced interoperability in Europe, very much in line with the ongoing focus of the Digital Decade Policy (Hoffmann & Solarte Vasquez, 2022).

Currently, the development of digitalization of courts in Finland is covered by the Strategy of the National Courts Administration, 2021–2025. One of the main objectives is to develop the competence and functioning of the judicial system and improving its quality through qualification development and the upgrade of digital services. The strategy also focuses on designing processes for digital environments that could serve the courts better. Finland has developed steadily, largely due to its exceptionally well-developed technical infrastructure and relatively inexpensive telecommunication services. It is perhaps the Nordic country that has taken the boldest steps toward preparing for further integration and cross-border cooperation so far.

In the rest of the cases the states have prepared by way of formalizing (institutionalizing regulating) and issuing laws that capture the current sociotechnical realities but leaving lots to be desired in terms of preparedness for technology leaps beyond their immediate borders. Some of the legal acts are broad or simple; they modify access and format requirements, enable electronic filing of some documents, including electronic signatures, and create environments for the exchange of electronic documents. However, they show clear indications of progress and demonstrate a political readiness to meet the current requirements and achieve cooperation, as it is visible in the Scandinavian context, where social and regulatory "interoperability" has been long in place.

4 Concluding Remarks and Directions for Further Research

The legal evolution and regulation of the digitization of the courts in the Nordic countries can be described as a gradual and court-centered process. It may be asked whether the existing technological capacities that these countries possess have been largely overlooked, if to contrast them against the EU vision and the mainstream policies of the neighboring and more advanced countries. This study suggests that the Scandinavian legal culture does not develop in a haste. Based on their historical background, and the way they advance cooperation capacities while staying up to date at the technical level, it can be expected that joining efforts in the direction of the EU vision on cross-border cooperation, particularly in judicial matters, will cause no difficulties.

The reform processes that permitted the advancement of e-justice schemes and also supported their functioning during the COVID-19 pandemic are found to be standard, be they optional or mandatory, digital and electronic communication, case handling portals, record-keeping in digital formats, video recording of proceedings, and legal information databases. The restrictions imposed brought again to light the Nordic countries' pragmatism: The problems that arose such as the courts lacking equipment to fully conduct hearings remotely were solved as they came. The judges worked uninterruptedly. The temporary solutions implemented in the early weeks of the pandemic are being replaced by more permanent and functional solutions. In criminal cases, the rule of law safeguards and procedural justice checks may be more complicated than in civil cases, because the proceedings cannot be discretionarily altered or improvised (Hirschl, 2011). In all likelihood, once consolidated, the value of digital work practices will be reassessed. Some serious resistance to change does not refer exclusively to the new ways of working in public sector organizational settings but to deeper transformation. Judges are more interested in delivering quality verdicts and justice, rather than a "fast settlement." Efficiency gains and better court management are not to come at the expense of effective (substantial) access to justice.[24]

It may be said that among the perceived shortcomings affecting digitization and judicial cooperation (in regard to other European states), their uneven development is the most apparent. Even though, automation and more advanced technologies were beyond the scope of the study, no legal or actual ground-breaking developments mark their institutional evolution so far (Bjerke-Busch & Aspelund, 2021). A greater multidisciplinary collaboration is expected in the public management levels to accelerate changes and reap the benefits of digitalization in the sector, but the constraints are many, starting from the guiding values of both the law and the ICTs. Artificial intelligence (AI) and automation are used for general governance purposes, as in most EU countries, in building and keeping records, routine and information sorting tasks (Kerikmae et al., 2020). But in the judiciary, the main stakeholders, decision makers, and agents continue to be the judges, legal experts, the public administrators, and the public. The courtrooms were long untouched by the general trends of digitalization, but access to the jurisdiction in broad sense has been technically ensured.

Presumably, the law will soon start to keep pace with the sociotechnical developments more decisively. The much expected "smart" legislation will provide a

[24] https://www.europarl.europa.eu/news/en/headlines/society/20201126STO92502/new-eu-rules-digitalisation-to-improve-access-to-justice

fuller range of opportunities for societal contentment as long as it meets the communities' legal needs without referring to specific technologies. While it may be seen as one of the foundational goals of the field of legaltech, good tech-neutral legislation is not to discourage the deployment of technology (Saarenpää, 2004; Pöysti, 2010). Until now, a common trend of swift enactment of laws or reforms in relation to issues that arise in conjunction with the growth of cutting-edge technology has been observed in other European countries and regions more acutely.

One more hypothesis on the "slow" digitization advancements of courts before the pandemic is that the Nordic countries do not experience the urgent need to codify or restructure their systems in order to uphold their well-established indicators of democracy, welfare, and social development.

A way of expanding the scope of our study is to follow up on these indicators, in regard to common decisions within the Nordic countries and inside Europe, to determine if their identity as a distinct legal subfamily is maintained or diminished. This in itself would be a consideration to put before the legislative body of each country, to guide a purposeful regulatory design. As Hjort (2021) has explained, the Nordic legal family is restricted in size, and if one follows the categories determined in comparative law theory, access to sources would be limited. If acceptable, every new law of significance would be subject to a cross-cultural consultation, even if at first glance this effort does not involve inter-Nordic contact, either in terms of people or goods. It should be borne in mind that when it comes to creating legislative responses to new issues arising from the growth of modern communities, a group of small countries with a common cultural background have shown to benefit if they cooperate. The involvement of all member countries early in the discussion also has a positive effect on the outcomes, as it was historically noted.

To end, Denmark, Sweden, Norway, and Finland have been at the top of the world e-government ranking (Joseph & Avdic, 2016). As e-justice is one of its elements, which rests on the digital capabilities of the courts, it is sensible to deduce that these countries, including Iceland, will meet the standards for the effectuation of regional policies on cross-border cooperation in digital matters after having accomplished their own domestic aims. Moreover, these countries have been and still are models for emerging economies to emulate. Looking at the experience of common evolution and cooperation, living standards, and prosperity in the Nordic countries, they seem capable to cope with the extended European challenges ahead.

References

Backer, I. L. (2007). The Norwegian reform of civil procedure. *Scand Stud Law, 51*, 41–75.

Bernitz, U. (2000). Nordic legislative cooperation in the New Europe-a challenge for the Nordic countries in the EU perspective. *Scandinavian Studies in Law, 39*, 29–42. https://scandinavianlaw.se/pdf/39-2.pdf

Bernitz, U. (2007). What is Scandinavian Law? concept, characteristics, future. *Scandinavian Studies in Law, 50*, 13–30. https://www.scandinavianlaw.se/pdf/50-1.pdf

Bjerke-Busch, L. S., & Aspelund, A. (2021). Identifying barriers for digital transformation in the public sector. In *Digitalization* (pp. 277–290). Springer.

Center for the Study of Democracy. (2011). Electronic tools for criminal justice in Finland – best practice case. In *E-Tools for criminal case management within selected EU member states* (pp. 93–104). https://www.dadinternational.org/images/PDF/4_Finland.pdf

Contini, F., Ontanu, A., & Carnevali, D. (2017). *Handle with care: Assessing and designing methods for evaluation and development of the quality of justice.* http://hdl.handle.net/1765/123477.

Council of Europe. (2018). *European judicial systems. Efficiency and quality of justice. CEPEJ Studies No.* 26.

European Commission. (2021). *2030 Digital compass: The European way for the digital decade.* The European Commission. Available from https://eur-lex.europa.eu/legal-content/EN/TXT/HTML/?uri=CELEX:52021DC0118&from=EN

European Commission. (2022a). *Digital Economy and Society Index (DESI) 2022. Thematic chapters.* The European Commission. Available from https://ec.europa.eu/newsroom/dae/redirection/document/88764

European Commission. (2022b). *eGovernment Benchmark 2022.* The European Commission. Available from https://digital-strategy.ec.europa.eu/en/library/egovernment-benchmark-2022

European Labour Court Judges Meeting. (2019). *Data protection and digitilisation.* XXVII Meeting of European Labour Court Judges. https://www.ilo.org/wcmsp5/groups/public/%2D%2Ded_dialogue/%2D%2D-dialogue/documents/meetingdocument/wcms_720830.pdf

Fredriksen, H. H., & Strandberg, M. (2017). Is E-justice reform of Norwegian civil procedure finally happening? *Oslo Law Review, 3*(2), 72–88. https://doi.org/10.5617/oslaw4079

Gras, I. O. (2021). Online courts: Bridging the gap between access and justice. *UCL Journal of Law and Jurisprudence, 10*, 24. https://doi.org/10.14324/111.444.2052-1871.1214

Hamulák, O. (2018). La carta de los derechos fundamentales de la union europea y los derechos sociales. *Estudios constitucionales, 16*(1), 167–186.

Hanseth, O. (2014). Developing pan-European e-Government solutions: From interoperability to installed base cultivation. In *The circulation of agency in E-Justice* (pp. 33–52). Springer.

Hasselbalch, O. (2017). The Nordic legal family and unification of laws. *Journal of Foreign Legislation and Comparative Law, 2017*(1), 43–46. https://doi.org/10.12737/24287

Hirschl, R. (2011). The Nordic counter-narrative: Democracy, human development and judicial review. *International Journal of Constitutional Law, 9*(2), 449–469.

Hjort, M. A. (2021). Sources of inspiration of Nordic procedural law: Choices and objectives of the legal reforms. In L. Ervo, P. Letto-Vanamo, & A. Nylund (Eds.), *Rethinking Nordic courts* (pp. 69–88). Comparative Perspectives on Law and Justice. https://doi.org/10.1007/978-3-030-74851-7_5

Hoffmann, T., & Solarte Vasquez, M. C. (2022). The Estonian e-residency programme and its role beyond the country's digital public sector ecosystem. *Revista CES Derecho, 13*(2), 184–204.

Hreinsson, P. (2004). Electronic administration in Iceland. *Scandinavian Studies in Law, 47*, 225–243. https://www.scandinavianlaw.se/pdf/47-11.pdf

Husa, J. (2015). Nordic law and development – See no evil, hear no evil? *Scandinavian Studies In Law, 60*, 15–33. https://scandinavianlaw.se/pdf/60-1.pdf

Husa, J., Nuotio, K., & Pihlajamäki, H. (Eds.). (2007). *Nordic law – Between tradition and dynamism.* Antwerp-Oxford.

Jensen, M. F. (2000). *The Danish administration of justice act.* Det Retsvidenskabelige Institut D, Københavns Universitet.

Joseph, S., & Avdic, A. (2016). Where do the Nordic nations' strategies take e-Government? *The Electronic Journal of e-Government, 14*(1), 3–17. https://avdic.se/onewebmedia/ejeg-volume14-issue1-article441.pdf

Juul-Sandberg, J. (2016). Reform and development of preparatory proceedings in the Danish civil justice system: Towards (even) more efficient courts. In *Current Trends in Preparatory Proceedings* (pp. 81–107). Springer. https://doi.org/10.1007/978-3-319-29325-7_4

Kalvet, T. (2012). Innovation: A factor explaining e-government success in Estonia. *Electronic Government, an International Journal, 9*(2), 142–157. https://doi.org/10.1504/EG.2012.046266

Karhu, J. (2010). *Introduction to Finnish law. Legal history: The past and the present of Nordic law.* University of Lapland. https://www.ulapland.fi/loader.aspx?id=cb85342a-8d27-4128-ba0a-b46dea5e9037

Keller, S. (2009). Welfarism. *Philosophy Compass, 4*(1), 82–95. https://doi.org/10.1111/j.1747-9991.2008.00196.x

Kerikmae, T., Troitiño, D. R., & Vasquez, M. C. S. (2020). La inteligencia artificial, futuro, reto y necesidad: presentación. In *Inteligencia artificial: De la discrepancia regional a las reglas universales: integración de percepciones políticas, económicas y legales* (pp. 25–34). Aranzadi Thomson Reuters.

Kischel, U. (2019). *Comparative law.* Oxford University Press.

Kramer, X. E. (2022). Digitising access to justice: The next steps in the digitalisation of judicial cooperation in Europe. *Revista General de Derecho Europeo, 56*, 1–9.

Kristjánsson, H. D. (2020). Iceland's rule of common sense ... and law? *Verfassungsblog: On Matters Constitutional.* https://doi.org/10.17176/20200420-182624-0.

Kujanen, K. (2004). *E-services in the courts in Finland.* http://www.weblaw.ch/jusletter/Artikel.asp?ArticleNr=3488

Letto-Vanamo, P., & Tamm, D. (2019). Nordic legal mind. In P. Letto-Vanamo, D. Tamm, & B. Gram Mortensen (Eds.), *Nordic law in European context* (pp. 1–19). Comparative Perspectives on Law and Justice. https://doi.org/10.1007/978-3-030-03006-3_1

Loebl, Z. (2019). *Designing online courts: The Future of Justice is open to all.* Kluwer Law International BV.

Lukianov, D. (2015). Legal families approach: Consistent patterns and trends. *Evropský politický a právní diskurz, 1*(1), 168–175. https://eppd13.cz/wp-content/uploads/2015/2015-2-1/25.pdf

Lupo, G., & Bailey, J. (2014). Designing and implementing e-Justice Systems: Some lessons learned from EU and Canadian Examples. *Laws, 3*(2), 353–387. https://doi.org/10.3390/laws3020353

Mäenpää, O., & Fenger, N. (2019). Public administration and good governance. In L. Ervo, P. Letto-Vanamo, & A. Nylund (Eds.), *Rethinking Nordic Courts* (pp. 163–178). Comparative Perspectives on Law and Justice. https://doi.org/10.1007/978-3-030-03006-3_10

National Courts Administration. (2021). *Strategy of the national courts administration 2021–2025* [Brochure]. https://tuomioistuinvirasto.fi/material/collections/20211214152444/7VY6rRDSz/TIV_Strategia_vihko_EN_verkkoon.pdf

Nousiainen, K. (2018). 2 Transformative Nordic Welfarism: Liberal and communitarian trends in. *Responsible Selves, 2.*

Nylund, A. (2021a). The past, present and future of Nordic Courts. In L. Ervo, P. Letto-Vanamo, & A. Nylund (Eds.), *Rethinking Nordic Courts* (pp. 291–308). Comparative Perspectives on Law and Justice. https://doi.org/10.1007/978-3-030-74851-7_16

Nylund, A. (2021b). Rethinking Nordic Courts: An introduction. In L. Ervo, P. Letto-Vanamo, & A. Nylund (Eds.), *Rethinking Nordic Courts* (pp. 1–17). Comparative Perspectives on Law and Justice. https://doi.org/10.1007/978-3-030-74851-7_1

Organisation for Economic Co-operation and Development. (2021). *Norway's strategic process to capitalise on the potential of new technology.* Organisation for Economic Co-operation and Development. https://www.oecd.org/development-cooperation-learning/practices/dynamic/dcd-best-practices/a5c76fbf/pdf/norway-s-strategic-process-to-capitalise-on-the-potential-of-new-technology.pdf

Petrén, S. (1968). Nordic and international lawmaking. *Scandinavian Studies In Law, 12*, 67–89. https://scandinavianlaw.se/pdf/12-3.pdf

Plesner, U., & Justesen, L. (2022). The double darkness of digitalization: Shaping digital-ready legislation to reshape the conditions for public-sector digitalization. *Science, Technology, & Human Values, 47*(1), 146–173. https://doi.org/10.1177/0162243921999715

Pöysti, T. (2010). Information Government in Practise: Functional gains and legal perils. *Scandinavian Studies in Law, 56*, 91–124. https://scandinavianlaw.se/pdf/56-5.pdf

Riekkinen, J. (2016). Evidence of cybercrime and coercive measures. *Digital Evidence and Electronic Signature Law Review, 13*, 50–66. https://doi.org/10.14296/deeslr.v13i0.2296

Rolén, T. (2021). Our system of law must remain steadfast, even in troubled times. *The Court Administrator, 9*, 15–18. https://www.iaca.ws/assets/The_Court_Administrator/The%20Court%20Administrator_Volume%209_Winter%202021.pdf

Saarenpää, A. (2004). E-government and Good Government: An impossible equation in the new network society? *Scandinavian Studies In Law, 47*, 245–273. https://scandinavianlaw.se/pdf/47-12.pdf

Scott, J. (2014). Extraterritoriality and territorial extension in EU law. *The American Journal of Comparative Law, 62*(1), 87–126. https://doi.org/10.5131/AJCL.2013.0009

Sheppard, B. (2015). Incomplete innovation and the premature disruption of legal services. *Michigan State Law Review*, 1797. https://doi.org/10.17613/y4by-hd89

Skogh, G., & Stuart, C. (1982). A contractarian theory of property rights and crime. *The Scandinavian Journal of Economics*, 27–40. https://doi.org/10.2307/3439777.

Sunde, J. Ø. (2021). The history of Nordic legal culture and court culture: The story of what should not have been, but still came to be. In L. Ervo, P. Letto-Vanamo, & A. Nylund (Eds.), *Rethinking Nordic Courts* (pp. 49–67). Comparative Perspectives on Law and Justice. https://doi.org/10.1007/978-3-030-74851-7_4

The Ministry of Justice. (2015). *The Swedish judicial system.* [Brochure]. https://www.government.se/49ec0b/contentassets/9ebb0750780245aeb6d5c13c1ff5cf64/the-swedish-judicial-system.pdf

Velicogna, M. (2011). Electronic access to justice: From theory to practice and back. *Droit et cultures. Revue internationale interdisciplinaire*, (61). https://doi.org/10.4000/droitcultures.2447.

Velicogna, M., Steigenga, E., Taal, S., & Schmidt, A. (2020). Connecting EU jurisdictions: Exploring how to open justice across member states through ICT. *Social Science Computer Review, 38*(3), 274–294. https://doi.org/10.1177/0894439318786949

Von Eyben, W. E. (1962). Inter-Nordic legislative co-operation. *Scandinavian Studies In Law, 6*, 63–93. https://scandinavianlaw.se/pdf/6-3.pdf

Voss, E., & Rego, R. (2019). *Digitalization and public services: A labour perspective*. Public Services International/Friedrich-Ebert-Stifung Bonn. http://hdl.handle.net/10451/40088

Waage, F., & Motzfeldt, H. M. (2022). *Digitalization at the courts*. Nordic Council of Ministers.

Wallström, Å., Engström, A., Salehi-Sangari, E., & Styvén, M. E. (2009). Public E-Services from the Citizens' perspective – Adopting a market orientation. *International Journal of Public Information Systems, 2009*(2), 123–134.

Zinchenko, O. (2017). Constitutions of Scandinavian countries: Comparative analysis. *Visnyk Natsionalnoi akademii pravovykh nauk Ukrainy, 88*(1), 58–68. http://visnyk.kh.ua/web/uploads/pdf/ilovepdf_com-58-68.pdf

Alina Hruba is a Ph.D. student at the Yaroslav Mudryi National Law University (Ukraine) and a visiting doctoral researcher at the University of Helsinki (Finland). Her current research focuses on the development of the rule of law in Nordic legal systems. She is currently working on her thesis "Implementation of the Rule of Law in the Nordic countries via Access to Justice."

Maria Claudia Solarte Vasquez , LLM, Ph.D., is Senior Lecturer of the Department of Law of Tallinn University of Technology and a researcher in the field of proactive contracting, dispute resolution systems, digital ecosystems management, and transaction usability and design for dispute prevention. She has worked on transitional regulatory pathways to automation, especially on the operationalization of legal values and methodologies in the adoption of technology-based products, services, interaction, processes, and systems.

Digital Sovereignty in the EU: Searching for Legal Mechanisms for Marking the Borders

Lusine Vardanyan, Hovsep Kocharyan, Ondrej Hamuľák, and Tanel Kerikmäe

Abstract In accordance with the principles of international law, each state has the exclusive authority to enact laws on its territory. However, digital transformation, in particular the dissemination of data, threatens the traditional foundations of sovereignty and jurisdiction. At the same time, in recent years, the EU has increasingly turned to the need to introduce the idea of digital sovereignty into its agenda. In this chapter, researchers analyse EU legislation, in particular GDPR (Regulation (EU) 2016/679 of the European Parliament and of the Council of 27 April 2016 on the protection of natural persons with regard to the processing of personal data and on the free movement of such data, and repealing Directive 95/46/EC (General Data Protection Regulation) (Text with EEA relevance). Available at: https://eur-lex.europa.eu/legal-content/EN/TXT/?uri=CELEX%3A32016R0679) and DSA, as well as EU law enforcement practice, in order to identify the legal mechanisms by which the EU is trying to realize its idea of digital sovereignty. Special attention is paid to the mechanisms of localization and extraterritoriality. The researchers conclude that EU data protection laws distinguish between the establishment of an extraterritorial effect and the localization of data.

The paper was prepared within the implementation of the project reg. no. CZ.02.2.69/0.0/0.0/19_073/0016713 "The Role of the EU in the Quest for Digital Sovereignty."

L. Vardanyan · H. Kocharyan
Faculty of Law, Palacký University Olomouc, Olomouc, Czech Republic
e-mail: lusine.vardanyan01@upol.cz; hovsep.kocharyan01@upol.cz

O. Hamuľák (✉)
Faculty of Law, Palacký University Olomouc, Olomouc, Czech Republic

Department of Law, School of Business and Governance, Tallinn University of Technology, Tallinn, Estonia
e-mail: ondrej.hamulak@upol.cz

T. Kerikmäe
Department of law, Tallinn University of Technology, Tallinn, Estonia
e-mail: tanel.kerikmae@taltech.ee

D. Ramiro Troitiño et al. (eds.), *Digital Development of the European Union*,
https://doi.org/10.1007/978-3-031-27312-4_14

1 Introduction

In recent years, the issue of digital sovereignty has been one of the most discussed and problematic topics on the EU's geopolitical agenda. In February 2020, the European Data Strategy was adopted, aimed at strengthening of the EU's leadership in the digital society. Despite the fact that the specified strategy mainly pursues economic goals, the issue of ensuring digital sovereignty also occupies here an important place. The emergence of the term "digital sovereignty" became a response to the new challenges facing the EU: a wide range of digital practices, the implementation of which inevitably leads to the transformation of many spheres of life, as well as raise concerns about the increasing economic and social influence of technology companies outside the EU (Gábriš & Hamuľák, 2021a, 2021b). By implementing the idea of digital sovereignty, the EU wants to assume the role of a regulatory actor in the field of digital technologies, for which legislative initiatives have already been submitted—the Digital Service Act[1] (hereinafter—*"DSA"*) and the Digital Markets Act[2] (hereinafter—the *"DMA"*).

Nevertheless, the implementation of digital sovereignty raises a number of both theoretical and practical issues, primarily because digital sovereignty sounds like an oxymoron, which binds together seemingly incompatible concepts: (1) *sovereignty*, which is based on territorial jurisdiction and which is (2) *boundless cyberspace*. In order to fulfil the role of a regulatory actor in the digital world, the EU must determine the boundaries within which it can extend its jurisdiction in cyberspace. This condition becomes decisive in the issue of the effectiveness of the implementation of the idea of digital sovereignty without creating tension in the field of international cooperation and avoiding duplicate jurisdictions. In this framework, the issues of the possibility of defining digital boundaries and the foundations of the legitimization of the power order come to the fore, and the following questions arise: What legal mechanisms do the EU use to mark its territory in order to realize its digital sovereignty? We will try to find answers to this question in our research.

[1] Proposal for a Regulation of the European Parliament and of the Council on a Single Market for Digital Services (Digital Services Act) and amending Directive 2000/31/EC COM/2020/825 final. Available at: https://eur-lex.europa.eu/legal-content/en/TXT/?uri=COM%3A2020%3A825%3AFIN

[2] Proposal for a Regulation of the European Parliament and of the Council on contestable and fair markets in the digital sector (Digital Markets Act) COM/2020/842 final. Available at: https://eur-lex.europa.eu/legal-content/en/TXT/?uri=COM%3A2020%3A842%3AFIN

2 Digital Sovereignty of the EU: An Indefinite Term with Great Regulatory Potential

2.1 Digital Sovereignty as One of the Dimensions of the Post-Westphalian Understanding of Sovereignty

The growth of data flows and widespread digital transformation have recently been seen as the forerunners of the post-territorial world, and flows and networks as a replacement for the traditional territorial order of states (Friedman, 2007). It is already clear that the collection, control, and use of data is a form of political, economic, and social power, which is problematized as a threat to a sovereign state. In this context, it is not surprising that the debate on data regulation has become more frequent, and the issue of cross-border data flows is becoming "increasingly securitised, territorialised and ultimately geopoliticised" (Lambach, 2019). Nowadays, numerous national initiatives have emerged to restore control over strategic data (data localization policies or changes in the connection architecture). This fundamental shift has led to a call for strengthening the role of the state, for strategic autonomy and digital borders, including in the context of the EU, which caused narratives about "digital sovereignty" to appear on the EU agenda in 2020.

However, despite the almost unanimous consent on the need to promote digital sovereignty for the EU and its significance for defining the new geopolitics of data flows, the term still remains vague, collective and unifying various concepts. An attempt to reveal the essence of "digital sovereignty" intuitively begins with an assessment of the possibility of applying the categories underlying sovereignty to cyberspace, because digital sovereignty logically should appear when adapting the concept of sovereignty to the technological world. Although here the whole complexity of such adaptation is revealed. On the one hand, the very issue of sovereignty is known to be one of the most complex in international law. Thus, according to M. Koskenniemi, sovereignty has no fixed content, subjecting it to a variety of meanings and uses (Koskenniemi, 2011), so there is practically no unambiguous definition of sovereignty. On the other hand, there are various conceptual approaches to cyberspace due to its "limitless" nature, primarily related to the problems of territorialization of such space.

Nevertheless, it is believed that modern international law is based on the Westphalian understanding of sovereignty, which means such a system of power organization in which the state has full supremacy within its territory, political independence in domestic and foreign policies and is legally equal in relations with other states.[3] As one can see, three categories are crucial for such sovereignty:

[3] In general, it can be argued that for the first time the idea of a community of sovereign states as the basis of the system of international relations appeared in the middle of the eighteenth century, i.e. a century after the conclusion of the Peace of Westphalia. This idea did not immediately become dominant and its final approval dates back to the middle of the twentieth century.

(1) the supremacy of power over a certain territory, (2) the territory itself and (3) state-centrism.

However in the digital space all those three categories are under strain, so the search for new definitions for sovereignty mainly revolves around the rejection of these categories. For example, J. Agnew put forward the theory of multiple sovereignties, which considers sovereignty as not necessarily related to a state or territory (Agnew, 2005). A similar position is also held by N. Tsagourias and R. Buchan, who point out the absence of an inextricable link between territory, on the one hand, and sovereignty and jurisdiction, on the other, linking the nature of sovereignty with the powers of the state, but not with the territory (Tsagouring & Buchan, 2015). These approaches reflect the essence of the idea of the post-Westphalian world, which, through fragmentation and transformation of sovereignty, changes the nature of international relations and develops power and competition between both states and non-state actors, where the distinction between internal and external sovereignty has lost its relevance. This idea negates the traditional notions of power as coming exclusively from sovereigns exercising independent and exclusive control over the territory.

A similar trend can be seen in the reflections of researchers who are trying to doctrinally define digital sovereignty. In particular, L. Moerel and P. Timmers define digital sovereignty as “the capabilities and capacities to decide and act autonomously on essential aspects of the longer-term future in the economy, society and democracy” (Moerrel & Timmers, 2021). In turn, L. Floridi defines digital sovereignty as “the control of data, software (e.g. AI), standards and protocols (e.g. 5G, domain names), processes (e.g. cloud computing), hardware (e.g. mobile phones), services (e.g. social media, e-commerce), and infrastructures (e.g. cables, satellites, smart cities), in short, for the control of the digital” (Floridi, 2020).

As one can see, the peculiar separation of digital sovereignty itself from the state also calls into question its separation from the principle of territoriality and the category of territory as a central element and a constructive feature of the state. Nevertheless, none of the doctrinal definitions of digital sovereignty can be considered exclusively acceptable. Thus, the specified definitions, as well as most doctrinal definitions on digital sovereignty, cannot be considered to reflect all aspects of the digital sovereignty, including its functional aspect. In particular, in the above definitions, digital sovereignty can be understood both as the goal set by the one who exercises power and as the quality of the one who exercises power.

In addition, it should be noted that digital sovereignty is an extremely contextual phenomenon, due to the fact that nowadays it is often said about the existence of various models of digital sovereignty, which are included by various states in their agenda. Therefore, it seems to us that it is impossible to fully disclose the content of digital sovereignty outside of connection with a certain state (Gábriš & Hamuľák, 2021a, 2021b). In this regard, in order to understand what meaning the EU puts into this concept, it is necessary to analyse how it is used in the EU documents.

2.2 *Digital Sovereignty in the EU Agenda: The EU's Claims for Regulatory Leadership*

In the context of the EU the discussion of sovereignty is particularly difficult because of its uniqueness as a subject of world politics. With the recognition of "digital sovereignty" as a derivative of "sovereignty" in relation to the EU, the discussion turns to the area of the legal nature of the EU, the question of the transfers of powers from the member states to the EU, while the constituent documents of the EU do not contain the term "sovereignty" at all. At the same time, the analysis of official EU documents shows that there is no unambiguous interpretation of the content of the term "digital sovereignty", as German Chancellor Angela Merkel also pointed out (Merkel, 2019). For example, there are different political contexts for its use: if the European Parliament talks about "*establishing*",[4] then the European Commission is about "*defending*",[5] and the European Council is about "*achieving*"[6] digital sovereignty. Besides, U. Von der Leyen uses it interchangeably with the term "technological sovereignty", defining it as "the capability that Europe must have to make its own choices, based on its own values, respecting its own rules" (Von der Leyen, 2020).

Moreover, serious concerns were expressed about the economic and social impact of non-EU technology companies and the ability of European regulators to enforce their laws. In this context, "digital sovereignty" means Europe's ability to act independently in the digital world and should be understood both in terms of defensive mechanisms and offensive tools to stimulate digital innovation. In this context, Ursula von der Leyen identified digital policy as one of the key political priorities of her tenure in 2019–2024 and promised that Europe should achieve "technological sovereignty" in critical areas. The European Council stressed that the EU needs to go further in developing a competitive, secure, inclusive, and ethical digital economy with world-class connectivity, and called for special attention to data security and artificial intelligence (AI) issues.

As one can see, the above definition of the EU's digital sovereignty reflects the ideas of the Lisbon Treaty, according to which the EU intends to play a significant role outside its territory, not only in relation to the external dimension of the single market but also in relation to the normative goals of global justice, such as human rights protection. In particular, according to Article 21(2)(b) of the TEU, one of the

---

[4]European Parliament. (2020). EU institutions establish common priorities for 2021 and until next elections. https://www.europarl.europa.eu/news/en/press-room/20201217IPR94201/eu-institutions-establish-common-priorities-for-2021-and-until-next-elections

[5]-European Commission. (2020). Statement by the President at "Internet, a new human right". https://ec.europa.eu/commission/presscorner/detail/en/statement_20_2001

[6]European Council. (2020). Remarks by President Charles Michel after the Special European Council meeting on. https://www.consilium.europa.eu/en/press/press-releases/2020/10/03/remarks-by-president-charles-michel-after-the-special-european-council-meeting-on-2-october-2020/

objectives of the EU in its external activities is to "consolidate and support democracy, the rule of law, human rights and the principles of international law". At the same time, Article 3(5) of the TEU refers, among other things, to the fact that the Union supports and promotes its values in relations with the wider world. With the joint reading of Articles 3(5) and 21 of the TEU, the EU becomes a "generator" of global rules and standards, which, in turn, creates prerequisites for the EU to realize its ambitions in the digital world as a regulatory actor. In the specified context, a wide range of interpretations of the EU's digital sovereignty can lead to the justification of a broad spectrum of extraterritorial measures, especially when digital sovereignty is associated with an even greater spread of the "Brussels effect". At the same time, through the term "digital sovereignty" the EU proclaims itself as a regulatory superpower, and it is the expansion of the "Brussels effect", coupled with value-oriented legitimization, that becomes the third alternative model of the EU, different from the approaches of the USA, China, and Russia. In this regard, the digital sovereignty of the EU can be defined as the ability of Europe to establish its own values and use its instruments and regulatory powers to promote the formation of global rules and standards in order to achieve strategic autonomy in the digital sphere.

3 The Principle of Territoriality in the Context of Data Flow Regulation

3.1 The Principle of Territoriality

The principle of territoriality is the main basis for the assertion of jurisdiction in international law (Ryngaert, 2015). It attributes jurisdiction to the state in all matters that are within its territorial sovereignty. This approach has also been approved by the international judicial practice in the *Lotus* case.[7] Thus, in the decision on this case, the Permanent Court of International Justice stated that "Now the first and foremost restriction imposed by international law upon a State is that (. . .) it may not exercise its power in any form in the territory of another State. In this sense jurisdiction is certainly territorial".[8]

However, in the context of cyberspace, of which data is an integral part, the situation becomes a little more complicated: data is stored in different places, disassembled, reassembled, and transferred to different places and without the knowledge of the users and various data components are stored in different places and can be accessed from several places at the same time. Due to the nature of the

[7] The case of the *S.S. "LOTUS"* (France v. Turkey). Judgment of 7 September 1927. PCIJ Series A—No 10. Available at: https://www.icj-cij.org/pcij/serie_A/A_10/30_Lotus_Arret.pdf

[8] Ibid. p. 45

data, determining territorial jurisdiction based on the location of the data may not be possible, which in turn challenges traditional spatial concepts of international law.

At the same time, the principle of territoriality does not limit the possibility of extending jurisdiction beyond its borders. In the same case of *Lotus*, the Court pointed out that although there is a restriction imposed by international law on a state about the non-exercise of power in any form on the territory of another state, but at the same time indicates that: "It does not, however, follow that international law prohibits a State from exercising jurisdiction in its own territory, in respect of any case which relates to acts which have taken place abroad, and in which it cannot rely on some permissive rule of international law. (...) Far from laying down a general prohibition to the effect that States may not extend the application of their laws and the jurisdiction of their courts to persons, property and acts outside their territory, it leaves them in this respect a wide measure of discretion (...)".[9] Thus, according to the approach in the Lotus case, both territorial and extraterritorial jurisdictions are possible, if there is no norm prohibiting extraterritorial jurisdiction in international law.

The second basic approach prohibits states from exercising jurisdiction unless there is a positive rule allowing them to do so. Most states and doctrines support this presumption against the exercise of extraterritorial jurisdiction (Storey, 2017; Dodge William, 2017; Clopton Zachary, 2014). Nevertheless, the permissive principle of jurisdiction outside the prohibiting rule to the contrary is a principle on which the EU still relies. In other words, the judicial practice of the CJEU tends to strengthen the first approach, i.e. the *Lotus* principle. In particular, in the cases of *Schrems I*[10] and *Schrems II*,[11] the CJEU emphasizes the supreme authority of national data protection authorities in relation to the European Commission, on the one hand, and in relation to states outside the EU, on the other.

It should be noted that the EU can also safely circumvent the restrictions provided for by public international law to legitimize its extraterritorial prescriptive jurisdiction. Firstly, the EU constitutional documents provide little concrete guidance on how international law should be applied within the framework of the EU legal order. Secondly, the CJEU did not use existing provisions that emphasize the EU's obligation to comply with international law. As a result, there is practically no indication of how the EU should exercise its powers and implement its policies and laws outside its borders. It follows from this that extraterritorial prescriptive jurisdiction can be exercised at the discretion of the EU, while there are no sufficiently clear restrictions for such discretion.

[9] Ibid. p. 46

[10] (Case C-362/14). *Maximillian Schrems v Data Protection Commissioner*. Judgment of the Court (Grand Chamber) of 6 October 2015. Available at: https://curia.europa.eu/juris/liste.jsf?num=C-362/14

[11] (Case C-311/18). *Data Protection Commissioner v Facebook Ireland Limited and Maximillian Schrems* Judgment of the Court (Grand Chamber) of 16 July 2020. Available at: https://curia.europa.eu/juris/liste.jsf?num=C-311/18

3.2 *Binary Territoriality/Extraterritoriality and Non-territoriality of Data*

Despite the fact that the EU has the latitude of discretion in the implementation of jurisdiction, nevertheless, it should be taken into account that the Union must respect international law in the exercise of its powers. The compliance with international law is presented as one of the principles that inspired the creation of the Union with such fundamental values as democracy, the rule of law, human rights, and fundamental freedoms. That is, the respect for international law should be considered as one of the fundamental constitutional principles of the Union. In this framework, the issue of ensuring digital sovereignty requires the EU to adhere to the binary of territoriality/extraterritoriality. However, international law reacts very slowly to changes in the modern world under the influence of digitalization: it only makes attempts to adapt existing legal regulation schemes that depend on classical categories defining sovereignty to cyberspace. Moreover, the steps that are being taken in this direction mostly involve the application of the principle of sovereignty to the material part of information and communication technologies,[12] while the question of the possibility of extending existing legal regulation schemes to data/information flows remains open (Finnemore & Hollis, 2016).

On the other hand, the attempt to adapt new relations to already existing concepts and categories helps to support the defining role of the state, including in cyberspace, positioning itself as the main regulator both unilaterally and within the framework of multilateral agreements. That is, the adaptation of the provisions of international law leads to the "territorialization" of cyberspace—the extension of such a configuration of power to it that acts in relation to territories. The territorial/extraterritorial binary existing in international law limits the models for regulating data flows: the EU can offer only two models of regulation—(1) the model of extraterritoriality and (2) the model of territorialization, aimed at overlapping digital borders with physical borders of the state. The nature of information/data (immateriality, divisibility, mobility, and interconnectedness) does not easily fit into this binary, which calls into question the possibility of applying jurisdictional rules (Daskal, 2015). The location of the data is the main issue for territorial jurisdiction. As noted by K. Eichensehr: "Data's intangibility poses significant difficulties for determining where data is located (and) access to data does not depend on physical proximity" (Eichensehr, 2017). The inherent "non-territoriality" of data is not a concept of international law: their movement across borders creates new geographies of private and state power that are not tied to traditional binary concept. The complex relationship of data to the territory does not fit into the clear framework of categorization of territoriality/extraterritoriality.

[12] Tallinn Manual on the International Law Applicable to Cyber Warfare, p. 25.

It is also necessary to refer to "container" understanding of the state, which prevails in international law (Storey, 2017; Taylor, 1994).[13] This understanding of the state can be found in the Montevideo Convention on the Rights and Obligations of States (1933). Despite the fact that the Montevideo Convention is a regional international treaty signed at the International Conference of American States, and states located outside the Western Hemisphere did not participate in its discussion, the criteria set out in it have become *de facto* a global standard (Hunziker, 2017). Thus, according to Article 1 of the Convention, the "state", as a subject of international law, must have the following qualities: (1) permanent population, (2) defined territory, (3) government, and (4) ability to enter into relations with other states. This definition has acquired the status of customary law and since its adoption it has been repeatedly confirmed in international judicial practice (Musgrave Thomas, 2000).[14] The requirement of a defined territory in Article 1 of the Convention refers to borders: it means a discrete territory separated from other territories. If this element in the definition of state in Montevideo Convention is combined with the element government, the territory becomes a "container" over which the sovereign exercises its supreme and exclusive power.

Nevertheless, even if the "container" understanding of the state is generally recognized in international law, the specific nature of data flows leads to the fact that the regulation of cyberspace begins to be regarded as extraterritorial. In any case, it seems to us that the regulation of data flows within the framework of an extraterritorial jurisdiction will be a kind of fiction. However, attempts to territorialize data by placing data in the locations of infrastructure, data subjects, or data controllers are also a fiction. Even if data is stored in a certain territory, this does not mean that the state has access to or control over this data, since control and management of data are carried out through networks of private entities. This shows that digital sovereignty cannot always be derived from territorial sovereignty.

[13] The "container" ideas about a territorial state are a solid shell separating it from other states, inside which there is an unsurpassed and hierarchically organized state power.

[14] For example, the Arbitration Commission of the European Conference in Yugoslavia in Opinion No. 1 declared that "the State is commonly defined as a community which consists of a territory and a population subject to an organized political authority; that such a state is characterized by sovereignty".

4 Extraterritoriality and Localization: As Necessary Elements of Digital Sovereignty in GDPR, the DSA, and the Judicial Practice of the CJEU

4.1 Localization

In the EU data flow regulation regime, various steps can be noticed towards the territorialization of data in order to assert jurisdiction over them. In particular, they are achieved through legal mechanisms in several areas, in particular: (1) by appointing a representative in the EU for data controllers located outside the EU; (2) by connection with the location of "data subjects" in the EU; and (3) by localization of data. The trend of restoring borders in the digital sphere is accompanied by attempts to regain control over data through data localization—reterritorialization of data by requiring their storage on servers and in data processing centres within national borders as an exercise of "data sovereignty". Recently, this trend has been accompanied by attempts of states to regain control over data and digital infrastructure.

Data control is the main driving force behind "data localization" laws, and several states, including the EU, have adopted data localization laws and taken initiatives to create national or regional digital infrastructures. In particular, in October 2019, a grandiose European project of the European data transmission infrastructure was launched. The initiative aimed at developing cloud services in Europe does not solve the problems that it was supposed to solve. Initially, Gaia-X was welcomed as a project providing common rules for the creation of European cloud services and data exchange market while respecting "European values" in the field of data protection, cybersecurity, and data processing (Troitiño, 2021). Senior French and German ministers trumpeted the birth of Gaia-X, a cloud data project that they promised would help Europe regain its "digital sovereignty" in the face of dominant foreign players such as Amazon and Google.[15]

In November 2021, the European Data Protection Board (hereinafter—*"EDPB"*) published two documents with instructions on when it will be allowed to transfer personal data to third countries outside the EU.[16] However, the experts came to the conclusion that the draft documents from EDPB have the effect of rigid localization of data, limiting data flows from the EU. In particular, as Th. Christakis writes: "The EDPB Guidance seems nonetheless to prohibit almost all such transfers when the

[15] https://www.techzine.eu/news/infrastructure/67696/infighting-threatens-the-future-of-europes-gaia-x-project/. See also: https://www.politico.eu/newsletter/digital-bridge/digital-bridge-trouble-in-the-european-cloud-privacy-in-the-u-s-washingtons-tech-wish-list/

[16] Recommendations 01/2020 on measures that supplement transfer tools to ensure compliance with the EU level of protection of personal data, European Data Protection Board. Available at: https://edpb.europa.eu/our-work-tools/documents/public-consultations/2020/recommendations-012020-measures-supplement_en

personal data is readable in the third country" (Christakis, 2020). In addition, there are technical obstacles to providing online services in a strict data localization mode.

Besides, data localization and reterritorialization can also be found in the case law of the CJEU. Thus, the CJEU demanded that Google removed search results from its website on the basis of the right to be forgotten and limit such removal to the EU territory, encouraging the use of geo-blocking technologies (*Google v. CNIL*).[17] Although it is in the context of restoring independence from foreign service providers that the concept of digital sovereignty is mentioned. However, data localization by itself does not guarantee sovereignty in terms of security and confidentiality of personal data (Komaitis, 2017). Territorial jurisdiction extends to the cyber infrastructure located on the territory of the state, to persons engaged in cyber activities on this territory, and to such activities in cyberspace that occur, terminate, or have a significant impact within state borders.

It should be noted that excessive desire to attract data to Europe and the creation of digital infrastructure in the EU can have the opposite effect. Digital sovereignty can lead to an acute struggle, which ultimately will not contribute to the development of the economy. In 2011, the Council of Europe warned of the potential adverse cross-border impacts on Internet access and use that could result from the exercise of national sovereignty. Digital borders will entail a ban on the activities of many cloud service providers.

4.2 The GDPR and the DSA: A Hybrid of a Territorial and Target Approach

EU law can be considered an illustration of how, through legislative acts, it is possible to bring one's jurisdiction into line with technology, using conceptual stretch and circumventing international norms, adjusting their interpretation to one's needs, often finding far-fetched connections. Some researchers consider the extraterritorial scope of the GDPR application "as a consequence of European digital sovereignty" (Irion, 2022). In particular, according to the regulation, any transfer of personal data outside the EU is presumed illegal, unless the European Commission considers that a third state adequately protects this personal data. This principle of adequacy can be called a kind of reflection of the principle of the subjective principle of territoriality. According to the GDPR, the data controllers and processors to whom this rule applies must appoint a representative in the EU. As one can see, this requirement creates a direct territorial connection.

[17](Case C-507/17). *Google LLC, successor in law to Google Inc. v Commission nationale de l'informatique et des libertés (CNIL)*. Judgment of the Court (Grand Chamber) of 24 September 2019. Jugdment ECLI:EU:C:2019:772. Available at: https://curia.europa.eu/juris/liste.jsf?language=en&num=C-507/17

However, the most common in recent digital laws should be considered the doctrine of consequences—a type of objective territorial jurisdiction in which the intention to produce economic consequences on the territory of a state is considered sufficient for the exercise of jurisdiction by that state. However, this doctrine is a very controversial legitimization for extraterritorial jurisdiction. Despite this, the EU uses it in its own interpretation, and the EU member states countries are increasingly exercising this type of jurisdiction. For example, Article 3(2) of the GDPR applies to controllers and data processors that are not established in the EU, but who process the data of individuals in the EU in two situations: (1) if it is related to the supply of goods and services in the EU and (2) if it is related to monitoring the behaviour of people in the EU.

In this context, J. B. D. Svantesson points out that the GDPR's provisions on monitoring the behaviour of EU residents fall under the doctrine of consequences (Svantesson, 2015). Nevertheless, it is somewhat difficult to prove in general the existence of a connection between the consequences that have occurred and the act of data processing. At the same time, the consequences are assumed not only economic, which is why the data protection legislation is criticized. Nevertheless, it should be pointed out that the GDPR reflects a combination of territorial and target justification.

The situation is different with the DSA. Thus, Article 1a of the DSA restricts the scope of the DSA, providing that the DSA applies to intermediary service providers that are offered to recipients who have their location or are located in the EU, regardless of the place of establishment of the providers of these services (Ivic & Troitiño, 2022). Moreover, the concept of "offering services in the Union" is understood as "enabling legal or natural persons in one or more Member States to use the services of the provider of information society services which has a substantial connection to the Union". Moreover, a substantial connection exists if the provider has an enterprise in the Union or when the enterprise (a) has a significant number of users in one or more member states or (b) focuses its activity on one or more member states.

However, the question remains, what does a significant number mean? The use of an evaluation criterion in a provision that defines the scope of application of a legislative act provides a loophole for creative interpretation. In addition, this paragraph does not at all reflect the intention of the enterprise itself to carry out its business activities in the EU; nevertheless, the EU is trying to extend its jurisdiction to companies that have decided to apply from its territory.

It should be noted that the spread of jurisdiction is very far-fetched and the criteria for evaluating communication do not meet the criterion of "targeting" as in Article 3(2) of the GDPR. Reterritorialization occurs through communication with the user's location in the EU, thus finding an alternative to solve the dilemma of regulating the digital society. By linking regulation to the location of the data subject or users, rather than the location of the data, GDPR and DSA expand their scope of application and they extend even to those organizations that are outside the EU. Due to the interconnectedness of global communications today, the data of a citizen of one country can be mixed with the data of a person who is not a US citizen, to the extent that data can be "accidentally" collected about several people at once, even unintentionally. This makes it difficult to regulate based on the location or identity of the user. Such an interpretation of this doctrine is fraught with too broad jurisdiction. There is a tendency towards extraterritorial coverage—regardless of whether such consequences are implied or have such consequences, the basis for jurisdiction remains controversial.

As we can see, the GDPR and the DSA strike at the principles of territoriality but at the same time do not abandon it. The element of territory, which was central to the traditional concept of sovereignty, is gradually losing its central role. The reference to the territory becomes just one of the various mechanisms that the EU uses to confirm its jurisdiction over the digital world. At the same time, the approach to data regulation in the EU contains both territorial and extraterritorial aspects.

4.3 Localization and Extraterritorialization in the Case Law of the CJEU

The analysis of the EU judicial practice shows that in order to solve issues related to the regulation of the data flow, the CJEU's judicial practice also tries to coordinate and use through extraterritoriality and localization. Thus, in the opinion of the *Salemink* case, the Advocate General argued that "(...) for EU purposes, the 'territory' of the Member States is the area (not necessarily territorial, in the spatial or geographical sense) of exercise of the competences of the Union",[18] calling the connection between the exercise of sovereignty and physical territory a conditional rather than a necessary truth. The EU territory as a non-physical space is a convincing idea. The principle of territoriality is certainly important in the implementation of EU jurisdiction in the cybersphere. For example, in the decision in the *Lindqvist* case, it was stated that EU legislation does not apply indiscriminately to the entire Internet.[19] The Court ruled that simply being in the EU and uploading personal data to a web page that can be accessed by anyone in the world with Internet access does not constitute the transfer of personal data to a third state (Hamuľák, 2018). Thus, the Court limited the scope of application of EU legislation. So, there is a transition from "territory" to "jurisdiction". Tendencies to the fact that "territory" is not physical or "jurisdiction" replaces "territory" may have significant consequences: it may open the way for the EU to expand its jurisdictional scope through creative interpretation or the use of other forms of initiation of jurisdiction.

[18] Opinion of Advocate General Cruz Villalón delivered on 8 September 2011. (Case C-347/10) *A. Salemink v Raad Van Bestuur Van Het Uitvoeringsinstituut Werknemersverzekeringen*. P.54. Available at: https://curia.europa.eu/juris/document/document.jsf?text=&docid=109265&pageIndex=0&doclang=en&mode=lst&dir=&occ=first&part=1&cid=24621

[19] (Case C-101/01). *Criminal proceedings against Bodil Lindqvist*. Judgment of the Court of 6 November 2003. Judgment ECLI:EU:C:2003:596. Available at: https://curia.europa.eu/juris/liste.jsf?num=C-101/01

The second approach can be seen, for example, in the case of *Google v. CNIL.*[20] In this case, the Court faces a choice between recognizing the global application and recognizing the non-universal application of the right to be forgotten, which will reduce the level of protection of this right, but will take into account the sovereignty of states. At first glance, the Court chooses the latter. The CJEU held that various third states either do not recognize the right to be forgotten or have a different approach to it and can resolve a dispute with the right to freedom of information in favour of the latter. The Court concluded that under EU law there is no obligation to remove links for all language versions of the search engine.

The Court essentially expresses respect for the right of other states to strike a different balance between the right to data protection and freedom of information. The CJEU notes that the GDPR was adopted on the basis of Article 16 of the TFEU, and its purpose is to guarantee a high level of protection of personal data throughout the European Union. The Court also recognizes that the concepts of rights recognized in the EU should not be unilaterally extended to the jurisdictions of third tates. At the same time, the Court tried to ensure the highest possible level of protection of the right to data protection while respecting international comity and legal diversity.

Besides, the Court narrows the scope of application of the GDPR, strengthening the territorial principle: although according to the GDPR protection should be applied to individuals, regardless of their nationality or place of residence, with respect to the processing of their personal data, the Court does not require the application of the GDPR to those who are outside the territory of the Union. The Court uses the concept of territorial consequences emphasizing the need to prevent harmful effects within the EU. Thus, the Court goes beyond territoriality to consider where the consequences are felt. Nevertheless, it is very standard that in the same case, the Court recognized the indirect recognition and the possibility of global application of EU law.

The CJEU clearly stated that the EU does not currently require the global application of the right to be forgotten; it also does not prohibit such practices. In addition, the CJEU judgement does not change the broad interpretation of Article 3(1) of the GDPR given in the *Google Spain* case.[21] Therefore, the position of the CJEU that the EU legislator does not grant the rights enshrined in the GDPR outside the territory of the EU member states is questionable. Thus, within the framework of the right to be forgotten, which develops almost exclusively as a result of the CJEU's judicial activism, in conditions when "digital imperialism" becomes the goal of various developed countries, the EU turns into a kind of tool for asserting its digital power far beyond its borders (Hamuľák et al., 2021). As we can see, the parallel

[20] (Case C-507/17). *Google LLC, successor in law to Google Inc. v Commission nationale de l'informatique et des libertés (CNIL).* Judgment of the Court (Grand Chamber) of 24 September 2019. Jugdment ECLI:EU:C:2019:772. Available at: https://curia.europa.eu/juris/liste.jsf?language=en&num=C-507/17

[21] (Case C-131/12). *Google Spain SL and Google Inc. v Agencia Española de Protección de Datos (AEPD) and Mario Costeja González.* Judgment of the Court (Grand Chamber), 13 May 2014. Available at: https://eur-lex.europa.eu/legal-content/EN/ALL/?uri=CELEX%3A62012CJ0131

application of legal mechanisms of localization and extraterritoriality in order to establish digital sovereignty can also be seen in the practice of the EU.

5 Conclusion

As one can see, the role of the territory in the implementation of the EU's digital sovereignty is ambiguous. The modern international order is still based on the territorial principle, but cross-border data flows ultimately challenge the traditional territorial understanding of sovereignty. The EU data protection laws simultaneously distinguish between the establishment of an extraterritorial effect and the localization of data, but at the same time, none of these options completely deviate from the principle of territoriality, and even through disputable fictions, they seek to substantiate the connection between the situation/fact and the territory. The EU's digital sovereignty presupposes the existence of both data localization laws and laws with extraterritorial effect. In this regard, the EU is trying to exercise jurisdiction in cyberspace in terms of the traditional concept of territoriality while simultaneously moving away from it and creating regulatory bindings to the location of users, data collectors, or infrastructure. The research shows that the use of these two legal mechanisms leads to a revision of the concept of territory in EU law.

Bibliography

Agnew, J. (2005). Sovereignty regimes: Territoriality and state authority in contemporary world politics. *Annals of the Association of American Geographers, 95*, 437–461.

Christakis, T. (2020). "Schrems III"? First thoughts on the EDPB post-Schrems II recommendations on international data transfers (Part 3). *European Law Blog*. Available from "Schrems III"? First Thoughts on the EDPB post-Schrems II Recommendations on International Data Transfers (Part 3) – European Law Blog.

Clopton Zachary, D. (2014). Replacing the presumption against extraterritoriality. *94 Boston University Law Review, 1*, 53.

Daskal, J. (2015). The un-territoriality of data. *125 Yale Law Journal 326 (2015) American University*, WCL Research Paper No. 2015–5, 73.

Dodge, William S. (2017). *Jurisdiction in the fourth restatement of foreign relations law. 18 Yearbook of Private International Law 143*. UC Davis Legal Studies Research Paper, 28.

Eichensehr, K. (2017). Data extraterritoriality. *95 Tex. L. Rev*. See Also 145 (2017) UCLA School of Law, Public Law Research Paper No. 17–24.

Finnemore, M., & Hollis, D. B. (2016). Constructing norms for global cybersecurity. *American Journal of International Law NY, 110*, 460.

Floridi, L. (2020). The fight for digital sovereignty: What it is, and why it matters, especially for the EU. *Philosophy & Technology, 33*, 369–378.

Friedman, T. L. (2007). The world is flat: A brief history of the twenty-first century. *Douglas and McIntyre (2013) Limited.*

Gábriš, T., & Hamuľák, O. (2021a). Pandemics in cyberspace – Empire in search of a sovereign? *Baltic Journal of Law & Politics, 14*(1), 103–123.

Gábriš, T., & Hamuľák, O. (2021b). 5G and digital sovereignty of the EU: The Slovak way. *TalTech Journal of European Studies, 11*(2), 25–47.
Hamuľák, O. (2018). La carta de los derechos fundamentales de la union europea y los derechos sociales. *Estudios constitucionales, 16*(1), 167–186.
Hamuľák, O., Vardanyan, L., & Kocharyan, H. (2021). The global reach of the right to be forgotten through the Lenses of the Court of Justice of the European Union. *Czech Yearbook of Public & Private International Law, 11*, 196–211.
Hunziker, R (2017). *Understanding the European Union through comparison to sovereign states.* Baylor University. Waco, Texas, 199.
Irion, K. (2022). *The General Data Protection Regulation through the lens of digital sovereignty*, EU Cyber Direct Research Seminar, Leiden University (online). Available from https://www.ivir.nl/publicaties/download/Irion-GDPR-and-Digital-Sovereignty-11Mar22.pdf.
Ivic, S., & Troitiño, D. R. (2022). Digital sovereignty and identity in the European union: A challenge for building Europe. *European Studies, 9*(2), 80–109. https://doi.org/10.2478/eustu-2022-0015.
Komaitis, K. (2017). The 'wicked problem' of data localisation. *Journal of Cyber Policy, Taylor & Francis Journals, 2*(3), 355–365.
Koskenniemi, M. (2011). What use for sovereignty today? 1. *Asian Journal of International Law, 61*, 61–70.
Lambach, D. (2019). The territorialization of cyberspace. *International Studies Review, 22*(3), 482–506.
Merkel, A. (2019). Rede von Bundeskanzlerin Angela Merkel zur Eröffnung des *14. Internet Governance Forums in Berlin.* Available from https://www.bundeskanzler.de/bk-de/aktuelles/rede-von-bundeskanzlerin-angela-merkel-zur-eroeffnung-des-14-internet-governance-forums-26-november-2019-in-berlin-1698264.
Moerrel, L., & Timmers, P. (2021). Reflections on digital sovereignty EU. *EU Cyber Direct, Research in Focus Series, 2021*, 33.
Musgrave Thomas, D. (2000). *Self-determination and national minorities.* Oxford Monographs in International Law, Oxford University Press, 2000, 235.
Ryngaert, C. (2015). *Jurisdiction in International Law (2nd Edition).* Oxford University Press.
Storey, D. (2017). States, territory and sovereignty. *Geography, 102*(3), 116–121.
Svantesson, D. J. B. (2015). Extraterritoriality and targeting in EU data privacy law: The weak spot undermining the regulation. *International Data Privacy Law, 5*(4), 226–234.
Taylor, P. J. (1994). The state as container: territoriality in the modern world-system. *Progress in Human Geography, 18*(2), 151–162.
Troitiño, D. R. (2021). La «Década Digital» de la Unión Europea: desarrollos e impactos sobre su ciudadanía y economía. *IDP: Revista d'Internet, dret i política, 34*, 1–14.
Tsagouring, N., & Buchan, R. (2015). *Research handbook on international law and cyberspace* (p. 672). Edward Elgar Publishing.
Von der Leyen, U. (2020). *Shaping Europe's digital future: Op-ed by Ursula von der Leyen.* Available from https://ec.europa.eu/commission/presscorner/detail/en/ac_20_260

A Multidimensional Understanding of EU's Digital Sovereignty

Pablo Martínez-Ramil, Haridian Bolaños-Frasquet, Ondrej Hamuľák, and Tanel Kerikmäe

Abstract The notion of "digital sovereignty" is at the core of contemporary geopolitical trends and regulatory initiatives aiming to shape the development of the digital realm in the EU's regulatory and academic environments. In this light, developing policy and legislation that could enhance the strategic autonomy of the region and its citizens in the digital field has become imperative. The lack of a consensus on the features of the concept, the ambiguous interchangeable terms that refer to it and the contrasting scenarios where it is used suggest that a multidimensional approach to address its meaning is necessary. This research, based on a systematic revision of the literature on digital sovereignty, delineates a roadmap toward three differentiated but interconnected dimensions, and the streams of understanding that result from them. Hence, it examines the concept as a policy goal in the international realm, as the EU's regulatory capacity and as an empowerment mechanism for the e-citizens. By exploring the interconnections between these dimensions, this article helps to conceptualize the cornerstones of an increasingly important notion.

The paper was prepared within the implementation of the project reg. no. CZ.02.2.69/0.0/0.0/19_073/0016713 "The Role of the EU in the Quest for Digital Sovereignty."

P. Martínez-Ramil · H. Bolaños-Frasquet
Faculty of Law, Palacký University Olomouc, Olomouc, Czech Republic
e-mail: pablo.martinezramil01@upol.cz; haridian.bolanosfrasquet02@upol.cz

O. Hamuľák (✉)
Faculty of Law, Palacký University Olomouc, Olomouc, Czech Republic

Department of Law, School of Business and Governance, Tallinn University of Technology, Tallinn, Estonia
e-mail: ondrej.hamulak@upol.cz

T. Kerikmäe
Department of law, Tallinn University of Technology, Tallinn, Estonia
e-mail: tanel.kerikmae@taltech.ee

D. Ramiro Troitiño et al. (eds.), *Digital Development of the European Union*,
https://doi.org/10.1007/978-3-031-27312-4_15

1 Introduction

In an interview with a leading Spanish newspaper, Thierry Breton (European Commissioner for Internal Market) conceded that the Digital Services Act (hereinafter DSA) was completed "just in time." The Commissioner was alluding to the recent purchase of Twitter by Elon Musk (Abril, 2022). Depicting himself as a "free speech absolutist," the CEO of Tesla had been repeatedly advocating for loosening the platform's content moderation policies (Espinoza, 2022), potentially clashing with the notion of "digital ecosystem of trust" that the European Union (hereinafter EU) intends to establish. Some already consider the acquisition to be "the first big test for the DSA" (Browne, 2022). However, the digital realm and the questions addressing its control transcend any single regulatory initiative or territorial border. Its constant fast-paced development, sped up by the COVID-19 pandemic, led to several clashes between big tech companies and sovereign states in the past few years. Floridi (2020, pp. 369–371) raised awareness on this matter in a very enlightening article, listing several confrontations between those who de facto control the digital world and those with the (de jure) legitimacy to claim sovereignty over it.

In essence, the purchase of Twitter is nothing more than just another bout in the quest for digital sovereignty, the notion at the core of these conflicts. Nonetheless, despite the attention received, the concept remains obscure. While some approach it from a traditional understanding of state sovereignty (López Velarde Campa, 2021, p. 15), others focus on geopolitical aspects to explore its growing relevance (Kleinhans & Lee, 2022; Creemers, 2020). While some define it as 'control' over the digital sphere (Vučić, 2021), others conceive it as a measurable capacity of the states (Kaloudis, 2021).

This research argues that the defining problem derives from the many facets of a very dynamic process as well as from the multiplicity of actors involved. Hence, it proposes instead a multidimensional approach to achieve a comprehensive overview. This will enable not only the incorporation of previous research works but also to map the hot spots, the stakeholders involved and their respective interests.

Thus, framing the EU's digital sovereignty as a policy goal (Roberts et al., 2021, pp. 2–3; Werthner, 2022, p. 244) reflects the paramount importance of the digital challenges resulting from geopolitical trends and an unregulated digital sphere. Understanding it as regulatory capacity (Werthner, 2022, p. 24; Shapiro, 2020, p. 12) evidences EU's proactive attitude toward shaping the future development of the digital realm beyond its territorial borders. Formulating it as the empowerment of e-citizens through the establishment of a digital ecosystem of trust illustrates the tensions between values, policy goals and EU's regulatory initiatives. Every highlighted dimension mirrors a different arena (international, European, and individual) where conflicts unfold. By exploring the interconnections between them, this article offers a detailed overview of the cornerstones of an increasingly important matter.

This conceptual paper introduces a clear stand on the issue from the viewpoint of the EU, serving as a roadmap toward a very fragmented discussion. Nevertheless, this research does not intend to provide a legal definition of the concept, for the many instances where it is used lack the consistency that such a conceptualization would demand. Instead, circumventing a purely legalistic perspective, this research claims that the notion rather refers to a multidimensional process. Here is where the real novelty lies: in the comprehensive analysis of the interactions between the different dimensions of the notion. This approach clears the way for the incorporation of disciplines other than the law to explain the checks and balances between the many actors in the field. Moreover, in light of the dynamic character of the concept, it smooths the traceability of its evolution and the mapping of the central points of conflict.

Above all, it should be noted that it is not possible to address in detail every relevant feature of the EU's digital sovereignty in one article. Consequently, this research aims to provide a comprehensive overview of this approach; an approach that will be furtherly developed in upcoming publications.

First, this article introduces some notes on the concept of digital sovereignty and its complex nature, justifying why a multidimensional approach constitutes the most suitable tool to examine the standpoint of the EU as well as the pitfalls and strengths attached to it. The text continues analyzing the international dimension of the EU's digital sovereignty, understood as a "policy goal." Special attention is paid to the struggle of the EU in light of geopolitical trends (attached to divergent worldviews), its dependence on third-party suppliers and its capacity as an economic power to influence the dynamics of the digital world. The latter element links the international dimension to the European one, for the EU's regulatory capacity, although limited within its borders, holds the potential to influence other regulatory regimes seeking benefits from its economic power (often known as the "Brussels effect"). Accordingly, the EU's regulatory capacity (understood as both proactive and reactive capacity) constitutes the second dimension. Its analysis takes into consideration both the legal and non-legal initiatives that shape the EU's Digital Ecosystem of Trust, its interpretation, and the rising tensions between them. Finally, the study of the empowerment of e-citizens through the aforementioned Digital Ecosystem accounts for the third dimension of "digital sovereignty." This dimension is highly interconnected with the previous ones and, as it is exposed below, it is a source of tensions between regulatory approaches.

2 Sovereignty in the Digital Age

To describe what lies behind the concept of EU's digital sovereignty, this research stems from previous works that understand the general concept of digital sovereignty as "control." However, it does so while acknowledging the material impossibility that imbues the notion of "absolute control" when addressing the digital realm. Hence, here it will be defined as the capacity to shape and exert influence over (i) the digital realm and (ii) its future development. An apparent follow-up question would necessarily address the concept of the "digital realm." Nonetheless, similarly to digital sovereignty, no consensus has been yet reached, often appearing next to terms of similar generic nature (such as "cyberspace" or "internet").

Floridi (2020, pp. 370–371) considered the digital realm to be composed of different elements, distinguishing "data, software (e.g., AI), standards and protocols (e.g., 5G, domain names), processes (e.g., cloud computing), hardware (e.g., mobile phones), services (e.g., social media, eCommerce), and infrastructures (e.g., cables, satellites, smart cities)." The elements listed here serve very well the purposes of this chapter, for they illustrate the diverse nature of the many elements that conform to the digital realm. While some are of analogic nature (e.g., hardware or infrastructures), others are purely digital (e.g., software). While some enable the participation of the users in the digital realm (e.g., standards and protocols), others directly constitute end uses (e.g., services, such as Facebook or Booking). The Panel for the Future of Science and Technology (hereinafter PFST), part of the European Parliamentary Research Service, acknowledged in a report the terminological confusion and proposed instead the physical elements sustaining the digital ones to be part of a different notion, namely technological sovereignty (PFST, 2021, pp. 2–4). Under this view, technological sovereignty would become a sine qua non without which there could be no digital sovereignty. Regardless of the approach employed, it remains clear that each one of these "segments" involves different stakeholders and poses different questions; all of them essential to the functioning of the digital realm.

Leaving aside China's "Great Firewall," the majority of the elements that constitute the digital realm, at least for most, know no borders. Therefore, it is not uncommon to see the notion of "digital sovereignty" appearing in the context of a fight, quest, or struggle among parts. For instance, in the aforementioned report, the PFST examined both challenges and risks that the EU is facing in the context of its technological sovereignty. Every single issue addressed there involves different actors, be it the increase of purchases of European start-ups by international companies -particularly the USA- (PFST, 2021, p. 34) or the EU's strong dependence in terms of access to critical raw materials from third countries (PFST, 2021, p. 6).

In light of the variety of elements sustaining the digital world and the unique nature of the EU (bearing in mind its international projection, its institutional organization and its multi-level governance system), a multidimensional approach emerges as the most suitable option to describe (i) the state of the "game board," (ii) the moves so far played by the EU, and (iii) the ones yet to take.

3 A Policy Goal, the International Dimension of EU's Digital Sovereignty

Charles Michel, the current president of the European Council, defined digital sovereignty as "a means to achieve autonomy" (European Council, 2021); the European Parliament as the "ability to act independently" (Madiega, 2020); and the president of the Commission as EU's capability "to make its own choices, based on its values, respecting its own rules" (Von der Leyen, 2020). The international dimension of the concept appears subtly attached to each one of these statements, for all of them imply the existence of hindering actors or scenarios where the EU is not autonomous, could not act independently or is unable to make its own choices.

This view is better clarified by the European Parliament resolution on Artificial Intelligence (hereinafter AI) in the digital age. The European Parliament (2022, para. 2) emphasized that "digital competitiveness and open strategic autonomy" have become policy goals in several states, as a result of the realization that "emerging technologies could affect the geopolitical power status of entire countries." Moreover, the parliament recognized the loss of a historic dominant role in the race for technological development, falling the EU behind the USA or China. The "fall" generates several risks, in light of the capacity of the digital realm to "deploy" its effects while blurring traditional analogue frontiers between states.

First and foremost, there is a risk of marginalization of European actors in the establishment of global technological standards, which might lead to technological developments carrying within its essence challenges to EU Values (e.g., the above-mentioned example of Twitter and the implications of different approaches toward "freedom of expression"). Second, the difficulties might increase when copying with the dangers possessed by digital tools, progressively becoming "an instrument of manipulation and abuse in the hands of some corporate actors as well as in the hands of autocratic governments for the purpose of undermining democratic political systems." And third, there is a risk of consolidating an increasingly disadvantageous European position in global digital markets. This obeys to the "high degree of market concentration" of many digital markets, allowing a small group of (mainly) US-based tech companies to attract the talent, commercialize ground-breaking technological innovations "and achieve extraordinary profitability" (European Parliament, 2022, para. 2–5). In sum, and in the words of the parliament, if no remedy is applied, the EU "will end up having to follow rules and standards set by others and risks damaging effects on political stability, social security, fundamental rights, individual liberties and economic competitiveness." (European Parliament, 2022, para. 6).

As highlighted, the most influential countries that shape the digital sphere are the United States (hereinafter USA) and China. Both represent divergent political and social models that, attached to their respective ideologies, impregnate their technological and digital R&D. The EU depends on both. On the one hand, it depends on China for manufacturing certain key enabling technologies (hereinafter KETs) such as raw materials employed in the construction of digital infrastructures and hardware

(Kleinhans & Lee, 2022). On the other one, it relies greatly on American software. To rephrase it, most of the IT technologies used in the EU come from third states (Kaloudis, 2021).

China, following its worldview, has relied on an interventionist approach to address digital matters. Examples are many, from the recent law on recommendation algorithms (that forced Chinese technology giants to share the details of their respective algorithms with the national regulators) (Kharpal, 2022) to the use of surveillance technology for the benefit of its regime (López Velarde Campa, 2021). The latter generated in some countries a debate concerning the possibility of establishing limits to Chinese technologies, such as banning Huawei from 5G technology (for it could potentially be used for espionage purposes) (Rühlig & Björk, 2020). On the other side of the political spectrum, the USA has systematically adopted a liberal approach in its relationship with technology, based on giving companies the freedom to innovate in the technological and digital sphere. However, this model is not free of dangers. Companies collect data from potential consumers and customers without any controls -as data became a monetizable good- and record sensitive data such as the one registering health and financial information (Bradford, 2012; Coyer & Higgot, 2020, pp. 43–48).

Both countries lead the technology race and, if nothing changes, will define technological global standards and the next decades' development of the digital realm. In other words, they will shape the impacts and changes that the digital world will promote in European societies.

The EU's responsiveness to this technological race is mostly limited to two lines of action. On the one hand, in the international sphere, the EU has, among other soft law initiatives, the capacity to enter into international agreements that affect its digital capacities. For instance, agreements celebrated to diversify the supply of raw materials considered KETs; or to ensure the legal transfer of personal data between states, like the upcoming Trans-Atlantic Data Privacy Framework. On the other one, its regulatory capacity enables the EU to determine its legal frameworks within the European borders. As it will be furtherly exposed, this regulatory capacity can be used to promote in certain instances legislative developments in third states.

4 A Legal Framework for the Digital Realm. The European Regulatory Capacity

The second dimension departs from the international sphere to hold its focus on the European one. It is mainly grounded in the EU's regulatory capacity. Its legal competencies are very well delimited in the EU primary law. A priori, regulatory instruments enacted within the EU should not deploy their effects beyond the EU's territorial border. Nevertheless, for years, the talk of the town has been the so-called "Brussels effect," the EU's ability to influence the international sphere either through its institutions or standards (Bradford, 2012). And the main tool that the EU has to protect its interests in the digital realm and face the challenges emanating from the international society is through the regulation of its internal market. Concretely, through the imposition of obligations or restrictions that companies operating in the market would need to meet.

Christakis (2020, p. 15) distinguished three main reasons that explain the external transposition of institutions, standards, and European regulatory frameworks beyond its borders. First, the economic strength of the EU (representing roughly a fifth of global GDP at market exchange rates) constitutes a great attractiveness for many actors interested in penetrating the internal market. Second, the development and establishment of stricter regulations than other markets have provoked a situation where, if a company intends to sell the same product in two different markets, it will tend to adapt the product to European standards (it is cheaper than developing the same product twice and wasting money "on having lots of different versions"). The combination of these two factors (the EU's economical attractiveness and its stricter regulatory frameworks) accounts for the third one, for it promotes the appearance of different stakeholders in third countries lobbying their respective governments to adjust their standards to the European levels. Under some conditions, the Brussels effect becomes some sort of unilateral regulatory globalization, where "regulations originating from a single jurisdiction penetrate many aspects of economic life across the global marketplace."

However, it must not be ignored that a large market is not enough to ensure the EU's capacity to exert such a level of external influence. It must be worthwhile for external operators to accept these standards, as well as impossible to evade them once accepted (Bradford, 2020; Bradford, 2012). That is why it is very problematic for the EU to strike the right balance in their newly established digital regulations, changing substantially after the lobbying of many interested parties during the legislative process. For instance, the rights-based approach originally proposed by the High-Level Expert Group on AI ended up shifting into "a language of risk-based AI, deemed to be rather innovation-friendly, with economic and human values remaining in tension with one another" (Antonov, 2022).

The Von der Leyen Commission entitled one of its six political priorities for the period 2019–2024 "A Europe fit for the digital age." The Commission aims to establish an EU digital ecosystem of trust, putting the technology "at the service of the people" while "empowering" individuals in a "human center, digital future" (European Commission, n.d.). Recent years have witnessed the early proliferation of the digital legal initiatives that are to give shape to the digital ecosystem of trust, which are, but are not limited to, the Digital Services Act (hereinafter DSA), the Digital Markets Act (hereinafter DMA), the General Data Protection Regulation (hereinafter GDPR), the AI Act, and most recently the Data Act and the Data Governance Act. The EU has found the legal basis for both their digital regulatory competencies and the necessity of the regulatory framework in the Treaty on the Functioning of the European Union Art 114, which enables the EU to establish measures that would ensure the functioning of the internal market.

Three main courses of action seem to emanate from the developing digital legal framework. First, they are imposing proactive and reactive obligations that constitute market barriers for companies entering or participating in the internal market. Second, they intend to facilitate European R&D through investment programs and the elimination of barriers to innovation. And third, seeking the empowerment of European e-citizens, the EU is adapting the legal frameworks to face the digital challenges for fundamental rights and implementing educational programs that seek to tackle the digital illiteracy that Europe suffers in comparison to China and the USA. The first two deserve a closer look in this section.

As Gábriš and Hamuľák (2021, p. 28) highlighted, regulatory authorities cannot tackle directly the cyberspace as the subject of a legal framework. Instead, they exercise their regulatory powers over the analogue legal and natural persons that, outside cyberspace, provide the services perceived as "essential or critical for the security and the interests of the sovereign." These are the entities that usually are the object of the sanctions delimited in the regulatory instruments. Thus, the proposed AI Act Art 71(3) contemplates specific fines for scenarios when the offender is a company, the proposed DSA Art 59 establishes a regime of sanctions for large digital platforms, and the proposed DMA Art 26 outlines the penalties to be imposed on the gatekeepers. All digital parties subject to penalties (such as gatekeepers or large digital platforms) have an analogical counterpart.

Hence, the measures set in the EU digital regulatory framework indirectly tackle the digital world by imposing obligations to the analogue parties that sustain it. Most of the established measures hold a proactive and a reactive side, for they are enacted as a reaction to digital developments while proactively shaping future ones. For instance, the AI Act Article 10 establishes a list of obligations related to Data governance to apply during the training process of algorithmic models. This article reflects the concerns that arose when several cases of algorithmic discrimination made it to the headlines in recent years (Martínez-Ramil, 2021, p.3) and proactively shapes the future development of these technologies by imposing in-design obligations.

The second course of action concerns the enforcement of the EU funding programs. In several reports and documents, the EU acknowledged that is falling behind the USA and China in terms of investment. The differences in AI technologies' investment illustrate this situation very well. While the estimated public yearly investment of the USA and China on AI is respectively 5.1 and 6.8 billion euros, the EU barely invests 1 billion euros annually (European Parliament, 2022, para.112). Aware of the situation, the Digital Compass 2030 acknowledges the necessity of increasing investments not only in the existing technologies but in the upcoming ones (such as the so-called quantum revolution). The under negotiations "Path to the Digital Decade," which seeks to concretize the plan and objectives defined in the Digital Compass into concrete measures, contemplates in its Art 12 the creation of multi-Country projects, another funding opportunity that adds up to the already existing ones (such as the Digital Europe Program or the Horizon 2020, among others). These funding programs seek to promote initiatives that would be increasing the digital competitiveness of the EU as well as the nurturing of European champions, companies that would function as a counterpart of the American GAMA or the Chinese BATX. In other words, these funding initiatives aim to promote European R&D to participate more actively in the shaping of the digital world.

In connection with the regulatory instruments, the role of the Court of Justice of the European Union (hereinafter CoJ) should be highlighted. As the supreme interpreter of EU Law, the CoJ has played a protagonist part in conjunction with the legislative power, amending the actions of the European Executive whenever they were not considered to be in accordance with it. Examples are many, such as the invalidation of the US-EU Privacy Shield (forcing the EU and the US to start negotiations again for a new data transfer agreement) in *Schrems II* or the recent confirmation of the 4 billion euro antitrust fine imposed on Google. Its role has been fundamental not only for the interpretation of the established digital legal framework but also for defining the content and limits of newly born digital rights, such as the right to be forgotten in *Google Spain* and *CNIL* (Hamuľák et al., 2021). Thus, the CoJ must not be overlooked, for the institution holds a decisive role when delineating the limits of the regulatory capacity (fundamental tool for the exercise of European digital sovereignty) of the EU (and therefore the potential influence of its measures over the digital realm).

Before closing this section, it should be emphasized that not all initiatives related to the EU's digital sovereignty are led by the EU itself. For instance, to reduce its dependence on structures placed in third states, France and Germany initiated in the year 2019 the project GAIA-X, a private-public joint initiative to provide Europe with data infrastructures to improve its autonomy. The EU only joined this enterprise a year later, under the auspicious of the Commission's president Von der Leyen. Hence, it is within the interest of the EU to support others' initiatives, as it does with funding programs promoting R&D on digital matters.

5 Beyond the Digital Ecosystem of Trust. The Empowerment of e-citizens

The third course of action mentioned above is the empowerment of EU e-citizens (Troitiño, 2021). It accounts for the third dimension of the EU's digital sovereignty, the one addressing the individual sphere of citizens. Two main issues are to be tackled by the upcoming regulatory instruments in this sphere: the digital illiteracy of the EU citizens (in comparison to China and the USA) and the pitfalls of the European fundamental rights protection in light of the challenges possessed by the digital realm.

The first one is not under debate. The European Parliament (2022, para. 83) acknowledged that "more than 70% of businesses report a lack of staff with adequate digital and AI skills as an obstacle to investment." As of 2019, estimations determined the number of ICT specialists in the EU as 7.8 million, far short of the necessary 20 million experts that the Commission estimates necessary. Moreover, it also recognized the constant brain drain of EU researchers, that are often hired by companies from third states and develop their careers outside the EU borders (European Parliament, 2022, para. 114). In the 2030 European Compass and the later proposed "Path to the Digital Decade," the goal is set for "80% of those aged 16–74 [to] have at least basic digital skills." This plan would be complementary to the Digital Education Action Plan 2021–2027, which would intend to address the problem at lower educational levels.

The second one is the most controversial point, the one generating a more heated discussion, for it has the potential to affect more stakeholders and therefore tilt the scales in a different direction. Part of the Digital Ecosystem of Trust that the EU intends to establish involves the empowerment of EU citizens. This empowerment partially comes from soft law instruments, like the European Declaration on Digital Rights and Principles for the Digital Decade, and partially through hard law, such as the translation of privacy rights to the digital realm in the GDPR or the EU Cookie legislation. The legal initiatives shall not be limited to the regulation of newly born rights (such as the right to be forgotten) but must also adapt analogue frameworks that fall short when dealing with digital tools, as it happens when cases of indirect discrimination caused by an AI system met the requirements for its justification (Martínez-Ramil, 2022). As a consequence, through the empowerment of E-citizens, the EU forces the actors developing the digital world to be mindful of the obligations that might emanate from the rights of the individuals. In other words, it could empower citizens, allowing them to bring an action before a Court if their rights are not respected.

These views fit well with the feelings of Europeans. A special Eurobarometer conducted in the year 2021 asked a sample of EU citizens on their views on digital rights and principles (Troitiño, 2022). Unsurprisingly, a large majority (82%) considered it useful for the EU "to define and promote a common European vision on digital rights and principles" (Eurobarometer, 2021). However, as emphasized above, this is not a straightforward path. In many cases, adopting the legal frameworks dealing with the protection of fundamental rights to the challenges possessed by digital technologies involves the adoption of measures "intrusive" enough to hinder innovation in certain fields. In addition, as highlighted many times in this chapter, the EU is lagging in both software and hardware, and a regulatory framework not particularly friendly toward innovation would discourage R&D investment in the EU. Strike the right balance between establishing a system of digital rights (sufficiently defined to ensure its stability and functioning) and a market attractive to R&D tech investors is not an easy task.

6 Conclusion

In this chapter, a multidimensional approach was drafted to illustrate the viewpoint of the EU before the fast-paced digital developments and their impacts on nowadays societies. Given the lack of a consensus on the notion of digital sovereignty, approaching it as "control" (understood as influence capacity) and addressing the tools that the EU possesses to exercise it constituted for the authors the most suitable approach to describe the position in which the EU finds itself.

All dimensions introduced here are interconnected, for is the regulatory capacity of the EU (European dimension) the one that impulses the external actions (policy goals) concerning third states (international dimension) while defining the borders of what will constitute the Digital Ecosystem of Trust. Concurrently, the legal framework of the Digital Ecosystem of Trust has the potential to establish global standards and influence regulatory developments in the international sphere (again, affecting the international dimension) and empower the European digital citizens with tools to exercise their rights in the digital sphere (involving the individual sphere of the citizens).

In this context, the added value of the EU lies in its representation of the midway between two opposite worldviews that determine the functioning of the digital realm. The establishment of a European Digital Ecosystem of Trust has the potential to illustrate that third path. However, the current state of the art and the position that the EU holds in relation to other superpowers could hinder this approach. This paper argues that it becomes necessary for the EU to resist the voices that call for greater liberalizations of the digital market to the detriment of the protections defined in the Charter of Fundamental Rights of the European Union. To create and sustain trust in a digital ecosystem, requirements of human agency and accountability must prevail. To empower people "in a human-centered digital future" the rights should not fall behind friendlier environments toward innovation.

Each one of the dimensions introduced here deserves its own article to address in detail the cornerstones of their functioning. That is why upcoming research must deal with each one of them separately. The intention of this article was nothing more than serve as a guide toward a complex process and to contribute to an increasingly important debate. One that will determine not only the relationship of Europeans with digital technologies but also the role that EU values will play within it.

List of References

Abril, G. (2022, April 27). El comisario europeo Thierry Breton avisa a Elon Musk: "En la UE hay una normativa que seguir." *El Pais*. Retrieved April 28, 2022, from https://elpais.com/internacional/2022-04-27/el-comisario-europeo-thierry-breton-avisa-a-elon-musk-en-la-ue-hay-una-normativa-que-seguir.html

Antonov, A. (2022). Steering complexity: A sketch of the EU's Contribution to AI Governance. Revista CIDOB d'Afers Internacionals, 131.

Bradford, A. (2012). The Brussels effect. *Northwestern University Law Review, 107*, 1.

Bradford, A. (2020). *The Brussels effect: How the European Union rules the world*. Oxford University Press.

Browne, R. (2022, April 26). Elon Musk's Twitter takeover sets him on a collision course with Europe. *CNBC*. Retrieved April 28, 2022, from https://www.cnbc.com/2022/04/26/elon-musks-twitter-takeover-sets-up-a-potential-clash-with-europe.html

C-311/18—Facebook Ireland and Schrems. Judgment of the Court (Grand Chamber) of 16 July 2020 Data Protection Commissioner v Facebook Ireland Limited and Maximillian Schrems.

Christakis, T. (2020, December). European digital sovereignty: Succesfully navigating between the 'Brussels effect and Europe's quest for strategic autonomy'. *Studies on Digital Governance*, Data Institute, pp. 1–103.

COMMUNICATION FROM THE COMMISSION TO THE EUROPEAN PARLIAMENT, THE COUNCIL, THE EUROPEAN ECONOMIC AND SOCIAL COMMITTEE AND THE COMMITTEE OF THE REGIONS. 2030 Digital Compass: the European way for the Digital Decade.

COMMUNICATION FROM THE COMMISSION TO THE EUROPEAN PARLIAMENT, THE COUNCIL, THE EUROPEAN ECONOMIC AND SOCIAL COMMITTEE AND THE COMMITTEE OF THE REGIONS. Digital Education Action Plan 2021–2027. COM(2020) 624 final.

Consolidated versions of the Treaty on European Union and the Treaty on the functioning of the European Union [2016] OJ C202/1 (TFEU).

Coyer, K., & Higgott, R. (2020). Sovereignty in a digital era. A report commissioned by the Dialogue of Civilizations Research Institute Berlin. Dialogue of Civilisations Research Institute of Berlin.

Creemers, R. (2020). China's conception of cyber sovereignty. In D. Broeders & B. van den Berg (Eds.), *Governing cyberspace: Behavior, power and diplomacy* (pp. 107–145). Rowman & Littlefield.

Espinoza, J. (2022, April 26). EU warns Elon Musk over Twitter moderation plans. *Financial Times*. Retrieved April 28, 2022, from https://www.ft.com/content/22f66209-f5b2-4476-8cdb-de4befffebe5

Eurobarometer. (2021). Digital rights and principles. September 2021–October 2021, European Commission. Retrieved September 1, 2022 from https://europa.eu/eurobarometer/surveys/detail/2270

European Commission. (n.d.). A Europe fit for the digital age. Retrieved August 20, 2022, from https://ec.europa.eu/info/strategy/priorities-2019-2024/europe-fit-digital-age_en

European Council. (2021, February 3). Digital sovereignty is central to European strategic autonomy—Speech by President Charles Michel at 'Masters of digital 2021 [Online event.]. Retrieved April 27, 2022, from https://www.consilium.europa.eu/en/press/press-releases/2021/02/03/speech-by-president-charles-michel-at-the-digitaleurope-masters-of-digital-online-event/

European Declaration on Digital Rights and Principles for the Digital Decade. COM (2022) 28 final.

European Parliament. (2022). Resolution of 3 May 2022 on artificial intelligence in a digital age (2020/2266(INI)). The European Parliament, para. 1–304.

Floridi, L. (2020). The fight for digital sovereignty: What it is, and why it matters, especially for the EU. *Philosophy & Technology, 33*, 369–378.

Gábriš, T., & Hamuľák, O. (2021). 5G and digital sovereignty of the EU: The Slovak Way. *TalTech Journal of European Studies, 11*(2), 25–47.

GAIA X. What we are. Retrieved August 29, 2022., from https://gaia-x.eu/

Hamuľák, O., Vardanyan, L., & Kocharyan, H. (2021). The Global Reach of the Right to be Forgotten through the Lenses of the Court of Justice of the European Union. *Czech Yearbook of Public and Private International Law, 12*(1), 126–141.

Kaloudis, M. (2021). Sovereignty in the Digital age–how can we measure digital sovereignty and support the EU's action plan? *New Global Studies, 6*, 1–25.

Kharpal, A. (2022, August 15). Chinese tech giants share details of their prized algorithms with top regulator in unprecedented move. *CNBC*. Retrieved August 20, 2022 from https://www.cnbc.com/2022/08/15/chinese-tech-giants-share-details-of-their-algorithms-with-regulators.html

Kleinhans, J. P., & Lee, J. (2022). Europe's dependence on Chinese semiconductor manufacturing. In T. Rühlig (Ed.), *China's digital power assessing the implications for the EU* (pp. 21–32). GGAP Report.

López Velarde Campa, J. A. (2021). *Derechos de la soberanía digital*. La Biblioteca.

Madiega, T. (2020). Digital sovereignty for Europe (Briefing PE 651.992; EPRS Ideas Papers). *European Internet Policy Review, 10* (3) | 2021 Parliamentary Research Service.

Martínez-Ramil, P. (2021). Is the EU human rights legal framework able to cope with discriminatory AI? IDP. *Internet, Law and Politics E-Journal, 34*, UOC, pp. 1–14.

Martínez-Ramil, P. (2022, May). Discriminatory algorithms. A proportionate means to a legitimate aim? *Journal of Ethics and Legal Technologies, 4*(1), 3–24.

Panel for the Future of Science and Technology. (2021). Key enabling technologies for Europe's technological sovereignty. European Parliamentary Research Service, Scientific Foresight Unit (STOA), European Parliament, PE 697.184, December 2021, pp. 1–96.

Proposal for a DECISION OF THE EUROPEAN PARLIAMENT AND OF THE COUNCIL establishing the 2030 Policy Programme "Path to the Digital Decade." COM (2021) 574 final 2021/0293(COD).

Proposal for a REGULATION OF THE EUROPEAN PARLIAMENT AND OF THE COUNCIL on contestable and fair markets in the digital sector (Digital Markets Act). COM/2020/842 final.

Proposal for a REGULATION OF THE EUROPEAN PARLIAMENT AND OF THE COUNCIL on a Single Market For Digital Services (Digital Services Act) and amending Directive 2000/31/EC. COM/2020/825 final.

Proposal for a REGULATION OF THE EUROPEAN PARLIAMENT AND OF THE COUNCIL LAYING DOWN HARMONISED RULES ON ARTIFICIAL INTELLIGENCE (ARTIFICIAL INTELLIGENCE ACT) AND AMENDING CERTAIN UNION LEGISLATIVE ACTS. COM/2021/206 final.

Roberts, H., Cowls, J., Casolari, F., Morley, J., Taddeo, M., & Floridi, L. (2021). Safeguarding European values with digital sovereignty: an analysis of statements and policies. *Internet Policy Review, 10*(3), 1–26.

Rühlig, T., & Björk, M. (2020). What to make of the Huawei debate? 5G network security and technology dependency in Europe. UI Paper, 1.

Shapiro, J. (2020). Introduction: Europe's digital sovereignty. In C. Hobbs (Ed.), *Europe's digital sovereignty: From rulemaker to superpower in the age of US-China rivalry* (pp. 6–13). Essay Collection, European Council on Foreign Relations.

Troitiño, D. R. (2022). The European Union facing the 21st century: The digital revolution. *TalTech Journal of European Studies, 12*(1), 60–78.

Troitiño, D. (2021). La «Década Digital» de la Unión Europea: desarrollos e impactos sobre su ciudadanía y economía. IDP, pp. 1–11.

Von der Leyen, U. (2020). Shaping Europe's digital future. European Commission. Retrieved April 27, 2022, from https://ec.europa.eu/commission/presscorner/detail/en/AC_20_260

Vučić, M. (2021). European Union's quest for digital sovereignty: policy continuations and strategy innovations. In K. Zakić & B. Demirtaş (Eds.), *Europe in changes: the old continente at a new crossroads* (pp. 99–115). Institute of International Politics and Economics.

Werthner, H. (2022). Geopolitics, digital sovereignty. . .What's in a word? In H. Werthner, E. Prem, E. A. Lee, & C. Ghezzi (Eds.), *Perspectives on digital humanism* (pp. 241–248). Springer.

Pablo Martínez-Ramil is a doctoral student at the Department of International and European Law, Faculty of Law, Palacký University Olomouc (Czech Republic).

Haridian Bolaños-Frasquet is a doctoral student at the Department of International and European Law, Faculty of Law, Palacký University Olomouc (Czech Republic).

Ondrej Hamuľák is a Senior Lecturer at the Faculty of Law, Palacký University Olomouc (Czech Republic) and Adjunct Professor in EU Strategic Legal Affairs, Tallinn University of Technology, School of Business and Governance, Department of Law (Estonia).

Tanel Kerikmäe is a professor at Tallinn University of Technology, School of Business and Governance, Department of Law (Estonia) and Senior Researcher at the Faculty of Law, Palacký University Olomouc (Czech Republic).

Digital Sovereignty or Sovereignty with Digital Elements?

Tomáš Gábriš and Ondrej Hamuľák

Abstract The notion of digital sovereignty suggests a need for a reassessment of the traditional notion of sovereignty. However, even in the case of bold libertarian expectations of freedom of cyberspace at its very beginning, the actual practice witnessed rather the traditional legal regulation entering cyberspace. Hence, one could similarly boil digital sovereignty down to being "only" the traditional sovereignty exerted over digital elements and digital content. The major difference in comparison with the traditional "analogue" sovereignty might then lie only in the fact of a need for faster and more flexible grasping of new digital phenomena. This already happens nowadays—via newly designed agile legal tools applied by a cascade of intermediaries—mostly public administration bodies and private gatekeepers.

1 Introduction

Digital sovereignty has become a buzzword recently. However, it remains unclear as to its extent, shape, and content. E.g., Pierre Bellanger defined it as follows: "*Digital sovereignty is the mastery of our present situation and our destiny as manifested and*

The paper was prepared within the implementation of the project no. 20-27227S "The Advent, Pitfalls and Limits of Digital Sovereignty of the European Union" funded by the Czech Science Foundation (GAČR).

T. Gábriš
Faculty of Law, Palacký University Olomouc, Olomouc, Czech Republic

Institute of State and Law, Slovak Academy of Sciences, Bratislava, Slovak Republic

O. Hamuľák (✉)
Faculty of Law, Palacký University Olomouc, Olomouc, Czech Republic

Department of Law, School of Business and Governance, Tallinn University of Technology, Tallinn, Estonia
e-mail: ondrej.hamulak@upol.cz

D. Ramiro Troitiño et al. (eds.), *Digital Development of the European Union*,
https://doi.org/10.1007/978-3-031-27312-4_16

directed by the use of computer technologies and networks" (*La souveraineté numérique est la maîtrise de notre présent et de notre destin tels qu'ils se manifestent et s'orientent par l'usage des technologies et des réseaux informatiques.*). (Bellanger, 2011). This rather poetic and abstract definition thereby broadly encompasses both the issue of individual digital autonomy (digital self-determination) as well as digital sovereignty of the states (and/or the European Union). But this still leaves the details of the definition and its practical manifestations unclear. In order to focus on digital sovereignty in detail here, let us distinguish in the very beginning between the individual and societal (etatist) concept of digital sovereignty—to use the notion of digital sovereignty solely in the meaning of sovereignty of states or of the EU respectively.

However, even with respect to this "narrower" concept of digital sovereignty being understood as independence and self-determination of the states and supra-states such as the EU, it still remains unclear whether one speaks here of a philosophical, political, or rather a doctrinal legal notion of sovereignty. In general, it might be claimed—with good reasons—that sovereignty is rather a philosophical and political concept, which manifests itself in a variety of legal tools and instruments being used in order to attain, enforce, and preserve sovereignty. This mirrors the standpoints which consider sovereignty to be best understood as a process, not a *status quo* (Bendiek & Stürzer, 2022, p. 2), which holds true especially in terms of daily practise of asserting sovereignty in cyberspace, by both legal and non-legal means. These tools are being drafted, introduced, and enforced at multiple levels—not only international, European, national, but also private levels, whereby various stakeholders are in fact being delegated the task to ensure the constant control.

Focusing here mostly on the European Union level, where the notion of digital sovereignty was coined, the European Commission initiated a variety of regulations and directives with a special focus on digital sovereignty, affecting a number of aspects and using an array of tools and intermediaries to reach the goals. One should mention here foremost the 2016 EU Network and Information Security Directive (NIS Directive), to be soon replaced by the NIS 2.0 Directive, extending in the near future beyond the sectors covered by the original NIS Directive so as to include postal services, waste management, large-scale food distribution, as well as central and state administrations—all in the effort to involve more sectors and more delegated entities ensuring the digital sovereignty of the EU in its practical aspects of network and information technology security.

In this line, in 2019, the European Parliament adopted the EU Cybersecurity Act, establishing in addition to NIS Directive also a general cybersecurity framework for information and telecommunication products and services, being entrusted to the EU Agency for Cybersecurity (ENISA). Additionally, the Commission also issued a recommendation for ensuring the cybersecurity of 5G networks in March 2019 and presented a "toolbox" on secure 5G networks in January 2020, which was supposed to be implemented by the EU Member States in their national legislation, aiming at the security of infrastructure and services related to 5G Networks.

Furthermore, the 2021 proposal of EU Artificial Intelligence Act (AI Act) intends to introduce a framework for AI technology products applied in the EU.

Moreover, the Digital Services Act (DSA) and Digital Markets Act (DMA) were passed in 2022, regulating the digital market anew, handling the issues of user data, hate speech and illegal content, hoping to be able to regulate the "gatekeepers" active in the digital services and digital market, which are already entrusted with numerous specific tasks and duties, ensuring the digital sovereignty of the EU in this field.

Finally, the data sovereignty of European citizens is also to be protected by the recently adopted Data Governance Act and the proposal of the Data Act, which clarify the use of data by public bodies and regulate under which conditions private data can be commercialized. (cf. Ibid.)

These are but the most recent examples of tools, by which the digital sovereignty of the EU is being asserted and hoped to be ensured. Still, the proper definition of the notion and its actual methodology remains undefined and unclear. Therefore, what we intend to do here is to offer fresh insight into the problem of defining digital sovereignty of the EU Member States and/or EU respectively, as well as to provide a suggestion for methodological instruments fit for regulation of this sector.

The greatest challenge thereby is that of stretching and applying the traditional concept of sovereignty to the new reality. Traditionally, sovereignty namely used to be applied to tangible assets (infrastructure) and persons, whereby the traditional concept of sovereignty might have sufficed even with regard to the modern technology infrastructure—as a sort of "technological sovereignty." However, it is more difficult to reconcile this perception with the idea of sovereignty over borderless intangible, digital phenomena (Burwell & Propp, 2020).

When searching for a solution, in fact, it might be suggested that even the traditional sovereignty has rather than being a power over tangible assets and persons actually been a toolbox applicable to enforce the sovereignty by solving the issues of jurisdiction and conflict of laws predominantly.

Our main idea and suggestion, therefore, is to shift the focus from philosophical and static definitions of sovereignty rather to the dynamic and evolutive perception of sovereignty as a box of legal tools constantly evolving in order to efficiently regulate and control the ever-changing (digital) phenomena. This perception reflects better the idea of digital sovereignty as being "*the ability to have control over your own digital destiny—the data, hardware and software that you rely on and create.*" (Fleming, 2021). The digital phenomena hence do not change much about the traditional "technologies of sovereignty," expressed in terms of legal rules and legal tools, being recognized as fluid borders stretching even beyond the analogue boundaries of an individual country.

This approach might indeed mean that instead of shaping digital sovereignty as a "new" concept, it might suffice to stick to the traditional concept of sovereignty, even being manifested by the same sort of tools and rules, mostly of legal nature. What is new and different is just the pace and the complexity of the phenomena to be grasped and regulated by these tools.

The phenomena of technology, market rules and social norms related to cyberspace are thereby sometimes even being perceived as separate normative systems which are in conflict with legal regulation of the respective phenomena. However, socio-legal methodology and socio-legal theory of law consider them as being only a part of the so-called material sources of law, i.e., forces and vehicles behind the big picture of legal machinery. There is nothing new about this theory or methodology claiming the need to take into account the actual reality to draft an efficient regulation. What is new here—to repeat it again—is just the pace and complexity of the phenomena to be regulated.

The problem we are facing here is thereby the same for legal regulators as well as for software developers. Just like in software development, lawyers also need to cope with the fact of an agile environment—and this may be done by using the agile methods both in software development as well as in law and legislation. Agile theory of legislation might thus greatly contribute to the efficient exertion of sovereignty over the digital assets—becoming instead of "digital sovereignty" rather an "agile sovereignty"—dynamically and elastically reflecting the ever-changing playground of various non-legal forces and needs.

The methodology for coping with these phenomena thereby clearly requires an approach different from the traditional normative or doctrinal legal methodology and legislation drafting. It requires a new methodology, possibly based on the models of agile management and of socio-legal studies, leaning onto thorough empirical knowledge of the phenomena to be regulated, and subsequent evaluation of the efficiency of regulation.

2 History and Recent Shapes of the Concept of Sovereignty

The term sovereignty is inherently an interdisciplinary concept. It is a subject of interest for constitutional lawyers, international lawyers, international relations scholars and political scientists.

From a historical point of view, however, it should be noted that the European Middle Ages originally did not know neither the concept of state nor that of sovereignty. Instead of sovereignty, the Latin terms *summa (plena) potestas, maiestas*, or *imperium* were used (Maiolo, 2007, p. 23). The term state appears for the first time only at the end of the fifteenth and the beginning of the sixteenth century in the works of Girolamo Savonarola, Niccolò Macchiavelli and Justus Lipsius (Ibid., p. 22). However, already in the Early Modern Age, the concept of dualism of economic power and political/legal power was also recognized by authors like Francisco Suarez, pointing to the fact of various normative forces being active behind the legal systems.

It was finally the French thinker Jean Bodin that is considered to be the author of the concept of sovereignty as we know it today. In fact, however, his earlier predecessor Bartolus of Saxoferrato also worked with the concept of sovereignty, but without explicitly using the term. Still, however, the question of the concept of sovereignty in the Middle Ages remains controversial among contemporary authors (Ibid., pp. 24–26), since there is no agreement on whether Bodin is the author or the sovereignty as a political and legal concept existed already before his times.

Bodin himself in his work *Les Six Livres de la République* (1576) defined sovereignty as "absolute and eternal power over the republic" and claimed that no one had defined it before him, but at the same time he acknowledged that the concept was already known to the Romans as *maiestas*, the Greeks as *kyrion archon* and *kyrion politeuma*, to Renaissance Italians as *segnoria*, and to Jews as *tomech shebet*. However, even Bodin, despite the introduction of the concept of sovereignty, did not abandon the continued use of the terms *maiestas*, *imperium*, or *summa potestas* as traditional notions for sovereignty (Ibid., pp. 82–84).

The theorists of "social contract" additionally brought about the idea according to which sovereignty in Bodin's understanding originates in the people, and based on their will the exercise of power is transferred onto state bodies. Thomas Hobbes thus claimed that the social contract creates a community, a state, where each individual is placed at the disposal of the sovereign (cf. Besterman, 1973, p. 17).

Unlike Hobbes, Locke took a different position on this issue. He abandoned the terminology of the social contract and introduced the idea of a tacit consent instead (Ibid., pp. 30–31). At the same time, Locke even deliberately avoided the concept of sovereignty. Unlike Hobbes, he namely recognized limited government, that is, two sovereigns—the government and the people (Ibid., pp. 35–38).

Finally, for the sake of completeness, we must also mention Rousseau and his concept of sovereignty, in which the sovereign is the people and sovereignty is not separable from the people—it is not transferable to representatives of the people nor to the monarch or any other state authorities. In such a case, the ideal solution is to have smaller communities that decide on common matters by daily voting.

Based on these starting points, in the twentieth century, the idea of the sovereignty of people gradually developed in two extreme directions—either towards the sovereignty of the parliament (in England, where "parliament can do everything"), or towards the sovereignty of the leader (who mystically could sense what the people, or the nation needs). However, any excess in reserving sovereignty to only one sovereign is at risk of abuse of sovereignty. According to Ellian, when the sovereign absorbs all law, law ceases to exist (Ellian, 2009, p. 211). Therefore, in post-war Europe (after 1945), a system of rule of law and power sharing as a special system of checks and balances were introduced, to limit the sovereign exercise of power at the internal level.

Still, as early as in the nineteenth century, while Jellinek proclaimed that sovereignty of the state is the internal power to determine its legal order, Gierke, known for his collectivist understanding, argued that the state as a collective is analogous to a person and has a sovereignty in relation to other states (Kurz, 1970, p. xi). This traditional teaching of the nineteenth century about the internal and external sovereignty of the state power and at the same time the even older teaching of the Enlightenment era about the sovereign people were taken over by legal scholarship of the twentieth century and still persist there until these days.

Nevertheless, in connection with the integration taking place on the European continent since the late twentieth century, the attention of representatives of European Union law, international law, constitutional law, political science, or international relations started focusing again more closely on the issue of sovereignty. The category of sovereignty at the turn of millenniums namely acquired a qualitatively as well as quantitatively novel dimensions and expressions than were not present in this form ever since the basic contours of the original concept of sovereignty were formed.

Specifically, the original concept of absolute sovereignty became clearly inapplicable in the changed globalized conditions of the contemporary world. Most experts namely agree that the traditional concept of sovereignty has been subject to two tendencies, at least since the end of the Second World War, that are eroding the traditional understanding of sovereignty (Šedivý, 1999, p. 100). On the one hand, it is the abovementioned globalization process and internationalization, which penetrates the framework of power structures and "attacks" the concept of sovereignty from above. On the other hand, there are decentralization processes taking place that shift the center of gravity of authority from the center to individual regional entities, but also non-state entities and even private entities, which might be considered an "attack from below." These opinions even claim that sovereignty is to be abandoned, because it was born in a world where nation states were the main players, centers of power and objects of interest, which no longer applies in the globalized world and in the world of strong non-governmental players (Lindahl, 2006, p. 53).

3 Reassessing Traditional Legal Concepts

The term and concept of sovereignty is not a proper legal term on its own. It is traditionally being perceived rather from a philosophical point of view, or a legal-philosophical point of view, which, however, results in its ontological and epistemological indeterminacy. Hence, it may be perceived as a catch-all and a catch-none term. In contrast to abstract philosophical notions, legal scholarship prefers to work with well-defined notions specified either in legal texts or in doctrinal teachings, which is an approach contrasting with (legal) philosophy (Bódig, 2021). Sovereignty may thereby also be understood in purely legal terms—as being manifested through specific legal instruments (legal norms) or legal practices following certain regulatory aims. In this sense, sovereignty can hence be perceived either as a blurry political concept that in fact undergoes constant policy changes, or it can be perceived procedurally/functionally, as a legal toolbox to reach the regulatory policy aims.

Regardless of the abovementioned two approaches, it is clear nowadays that it is not possible to speak about absolute independence and separation from other power-sources inside and outside. Absolute internal and external sovereignty was in fact an ideal (but also a legitimizing tool) of an absolutist state (Mrva, 2017, p. 8) and subsequently of the normativist legal concept of a state, which equated the state with the law. However, sociologically, it was always recognized that outer and inner social and economic forces exert important influence on law, its creation, daily implementation and its application by the respective authorities. An absolute isolation from other power-centers, other normative systems and driving forces was not possible and is impossible yet more nowadays—especially due to closer interconnections, globalization, and the role of private players.

The concept of sovereignty has additionally been facing even more challenges since the end of the twentieth century, especially with regard to the "borderlessness" of emerging cyberspace. The concept of sovereignty therefore occupied a central position in discussions about the normative architecture of cyberspace (Schmitt & Vihul, 2017, pp. 213–218), with a desired state control of a "national" slice of cyberspace.

In the first place, these attempts took form of classical "analogue" sovereignty assertion over the physical cyber infrastructure (hardware and persons using the infrastructure). Gradually, the intangible assets were to be regulated too—via the regulations of jurisdiction and conflict of laws, but also by specific legal regulations of geo-blocking, blocking of cross-border online gambling, or the so-called blocking laws in procedural law (Troitiño et al., 2020) Subsequently, focus has shifted to the issues of cyber security and financial control of cyberspace, where we find ourselves at currently.

The cyber-paternalistic views of cyberspace thus operated with an effort to subordinate cyberspace to state sovereignty, being manifested in the large juridification of cyberspace by EU law as well as by national legal systems. However, despite the efforts to create "borders" in cyberspace, it still remains primarily a space without borders, since all technological as well as legislative solutions pursuing the goal of erecting absolute and impermeable e-borders in cyberspace are failing. The existing imperfect borders, especially legal ones, might only be perceived as fluid, "liquid" borders that do not represent "traditional" analogue state borders, but rather only functionally aim to ensure the exercise of national sovereignty in the cyberspace. Concept of borders, just like the concept of sovereignty, hence apparently witnesses a turn to a more functional paradigm as well.

Clearly, the understanding of legal concepts of sovereignty and borders evolve too—instead of a static understanding of the concepts, new dynamic and agile paradigms seems to be embraced, taking into account the functional aspects and the forces pushing the evolution forward. These forces were thereby traditionally called "material sources of law", being a trigger and a reason to enact new legal rules in order to react to the challenges brought about by the non-legal externalities. Material sources of law are thereby a very broad category—comprising all the social, moral, ideological, philosophical, cultural, technological, economic, and various other aspects of everyday reality which require legal regulation or which influence or challenge the existing legal regulation. In fact, one can either call these external phenomena by the umbrella term of "material sources of law," or give them their individual proper names, such as Lessig did when describing the regulatory system of cyberspace, allegedly consisting of four layers: "law," "social norms," "market," and the "code" (Lessig, 2006)—being in fact just another denotation for societal changes, economic development, and technological progress. Often these elements go hand in hand and lead together to a substantial social change, meaning that a large group of population performs activities and enters into relations that are different from activities and relations of their predecessors (Vago, 1981, p. 285).

An example *par excellence* of such a change may be the current technological revolution in its multiple aspects. The law, willy-nilly reflecting the changing society, has to cope with this evolution and react. Should it ignore the new phenomena, the public order as well as the authority of law might get endangered. The main issue here is, however, how quickly the law is able to grasp and cope with the new social phenomena. E.g., back in the days, Jeremy Bentham (1748–1832) argued for the need of a quick response by legislators to the changing situation. On the other hand, Friedrich Carl von Savigny (1779–1861) opposed it. He claimed that an accepted custom and legal practice are able to cope with the new situations satisfactorily, albeit more slowly (Vago, 1981, p. 285), which would mean leaving it up to the judicial practice and jurisprudence to cope with the new situations. Finally, however, the former approach—the one advocated for by Bentham—prevailed in continental Europe. Still, the ever-faster pace of technological development challenges the ability of legislation to fully and satisfactorily regulate the emerging issues. Instead, the role of judicial practice (or administrative practice) and jurisprudence rises in situations where there are "gaps" in the law. Still, legal experts are lacking insight and knowledge of the complex technological phenomena they are supposed to regulate. The empirical knowledge of modern technologies in all their aspects, including the predictions of their evolution, is a necessary precondition to successfully master the regulatory task in this area.

However, the need to examine the complex societal factors or forces and their interconnection with law arises not only with regard to modern technology, but also concerning other driving forces. Probably that is why the new schools of legal thought are constantly emerging—such as Law & Economics, Behavioral Law & Economics, or the older ones are being reinforced again—such as various Socio-Legal schools, having their roots in the early 1900s. It is namely these theories and their methodologies that help us to better grasp the reality of law in a complex society. What may be recommended in this context is therefore to try to design a specific methodology suited for regulating the empirically very thick and agile environment of modern technologies—calling for a sort of techno-legal methodology.

4 Techno-Legal Methodology for Agile Law

Modern technologies represent the area where the quickest and greatest transformations can be witnessed. The tension between legal and non-legal normative systems and forces is mostly visible in this area as well. This naturally increases demands on traditional legal concepts, such as the abovementioned concepts of sovereignty or state borders.

Law, especially in situations where it faces agile development and requires agile approaches in theory and practice, apparently needs a new "agile methodology," designed to fully meet the needs of efficient legal regulation of the new phenomena. This novel methodology could possibly be built on the combination of empirical socio-legal methodology and agile methodology of software development. To introduce the two approaches, let us start here with the socio-legal scholarship first (see, e.g., McConnachie, 2020) which was challenging the traditional doctrinal legal scholarship already since the turn of the nineteenth and twentieth centuries. It has always sought to present a more complex understanding of "*how legal rules, doctrines, legal decisions, institutionalised cultural and legal practices work together to create the reality of law in action*" (O'Donovan, 2016). This is thereby in clear contrast with the doctrinal research which is obviously shaped by an "authority paradigm," i.e., a focus on authoritative sources and the normative function of rules, while identifying their prescriptive content rather than their actual effect, and disregarding the background forces and actual ability to regulate the object of regulation (Ibid.). Doctrinal approach is hence based on the belief that law may be projected as an autonomous instrument of social intervention, independent from other forces and normative systems (Ibid.).

The doctrinal (dogmatic) approach thereby proved to be a workable concept in the twentieth century, in a stable environment of a pool of traditional problems, with pre-solved and pre-fixed answers to prototypical situations. The focus of such legal scholarship is then naturally on hermeneutics and legal argumentation in the process of application of prototypical law onto prototypical facts. However, this approach proves not workable in a dynamic context, in complex environment, with numerous unsolved new problems (new, non-prototypical situtations) that are constantly emerging in practice, creating thus an unstable, ever-changing facts-base onto which the law should be applied or for which new legal rules should be designed. Empirical knowledge and specific expert knowledge are necessary in such situations in order to grasp the facts themselves, before applying the law or before suggesting new legal tools to tackle the new situations. Unfortunately, in most cases, it is not possible to combine the detailed technology knowledge with expert legal knowledge in one person. Therefore, close cooperation between technology experts (and/or other stakeholders) and legal experts is necessary in the process of drafting, application, and evaluation of outcomes, using evidence-based approach to assess practical effects of the legal tools introduced.

This approach is in fact well-known and applied already for decades under the headline of so-called agile methods in software development management. Agile methods started to evolve already in the early 1980s and their main principles were summarized in the Agile Manifesto of 2001. From the field of agile software development, this methodology was then transferred even to legal practice of agile software contracting (Nuottila et al., 2015), where there is an increasing demand for proactive coordination and flexible adaptation to changes.

Still, further promising extension into legal scholarship and legal practice, going beyond contracting to another type of rule-setting, namely to legislation, is yet to be developed and tested. Building on the experience with regard to agile contracting, legislation is namely also "*... designed by lawyers to protect different parties against risks and to get commitments from others; commitments that the law will* *enforce at the court in the worst case scenario.*" (Ibid.). Similar to agile software development, where the aspired end result cannot be precisely defined beforehand, software contracts and potentially also the legislation require nowadays increased flexibility. This means stronger demands for proactive coordination and flexible adaptation to changes, particularly in the context of a project characterized by complexity and uncertainty (Ibid.).

Just as in contract drafting, in the world of legislation too, it was often assumed that the project or product requirements were all known in the very beginning of a project. This, however, proved wrong or at least inefficient both in terms of software development as well as in terms of software contract drafting (Troitiño, 2022). Because of these challenges that the traditional model was facing, a newer, spiral model of management was developed. The spiral model started to change software development into an iterative and more responsive direction, in that it divided a project into smaller cycles. Each cycle could have been assessed separately upon its completion, which made the system more responsive and interactive, reflecting the actual needs of the customer (Ibid.).

Finally, the latest cry in this evolution is the proper agile methodology. The Manifesto for Agile Software Development from 2001 is based on four main principles of agile methodology, which may be reformulated for the needs of a lawyer as follows:

- Dynamic interactions should prevail over static processes and tools
- Working outcome prevails over the complex wording
- Collaboration prevails over formal top-down regulation, and
- Responding to changes over following a plan is a natural part of the plan

Speaking in software development terms again: "*Rather than starting out by defining the perfect requirements document,*" it is recommended "*developing release after release as quickly as possible, testing them on users, and modifying them to truly meet user needs.*" (Hassett & Burke, 2020). This means that in practice, software technology firms:

- Develop programs as quickly as possible
- Try them out on users
- Make changes based on feedback
- Then try them out again. (Ibid.)

This new approach thus promotes a more open, dynamic, and flexible approach, being an idea very much welcome also in the field of contract law and possibly even the field of legislation drafting. This novel approach hence consists foremost in the idea of an increased cooperation with stakeholders, as well as in the idea of encouraging changes rather than discouraging them, and by focusing on responding to change rather than on following a plan.

Possibly, this experience could similarly apply to other formats of project management—including that of legislation drafting. In fact, already in 2018, this was suggested by M. Lokin in her PhD. dissertation drawing heavily on the agile software development methods (Lokin, 2018). Lokin thereby embraced the idea that just like mainstream contract drafting moved to a rather proactive drafting, this could be successfully applied also in the field of legislation drafting. In both sectors, the agile methods namely promote the idea of "*continual refinement of the product and project practices*" (Nuottila et al., 2015), which is basically already the way how legislation in general works even nowadays, having abandoned the traditional stability of legal norms and having accepted the fact of "legislative avalanches" present in all EU Member States. However, there is more to agile than just the constant and fast legislation changes: "*in agile, policy or product designers need the input of users, citizens, or customers because their own experiences are not as useful or representative*" (Mergel et al., 2020, pp. 6–7). Additionally, "*agile assumes situations are fluid and change over time. As new information, constraints, or opportunities emerge, agile drives practitioners to revise and update early working versions to improve processes or services.*" Agile thereby even assumes failures—believing that agencies that experience failure in first iterations are better equipped to improve. This helps in the end to ensure that products and services are duly usable and fit for the purpose. Agile thus emphasizes bottom-up change over the top-down direction—meaning it recommends taking a close (empirical) look at the actual situation and the actual players, who are fully involved in the process of testing and iterations—often even with some degree of responsibility within the partial project cycles. Thus, stakeholders are themselves becoming involved in the regulation, being managers of the actual exertion of the tools in practice.

5 Agile Sovereignty as a "Sovereignty Cascade"

Moving back to the idea of digital sovereignty, it was already stated that some recent views question the concept of sovereignty as such (especially the sovereignty of states), arguing that sovereignty is to be abandoned as a historical concept, because it was fit only for the world of the Westphalian system (created after 1648), where states were the major players. In contrast, in the globalized world of today, more and more power and competences are gained by intergovernmental and non-governmental organizations or even private entities.

Still, instead of abandoning the concept of sovereignty completely, it might be wiser to "only" reconsider the concept as suggested *supra*, i.e., in more "functional" terms. Sovereignty thus could be perceived again "only" as a "supreme power" (*summa potestas*), i.e., a "power to exert control," albeit so far, instead of controlling the cyberspace properly, the regulatory authorities have apparently chosen a milder "intermediate" regulatory approach in which instead of control of cyberspace, they are in control of entities providing services perceived as "essential" or "critical" or acting as "gatekeepers," i.e., access points to the digital content and services. These are namely the entities that can be effectively "tamed" by sanctions imposed by the sovereign. This solution thus represents a relatively simple way how to control the provision of crucial services and products within the scope of competences of a traditional sovereign, while at the same time giving up on the hopes to control each and every move in the cyberspace (which would be a totalitarian approach not acceptable even in the analogue world).

The EU and any other sovereign thus currently effectively asserts its "sovereignty" in cyberspace only with respect to specific assets and persons (entities) in its direct scope of control (jurisdiction). This is in line with the notion of a "layered sovereignty," coined by Przemysław Roguski—meaning a distinction between the physical layer of cyberspace, the logical and social layers (Roguski, 2019). Roguski argues that while the physical layer is covered by state sovereignty by virtue of the principle of territoriality, the logical and social layers of cyberspace may be open to the exercise of state authority based on a criterion of proximity, i.e., whenever the state can establish a genuine link with the digital objects or online personae over which authority is to be asserted.

In other words, this means that sovereigns are controlling major cyberspace entities that fall in their competence in the analogue world. In fact, these persons are thereby being involved in the agile exercise of sovereignty, entrusted with certain patrolling tasks, such as those of data processors in case of data protection, or the operators of essential services in case of legal regulation of cybersecurity. Tropina and Callan in this respect speak of co-regulation (Tropina & Callanan, 2015), which could be situated on a continuum between the state-controlled regulation and self-governance. The sovereign is thus relying on cooperation with often private entities, which are being delegated the task of day-to-day management of EU (or State) sovereignty in the digital world. Their cooperation (or coordination) takes place under the supreme control of the traditional sovereign to whom they are responsible in the analogue world. The state as a traditional sovereign thus still has the upper hand, albeit it acts more as a coordinator rather than a co-regulator.

This is clearly the point where the traditional idea of sovereignty possibly turns into a new concept of sovereignty—shared between the state (or EU) authorities and non-state authorities acting as intermediaries of sovereignty, in a cascade going top-down from the EU to the Member States and to the private entities identified as operators or gatekeepers. What seems to be a new system of a "sovereignty cascade" is in fact to be perceived as a flexible agile management methodology, which allows for fixing the deficiencies at various levels of regulation—all the way from the top (EU legislation) to lower levels of national legislation and down to the actual exercise by the respective intermediaries in their constantly repeated iterations of law being applied in practice.

6 Conclusions

Pluralist contemporary streams of legal thought generally share a common trait—they describe and eventually criticize law while using non-legal vocabulary and non-legal approaches, offering a primarily external view on law. Although this approach is much welcome after a long period of internal doctrinal closeting of the legal scholarship, at the same time it is necessary to draw attention to the limitations and shortcomings of this approach, which gets too far-fetched from the actual functioning of law and from its normative use in everyday practice. External approaches to law and jurisprudence without transforming their conclusions into the creation of "new dogmatics," or "new legal science" are often perceived as useless. It is hence up to legal scholars to translate their findings and conclusions into legal language and to transpose them back into jurisprudence and law as such—in search of a new form of "neodogmatic jurisprudence." This could thereby be successfully attempted with regard to external viewpoints and external driving forces behind law traditionally being an object of socio-legal studies, researching law "in context." However, descriptive approach of socio-legal studies was predominantly used "only" as a tool of evaluation of law and of its functioning in practice. Law was and still is primarily drafted and applied by doctrinally trained lawyers who search for prototypical solutions to prototypical situations. However, with the ever-faster pace of evolution of external driving forces from among which the modern technology is the most visible, doctrinal legal scholarship is not in the position to offer prototypical solutions to prototypical situations any more—due to the complexity and "blackbox-nature" of technological problems. Traditional legal tools applicable to traditional situations thus fail to be straightforwardly applicable onto the agile environment of modern technologies. In this context, law, including legal tools through which the sovereignty of states is manifested, need reassessment based on the actual empirical research of the environment to be regulated as well as based on a proper evaluation of law functioning in an agile environment. New methodological approaches are hence necessary both in drafting the law as well as in its application and everyday use in practice—possibly drawing from both empirical socio-legal studies and from the agile software development, forming a sort of a new "techno-legal methodology" of agile and functional law. This new methodological toolbox should thereby involve tools for closer cooperation between lawyers and affected stakeholders, in all the phases—in drafting the law, in its application by state authorities, but also in the daily management often left over to the stakeholders involved in the exercise of the regulatory power cascade.

References

Bellanger, P. (2011). *Qu'est ce que la souveraineté à l'âge du numérique? Qu'avons-nous abandonné? Que peut-on reconquérir?* http://archives.lesechos.fr/archives/cercle/2011/08/30/cercle_37239.htm

Bendiek, A., & Stürzer, I. (2022). *Advancing European internal and external digital sovereignty: The Brussels effect and the EU-US Trade and Technology Council.* https://www.swp-berlin.org/publications/products/comments/2022C20_EuropeanDigitalSovereignty.pdf

Besterman, Th. (1973). The influence of Hobbes and Locke in the shaping of the concept of sovereignty in eighteenth century France. In *Studies on Voltaire and the eighteenth century. Vol. CI.* The Voltaire Foundation.

Bódig, M. (2021). *Legal doctrinal scholarship: Legal theory and the inner workings of a doctrinal discipline.* Elgar.

Burwell, F. G., & Propp, K. (2020). *The European Union and the search for digital sovereignty: Building "Fortress Europe" or preparing for a new world?* https://www.atlanticcouncil.org/wp-content/uploads/2020/06/The-European-Union-and-the-Search-for-Digital-Sovereignty-Building-Fortress-Europe-or-Preparing-for-a-New-World.pdf

Ellian, A. (2009). De staat van uitzondering: Soevereiniteit in moderne tijd tussen norm en beslissing. In *Soevereiniteit en recht: Rechtsfilosofische beschouwingen*. Boom Juridische uitgevers.

Fleming, S. (2021). *What is digital sovereignty and why is Europe so interested in it?* https://www.weforum.org/agenda/2021/03/europe-digital-sovereignty/

Hassett, J., & Burke, E. (2020). *Why the agile approach is so important to law firms*. http://nextgenlpm.com/wp-content/uploads/2020/07/Agile-LPM-from-Of-Counsel.pdf

Kurz, H. (1970). *Volkssouveränität und Staatssouveränität*. Wissenschaftliche Buchgesellschaft.

Lessig, L. (2006). *Code V.2*. Basic Books.

Lindahl, H. (2006). Sovereignty and symbolization. In *Relocating sovereignty*. Ashgate.

Lokin, M. H. A. F. (2018). *Wendbaar wetgeven: de wetgever als systeembeheerder* (Agile law making: The legislator as system administrator). Boom Juridisch.

Maiolo, F. (2007). *Medieval sovereignty: Marsilius of Padua and Bartolus of Saxoferrato*. Eburon.

McConnachie, K. (2020). *Routledge handbook of socio-legal theory and methods*. Routledge.

Mergel, I., Whitford, A., & Ganapati, S. (2020). *Agile: A new way of governing*. Universität Konstanz OPAS-Plattform Serie. https://www.researchgate.net/publication/340684906_Agile_A_new_way_Governing

Mrva, M. (2017). *Suverenita štátu a Európska únia* (Sovereignty of state and the European Union). Heuréka.

Nuottila, J., Kujala, J., & Nystén-Haarala, S. (2015). *Flexible contracting in software project business: What we can learn from agile methods in the software industry*. http://www.diva-portal.org/smash/get/diva2:1050437/FULLTEXT01.pdf

O'Donovan, D. (2016). *Socio-legal methodology: Conceptual underpinnings, justifications and practical pitfalls*. https://www.researchgate.net/publication/313640633_Socio-Legal_Methodology_Conceptual_Underpinnings_Justifications_and_Practical_Pitfalls

Roguski, P. (2019). Layered sovereignty: Adjusting traditional notions of sovereignty to a digital environment. In *11th International Conference on Cyber Conflict: Silent Battle*. NATO CCD COE Publications.

Schmitt, M. N., & Vihul, L. (2017). Sovereignty in cyberspace: Lex Lata Vel Non? *American Journal of International Law, 111*, 213–218. https://www.cambridge.org/core/journals/american-journal-of-international-law/article/sovereignty-in-cyberspace-lex-lata-vel-non/6C6FD06E2B02B72224DF7127483A33F0

Šedivý, J. (1999). Suverenite státní a nadstátní. Má evropský občan budoucnost? In *Demokracie a ústavnost*. Karolinum.

Troitiño, D. R. (2022). The European Union facing the 21st century: The digital revolution. *TalTech Journal of European Studies, 12*(1), 60–78.

Troitiño, D. R., Kerikmäe, T., De la Guardia, R. M., & Sánchez, G. Á. P. (Eds.). (2020). *The EU in the 21st century: Challenges and opportunities for the European integration process*. Springer.

Tropina, T., & Callanan, C. (2015). *Self- and co-regulation in cybercrime, cybersecurity and national security*. Springer.

Vago, S. (1981). *Law and society* (5th ed.). Prentice Hall.

Tomáš Gábriš Senior researcher at the Faculty of Law, Palacký University Olomouc (Czech Republic) and researcher at the Slovak Academy of Sciences, Institute of State and Law, Slovak Republic.

Ondrej Hamuľák Senior Lecturer at the Faculty of Law, Palacký University Olomouc (Czech Republic) and Adjunct Professor in EU Strategic Legal Affairs, Tallinn University of Technology, School of Business and Governance, Department of Law (Estonia).

EU Soft Power: Digital Law

Lilla Nóra Kiss

Abstract The world is facing a technological revolution where the internet is transforming our way of life in every dimension. The European Union (EU)—as the European Commission (EC) has identified its six priorities—aims to take the leading global role in several policy areas to "build a stronger Union in the world." The spread of the internet revolutionizes several sectors and industries and affects innovation, competitiveness, efficiency, security, and—among others—sustainability. The continuous innovation and the ubiquitous digital single market (DSM) open new possibilities for the European Union to become "fit for the Digital Age." The Tech revolution requires smart legislation which is able to foster innovation. Designing a good legal architecture is an opportunity for the European Union, on the one hand, to harmonize the 27 Member States (MSs) national digital markets into a single one and, on the other hand, to use its market power to influence digital laws all over the world. Extraterritoriality is a key feature of European Union law that is based on the European market power. Providing access to the world's largest single market for third-state actors is an advantage for which multinational corporations tend to comply with European standards and apply them outside the EU as well. Therefore, designing the digital laws is a power for the European Union enabling to export continental solutions outside the borders of the European integration. Exporting standards via corporate compliance is a soft power for the European Union to influence third countries' legal systems without signing international agreements with those countries to do so. This gives opportunities for the EU to become a world leader in digital legislation, but creates new challenges for it as well.

L. N. Kiss (✉)
Antonin Scalia Law School, George Mason University, Arlington, VA, USA

Freedom and Identity in Central Europe (FICE) working group, Washington, DC, USA
e-mail: lkiss@gmu.edu

D. Ramiro Troitiño et al. (eds.), *Digital Development of the European Union*,
https://doi.org/10.1007/978-3-031-27312-4_17

1 Introduction: The Development of EU Digital Law

The information society and then the network society has significantly changed our way of life in every dimension and enlarged the role of the states in Europe. The information and technology-based solutions dominate the B2B (business to business), C2C (citizen to citizen), B2C (business to citizen) and relations, and therefore, a significant need for high-quality G2B and G2C (government to business and citizen) relationships appeared. In these relations, the role of the state is to enable innovation for the private sector on the one hand and to increase its own competitiveness on the other. The latter turned the state's role from being a "night watchman-state" (Nozick, 1974) or "minarchy" (which is a faux libertarian capitalist approach to minimum statehood) to an active "service provider state" (Budai, 2009).

The process has also affected the position of the European Union (EU) intending to take leadership as a global economic player. Therefore, from the early 1980s, the European Communities and then the European Union triggered programs to incent continent-wide development in information-communication technologies and competitiveness. In 1984, based on the proposal of the European Commission, the Council of the European Communities decided to set up a "European Strategic Program for Research and Development in Information Technology" (ESPRIT) (European Communities, 1984), then the Research in Advanced Communications for Europe (RACE) in 1985 (European Communities, 1985), and several research and technology programs in areas of high interest (e.g., DRIVE—for delivery on public roads, AIM—for health care, DELTA—for distance education). The benchmark step in the European level innovation was the acceptance of the so-called Bangemann-report ("Europe and the global information society") in 1994. The report urged European legislative steps to set the agenda for data protection, intellectual property protection, and network security to achieve a leading role in determining the rules for the information society. The report highlighted the role of competition among legislations willing to adapt to the changing circumstances and raised the awareness of European decision-makers that those legislations will dictate the innovation-technology agenda who can adjust their legislations to these technological developments. The direct consequence of the Bangemann-report was the European Commission's action plan titled "Europe's way to the Information Society. An Action Plan—Communication from the Commission to the Council and the European Parliament and to the Economic and Social Committee and the Committee of Regions" in 1994 (European Commission, 1994). The process continued with the e-Europe Action Plan in 2002 (European Commission, 2000) which intended to encourage developments in the public sector (foster e-commerce, e-health care, e-public administration, etc.), then with the Common List of Basic Public Services (CLOBPS) in 2002 which included a list of those public services that the Member States shall provide electronically for their citizens (e.g., income tax services, vehicle registration services, reporting to the police, etc.). The next in this process was the adoption of the eEurope 2005 Action Plan in 2002, which was special since it involved the 2004 "class" of acceding candidate states, including the post-socialist countries in the Central and Eastern European area. The interstate dialogue supported the mutual understanding of states regarding their institutional, social, and legal differences. As a result of these dialogues, the European Parliament and the Council adopted their decision on interoperability solutions for the replacement of the European public administrations (ISA) (European Parliament and Council, 2009). The decision was a product of the EU's i2010 eGovernment Action Plan. Under the i2010 Action Plan umbrella, the EU has set for a complex pack of solutions to determine the requirements for electronic services in the EU based upon the eEurope2002 and eEurope2005 but stepped further from them. The i2010 intended to introduce a consumer and competition-friendly policy.

One significant element of the process was the 2006/123/EC directive which set the rules for services on the internal market. This declares the tasks of the states as service providers, including simplifying the procedures, informing the citizens in a wider range, providing electronic services, etc. The process led to the "Digital Agenda for the EU" (European Commission, 2010) which was a priority of the Europe2020 strategy. The digital agenda addressed to apply of infocommunication-technology tools all over Europe by 2020. Therefore, the digital agenda aims to create sustainable economic and social benefits through a single digital market based on high-speed, super-fast internet and interoperable applications. The strategy intended to create a digital single market (DSM) modeled on the single market, which allows transnational services in the EU, among others. The Digital Single Market strategy was adopted in May 2015 by the Jean-Claud Juncker-led European Commission. This is a real benchmark in the process. It has already achieved some of its objectives, for example ending roaming charges (European Parliament and of the Council, 2012) in the EU, which happened already back in 2017 and banning unjustified geo-blocking (European Parliament and Council, 2018) in the internal market in 2018 (Hamuľák et al., 2021).

It includes innovative proposals for both the private and public sectors—but sets objectives for the state on how to provide its services. In the end, the DSM intends to reach also a European Administrative Area by harmonizing several elements of the Member States' public administration. The latter means, for example, achieving similar or uniform identification processes for users, and harmonized rules for e-signature and authentication. The DSM is significant because it includes general and specific objectives for multiple sectors and actors affected by digitalization. For example, it generally aimed to remove virtual borders, boost digital connectivity, and make it easier for consumers to access cross-border online content and services. In addition, the DSM sought to fit the EU's single market for the digital age by moving from 28 (since the Brexit, 27) national digital markets to a single one and then opening up digital services to all citizens and strengthening business competitiveness in the digital economy (Laitinen, et al., 2020). The current European Commission (2019–2024) led by Ursula von der Leyen is highly committed to continuing the massive digitalization process, which leads to a 'Europe fit for the digital age. The objectives can be translated into 20 new legislative and non-legislative initiatives. These Ursula von der Leyen's Commission-driven initiatives are added to the 24 ongoing files from the Juncker administration (2014–2019) (European Parliament, 2022).

The two significant specific strategies adopted under the DSM by the current Commission are the Digital Services Act (DSA) (European Parliament and Council, 2020) and the Digital Markets Act (DMA) (European Parliament and the Council, 2020). The Digital Services Act proposes to upgrade liability and safety rules for digital platforms, services, and goods and take further steps towards completing the digital single market. The Digital Markets Act addresses the negative consequences arising from certain behaviors by platforms acting as digital "gatekeepers" to the single market. Regarding the Digital Markets Act, the institutions have already achieved a "political agreement on rules to ensure fair and open digital markets" in March 2022. The institutions reached a political agreement regarding the Digital Services Act in April 2022. The proposals are ambitious, but if the EU intends to lead the digital legislation all over the globe, it has to dictate its standards and use its market power to influence European jurisdictions (Troitiño et al., 2020). By 2030, the EU has ambitious goals to achieve in the digital sphere. These are the digital targets of the EU aiming to strengthen its digital sovereignty.

Designing an excellent legal architecture is an opportunity for the European Union to increase its competitiveness via harmonizing the 27 Member States (MSs) national digital markets into one. At the same time, harmonization is based on the EU's competences in regulating the single market. Practically saying, the EU can use its market power to influence digital law all over the world. Access to the world's largest single market for third-state actors is an advantage for which these corporations are meeting European standards. Furthermore, since extraterritoriality is a feature of EU law, harmonizing and, in some cases, unifying rules for the single market is influencing third parties, such as the USA or China. Therefore, designing the digital law is a soft power for the European Union which enables to export of continental solutions to other parts of the globe. The next chapter analyses the opportunities and challenges for the EU to become a world leader in digital legislation.

2 Digitalization and European law: Opportunities and Challenges

2.1 Digitalization as an Opportunity

International organizations and supranational alliances usually draw ambitious goals requiring close cooperation of its members, transfer of certain state powers (competences), and harmonizing some policy areas. The European Union is *sui generis* in the line of international organizations for multiple reasons, one of which is that its powers are limited by the Treaties. Still, it uses its soft power very effectively in order to bring closer cooperation among its Member States in multiple policy areas. Challenges (for example, the competition with the USA, Japan, and China) and crises (such as the economic crisis, the COVID-19 pandemic, and the energy crisis) create opportunities for the EU to co-apply its hard and soft powers and achieve cooperation even in areas where otherwise it would not have the competence to do so (e.g., direct taxation).

In the introductory chapter, I presented the main pillars of the development of the EU's soft power in digitalization, but did not highlight the political and social environment of each proposal (Hamulák, 2018). However, if we interpret several consumer protection-related legislative proposals, we can understand that the 2008 economic crisis significantly thematized the detailed rules of e-commerce and contracts. While, during the COVID-19 pandemic, the digital content-related concerns got higher focus to protect minors in the digital age and support reaching quality information while our personal data is in security. Third, Russia's current invasion in Ukraine highlighted the relevance of cybersecurity. The digitalization is a special policy area since it reaches all other areas of law. Due to the nature of the internet, it is impossible to just "regulate a bit" - those questions where the EU has powers - and avoid regulation, where the EU lacks those powers. This would lead to discrimination and loopholes which would not serve European or national interests.

Digitalization is a soft power. Soft, because the EU does not literally need to have competence to propose legislation (e.g., in the field of direct taxation). Soft, because the EU may take relevant actions within the framework of sectoral and horizontal policies, such as: industrial policy (Article 173 of the Treaty on the Functioning of the European Union (TFEU)); competition policy (Articles 101, 109 TFEU); trade policy (Articles 206, 207 TFEU); the trans-European networks (TENs) (Articles 170, 172 TFEU); research and technological development and space (Articles 179 and 190 TFEU); energy policy (Article 194 TFEU); the approximation of laws for improving the establishment and the functioning of the internal market (Article 114 TFEU); the free movement of goods (Article 26 and Articles 28–37 TFEU); the free movement of people, services and capital (Articles 45 and 66 TFEU); education, vocational training, youth and sport (Articles 165 and 166 TFEU); and culture (Article 167 TFEU).

The EU can use its market power and legislative competence in other areas to define certain concepts. For example, determining who is a minor in the digital sphere is not always obvious. If we consider that in the physical space (or in different legal frameworks) minor or infant can be someone under 18, 16, 14, 12, we can see that defining concepts can be very confusing. Another example is handling disinformation and designing a legal framework to fight against it while providing the free speech in the online sphere and avoiding unjust censorship—all at the same time. However, seemingly, the EU manages these challenges well and turns them into opportunities to evolve its soft power and influence the Member States' and the third parties' legislations, too.

Digitalization is an opportunity to increase competitiveness and to strengthen its digital sovereignty. The former concept is quite understandable, it refers to the availability of alternatives and thriving for better supplies, while consumers can benefit and the economy flourishes. However, the latter concept is less clear. Digital sovereignty is challenged by the digital transformation and the global technical infrastructure of the internet because the traditional principles of territoriality and state hierarchy appear opposed to the diffuse, flexible, forever-shifting constellations of global digital networks. (Pohle, Julia Berlin Social Science Center (WZB), Germany Thorsten Thiel, Research Group, 2020). Digital sovereignty is, therefore, a status where the states (or the EU) can defy keeping legal governance and control over their citizens' data (Hamulák et al., 2021) while they are fulfilling their roles as states but allowing their citizens and private corporations to enter into business relations. Digitalization is, therefore an opportunity to set rules to keep control over personal data (GDPR) (European Parliament and Council, 2016), over the capacity for innovation, and over the ability to shape and enforce legislation in the digital environment. (Várkonyi Gültekin & Sulyok, 2019) This also means that an effective and comprehensive legislative pack can design to enhance Europe's strategic autonomy in the digital field. This requires the Union to update and adapt a number of its current legal, regulatory and financial instruments, and to promote more actively European values and principles in areas such as data protection (Maksó, 2017), cybersecurity and ethically designed artificial intelligence (AI).

The EU grabbed this opportunity, and under the Digital Single Market umbrella, it has already translated its ambitions into 20 new legislative and non-legislative initiatives—which will extraterritorially impact the jurisdictions outside the EU when their private actors intend to step into the EU's digital single market. Ursula von der Leyen's Commission added to the 24 files which were ongoing from the Juncker administration (2014–2019). The files include legislative proposal for a Data Act, for digital levy, for a trusted and secure European e-ID, an initiative on improving the working conditions of platform workers, communication on updating the new industrial strategy for Europe, review of the competition policy regarding digital goods and services, a legislative proposal on leveling the playing field for foreign subsidies and action plan on synergies between civil, defense, and space industries.

The pivot pack of proposals is the Digital Services Act and the Digital Markets Act. The Commission published the Digital Services Act (DSA) proposal on 15 December 2020.

Under the Digital Services Act, binding EU-wide obligations would apply to all digital services connecting consumers to goods, services, or content, including new procedures for faster removal of illegal content and comprehensive protection for users' fundamental rights online. The new framework intends to rebalance the rights and responsibilities of users, intermediary platforms, and public authorities. It is based on European values (the respect for human rights, freedom, democracy, equality, and the rule of law). The proposal complements the European Democracy Action Plan (European Commission, 2020a) aiming to make democracies more resilient. (European Commission, 2020b) The Digital Services Act proposes to introduce a series of new, harmonized EU-wide obligations for digital services, such as: rules for the removal of illegal goods, services, or content online; safeguards for users whose content has been erroneously deleted by platforms; new obligations for very large platforms to take risk-based action to prevent abuse of their systems; wide-ranging transparency measures, including on online advertising and on the algorithms used to recommend content to users; new powers to scrutinize how platforms work, including by facilitating access by researchers to key platform data; new rules on traceability of business users in online market places, to help track down sellers of illegal goods or services; an innovative cooperation process among public authorities to ensure effective enforcement across the single market. (Complex Discovery, 2020) Therefore, the DSA provides a wide variety of rules in several policy areas which supports the EU's intention to harmonize its market rules further and increase the compliance level among them. Moreover, under the DSA, platforms that reach more than 10% of the EU's population (approx. 45 million users) are considered to be "systemic" in nature and thus are subject not only to specific obligations to control their own risks, but also to a new oversight structure.

The Digital Markets Act is a regulation that the European Commission proposed that the EU Council has just adopted in July 2022. The DMA intends to ensure a higher degree of competition in the European Digital Markets by preventing large companies from abusing their market power and by allowing new players to enter the market. Once it enters into force, it will establish a list of obligations for designated Gatekeepers and, in case of non-compliance, there will be enforced sanctions mechanisms, including fines of up to 10% of the worldwide turnover. It targets the largest digital platforms operating in the European Union. The extraterritorial impact of the DMA will be tangible once it enters into force, due to the fact that the largest platforms in Europe are not Europeans, but mostly US corporations. Therefore, DMA targets the Big Tech corporations (Google, Amazon, Meta, Apple, and Microsoft).

Therefore, the laws related digital content and activities provide a good opportunity for the EU as a global player to influence not just the markets, but also the legislation of third countries.

2.2 *Digitalization as a Challenge*

Digitalization is not just an opportunity but also a challenge for states and the EU. First, it challenges the EU's ability to harmonize the 27 national jurisdictions in the field of digitalization where almost every sector and industry is involved in and affected by the innovation. Secondly, digitalization tests the EU's ability to use market power to make its tech and internet standards extraterritorial (in third countries) applicable. Thirdly, digitalization is challenging as it tests if the EU can become a leader in the global sphere. Finally, the digital revolution is a challenge that helps decide if the EU can become a more robust and closer confederation of sovereign nation states than it is now. The question here is if the digital revolution and its legal and economic consequences can result in a more harmonized and, in some cases unified alliance of European states.

The first challenge is questioning the EU's ability to harmonize the national laws in the field of digitalization. Currently, the EU consists of 27 sovereign states, and the Treaties determine the share of powers between the EU and the Member States. Even if the proper functioning of the single market is a good reason to adopt

European laws, some policy areas are solely in the hands of the Member States, which can result in competency collisions. Therefore, the EU faces the challenge to bring its Member States' different interests and legislative traditions under its jurisdiction. A harmonized tech-policy all over the EU (Kerikmäe, 2014) could be a good approach in the long term, however, compliance in the short term is costly. The Member States may want to foster cooperation on those areas only where they see the immediate benefits for themselves.

Moreover, the EU has to consider all national and regional circumstances, the differences between the rules and the societies, too. For example, in some regions, the level of digital illiteracy is relatively high while in other places the rest of the citizens uses internet and infocommunication tools everyday. In addition to the digitally illiterate, catching up with the elderly is also difficult, as there is a significant generational gap between digital natives and those who are not. Therefore, the harmonization of the Member States' legislative material is not just a challenge from a legal point of view, but a political one, too. All Member States have different interests and finding a good compromise usually takes time. Time is a hard factor when it comes to tech issues, since there a realistic threat that the law will become outdated by the time the legislators finally accept it. Therefore, the rules should be long standing when it comes to harmonizing national jurisdictions.

Secondly, digitalization tests the EU's ability to use market power to make its tech and internet standards extraterritorial (in third countries) applicable. The EU is the largest single market in the globe with over 450 million consumers. The market power reaches a continental limit in the European Economic Area, which can influence third country actors to meet European standards if they want to enter the market. This is particularly true when it comes to digital platforms' operation. For example, when the EU adopted the GDPR, it extraterritorially influenced third countries, such as the USA. All actors who are settled in European jurisdictions had to adjust their data protection rules to the GDPR.

Thirdly, digitalization is a challenge as it tests if the EU can become a leader in the global sphere. If the second challenge is addressed properly and the EU's digital law is extraterritorially applied, the EU can reach its goals in becoming a global leader in tech. As tech is related to all sectors and industries, the harmonization of national laws in the EU may affect actors outside the EU and as a spillover, the EU would become a leader in more fields.

Finally, the digital revolution is a challenge that helps decide if the EU can become a stronger and closer confederation of sovereign nation states than it is now. The conference on the future of the European Union was just finished in spring 2022. Several recommendations and conclusions address digitalization, meaning that citizens are interested in this modernization process all over the EU. The big question in Europe is whether it can move towards a closer (more federation-like) union or the maximum level of cooperation has already been reached. The different levels of collaboration (two- and multispeed Europe) in the EU can work, but not in the dimension of digitalization. High internet standards and harmonized digital goods and services rules should be on the same page within the EU. Otherwise, it will not allow the proper function of the single market.

3 Concluding Thoughts

Digitalization, on the one hand, is the root of innovation, and the private sector can make it flourish if the different rules enable them. On the other hand, where there are too many and too strict rules, private actors are discouraged from entering the market or competing with each other. Although, without harmonized rules, specific digital economy sectors cannot operate appropriately since the differences among the national jurisdictions just fragment the market. Harmonized rules could encourage smaller and medium-sized corporations to step out from national markets and enter the single market as harmonization guarantees having similar standards as in the home country. On the other hand, the compliance is a big burden and it is particularly true when it comes to the coherence between the existing regulations (e.g., GDPR) and the newly adopted laws, besides harmonizing the national jurisdictions. This also means that EU digital law shall respect the principle of subsidiarity where it is applicable. That is why the EU's digital law shall consider the following.

A high emphasis should be placed on ensuring that all sectors and regions of the European Union benefit from the digital policy in question. This means that the regional and national differences should be considered when it comes to harmonizing the rules. Moreover, with regard to international aspects, more emphasis should be placed on cooperation with third countries. Europeans shall consider third country actors as several big corporations (including the Big Tech) are not Europeans. Their withdrawal from the single market would be a huge loss, therefore, digital laws shall consider their legitimate interests and find a balance between public interests (regulated sectors, high competitiveness, efficiency, sustainability, and accountability) and private interests (profit-maximizing, access to markets and compete). Furthermore, considering third states means avoiding antagonizing the potential external relations, e.g., in the transatlantic sense, and putting other regions on the map, including emerging regions.

To address the challenges of the different level of development in the Member States, the EU's digital law shall emphasize that four cardinal points stated in the Digital Agenda (the development of digital capacities and digital infrastructures, as well as the further digitization of businesses and public administration). It is important that the targets for each Member State are set with maximum regard to their national specificities, competences and current development level. The digital laws shall avoid further deepening the digital divide between Member States and among citizens. With regard to the objective of digital infrastructures, the principle of technology neutrality shall remain a top priority for the EU and the Member States along with the usage-focused regulation. In addition, EU institutions shall consider the compliance with and the coherence of the existing rules, otherwise the contradicting legislation may discourage competitors and the law of unintended consequences will apply.

Digital services, like their offline counterparts, should operate in a transparent, accountable manner, respecting fundamental rights and the rules of competition law. In addition to taking effective action against illegal content, the EU shall attach great importance to ensuring the maximum guarantee of the right to freedom of expression on online platforms. The platforms must be transparent and provide users with an effective redress procedure. In this regard it is of utmost importance that the DSA is without prejudice to the right of the recipients to initiate proceedings against the decision of an online platform before a court of the country where they are established, in accordance with the national law of that country.

The digital law is a very complex area affecting all sectors and industries globally. This challenges the role and position of states and international actors, such as the EU. Keeping up with the competition is a national interest, similarly to ensuring the security and the coherence of public and private services in the online sphere. Data protection, consumers' rights, human rights and the balance between legitimate interests shall be considered when it comes to applying digital laws. The EU has achieved a lot during the last decades; recently, a significant fasten up with digitalization is tangible. The digital law is a soft power to harmonize national legislations, bring the Member States closer to each other, and influence third countries via market standards. As a result, to support the EU's ambitions to become a global leader in tech revolution via its legislation and market.

Lis of References

Budai, B. B. (2009). *Az e-közigazgatás elmélete*. Akadémiai Kiadó.

Complex Discovery. (2020, December 15). *Complex discovery*. Source. https://complexdiscovery.com/the-digital-services-acttransformational-digital-regulation-from-the-european-commission/

European Commission. (1994). *Europe's way to the information society. An action plan—Communication from the Commission to the Council and the European Parliament and to the Economic and Social Commitee and the Committee of Regions'*. Brussels: COM_1994_0347_FIN.

European Commission. (2000). *eEurope 2002, An information society for all, action plan.*

European Commission. (2010). *Communication from the commission to the European Parliament, the European Council, the Council, the European Economic and Social Committee and the Committee of the Regions.*

European Commission. (2020a). *Communication on the European democracy action plan.*

European Commission. (2020b). *European Commission*. Source. https://ec.europa.eu/info/strategy/priorities-2019-2024/europefit-digital-age/digital-services-act-ensuring-safe-and-accountable-online-environment_en

European Communities. (1984, March 09). *ESPRIT: European programme for research and development in information technologies*. L 67–1984-03-09: Official Journal (OJ).

European Communities. (1985). *RACE: Research in advanced communications for Europe*. JOC_1985_249_R_0002_01.

European Parliament. (2022). https://www.europarl.europa.eu/legislative-train/. Date of downloading: 2022. 09, source: Legislative Train Schedule: https://www.europarl.europa.eu/legislative-train/

European Parliament and Council. (2009). *Decision on interoperability solutions for European public administrations (ISA).*

European Parliament and Council. (2016). Regulation (EU) 2016/679 (General Data Protection Regulation).

European Parliament and Council. (2018, February 03). *Regulation (EU) 2018/302 of the European Parliament and of the Council on addressing unjustified geo-blocking and other forms of discrimination based on customers' nationality, place of residence or place of establishment in the EU.* OJ L 60I.

European Parliament and Council. (2020). *Regulation on a Single Market For Digital Services (Digital Services Act) and amending Directive 2000/31/EC.*

European Parliament and of the Council. (2012). *Regulation (EU) No 531/2012 of the European Parliament and of the Council of 13 June 2012 on roaming on public mobile communications networks within the Union (recast) Text with EEA relevance.*

European Parliament and the Council. (2020). *Proposal for a regulation of the European Parliament and of the Council on contestable and fair markets in the digital sector (Digital Markets Act).*

Hamulák, O. (2018). La carta de los derechos fundamentales de la union europea y los derechos sociales. *Estudios constitucionales, 16*(1), 167–186.

Hamulák, O., Andraško, J., & Mesarčík, M. (2021). El desarrollo digital de la Unión Europea: aspectos de gobernanza de datos de la movilidad cooperativa, conectada y automatizada. *IDP. Revista de Internet, Derecho y Política, 2021, 34*, 1–16.

Hamuľák, O., Kiss, L. N., Gábriš, T., & Kocharyan, H. (2021). "This Content is not Available in your Country" A general summary on geo-blocking in and outside the European Union. *International and Comparative Law Review, 1*, 153–183. https://doi.org/10.2478/iclr-2021-0006

Kerikmäe, T. (2014). *Regulating eTechnologies in the European Union normative realities and trends*. Springer.

Laitinen, E., Troitiño, D. R., & Kerikmäe, T. (2020). European union and Great Britain: After brexit, who wins the break-up? In *The EU in the 21st century* (pp. 103–116). Springer.

Maksó, B. (2017). Concepts and rules in the general data protection regulation. In T. Kékesi (ed.), *MultiScience—XXXI. microCAD International Multidisciplinary Scientific Conference Miskolc* (old.: 1–7). Miskolc.

Nozick, R. (1974). *Anarchy, State and Utopia*. Basic Books.

Pohle, Julia Berlin Social Science Center (WZB), Germany Thorsten Thiel, Research Group. (2020, December 17). Democracy and digitalisation. *Internet Policy Review*. https://doi.org/10.14763/2020.4.1532. https://policyreview.info/concepts/digital-sovereignty.

Troitiño, D. R., Kerikmäe, T., De la Guardia, R. M., & Sánchez, G. Á. P. (Eds.). (2020). The EU in the 21st century: Challenges and opportunities for the European integration process. https://doi.org/10.1007/978-3-030-38399-2.

Várkonyi Gültekin, G., & Sulyok, M. (2019). *Life after the GDPR: Good data protection rules and prospects for the future*. Szegedi Tudományegyetem Állam- és Jogtudományi Kar.

Lilla Nóra Kiss, PhD Visiting scholar and adjunct faculty at Antonin Scalia Law School, George Mason University, co-founder of the Freedom and Identity in Central Europe (FICE) working group

Automated Vehicles and New Transportation Services: Exploring Selected Legal Issues

Jozef Andraško and Matúš Mesarčík

Abstract New forms of public transportation services are constantly introduced in the context of automated mobility. However, these services do not only raise comfort, foster safety, and reduce costs, they also present new legal challenges within regulatory landscape especially concerning the issue of remote driving from the point of requirements on a remote driver and processing of personal data.

The article starts with an explanation of the basic notions of automated mobility and the inclusion of automated means of transport for the public. Different levels of automation are explained. New transportation services are explored from the point of view of remote drivers. The Discussion contains several recommendations towards regulation of remoted driving.

1 Introduction

In March 2018, a woman pushing a bicycle laden with shopping bags was hit by a Volvo XC 90. The woman later died from her injuries in the hospital. This case would probably be one of the many fittings to unfortunate pedestrian death statistics. However, Volvo XC 90 was a test vehicle equipped with a development automated driving system provided by Uber. During the accident, the operator was present in the vehicle and reacted slowly to the approaching pedestrian. The investigation showed that the automated driving system classified the woman pushing a bicycle laden with shopping bags as an unknown object, a vehicle and a bicycle (The

The paper was prepared within the implementation of the project no. 20-27227S 'The Advent, Pitfalls and Limits of Digital Sovereignty of the European Union' funded by the Czech Science Foundation (GAČR).

J. Andraško (✉) · M. Mesarčík
Faculty of Law, Institute of IT & IP Law, Comenius University in Bratislava, Bratislava, Slovakia

Faculty of Law, Palacký University Olomouc, Olomouc, Czech Republic
e-mail: jozef.andrasko@flaw.uniba.sk; matus.mesarcik@flaw.uniba.sk

D. Ramiro Troitiño et al. (eds.), *Digital Development of the European Union*,
https://doi.org/10.1007/978-3-031-27312-4_18

National Transportation Safety Board, 2018). The operator of the vehicle sitting in the car responded slowly as she was not paying attention and watching a streaming video (Chesterman, 2021, p. 32).

The case shows risks associated with cars equipped with automated driving systems and considerations regarding the role of an operator ('driver'). Issues of liability and accountability of the operator and developer of the automated system naturally emerge and have been subject to considerable academic debate (see, e.g., Chesterman, 2021; Ebers, 2022). However, automated driving systems are not only used in personal vehicles, but also in new public and private mass transportation services. Also, the operator or driver of such vehicle might be located in another part of the world. These are subjects of our considerations in this article.

The case mentioned above in the sphere of public transport services shall not be considered as a science fiction scenario, as the European Union (EU) is heavily investing in the area of smart and digital mobility. In the words of Adina Vălean, the EU Commissioner for Transport: 'Digital technologies have the potential to revolutionize the way we move, making our mobility smarter, more efficient, and also greener' (The European Commission, 2022b). The benefits of implementing automated systems in new forms of transport include sustainability and reduction of carbon footprint, reducing energy consumption, greater efficiency, and safety on roads (The European Commission, 2022a). However, such transport systems are composed or directly dependent on several technologies including artificial intelligence (AI), excellent connectivity, and semiconductors (The European Commission, 2022b). The functioning of public transport with these technologies is also highly dependent on cooperative Intelligent Transport Systems (C-ITS). The EU has already adopted and presented a European Strategy on Cooperative Intelligent Transport Systems. Additionally, the European Commission adopted a delegated regulation on specifications for the provision of C-ITS although being objected by the Council and thus not entering into force in 2019. Furthermore, risks associated with cybersecurity, privacy, or data protection, accountability, and responsibility have been subject to considerable debate.

In this chapter, we are narrowing our research within the intelligent mobility in two manners. First, our focus is solely on new public road transportation services implemented with automated driving systems. Second, after an explanation of the fundamentals of automated mobility, we are emphasizing the issue of remote driving and a remote driver from the point of view of regulatory requirements and data protection issues. Our contribution lies in exploring the issue from the point of view of new transportation services, especially in the public domain, and the narrow focus on remote driving.

The chapter presents a systematic analysis of regulatory challenges within new forms of transport deployed with automated driving systems focusing on the issue of remote driving. To fulfil our aim, we conducted desk-based legal research, using various legal sources, including strategic documents on the level of the EU, legislation and interpretative guidance in the EU, and selected national legal systems together with research of literature. National legal systems were selected due to the advanced regulatory development on the topic. Additionally, we will illustrate regulatory challenges and our proposals on the fictional use-case scenario.

This chapter proceeds as follows: In Sect. 2 we provide the reader with the elaboration on new forms of automated transport. Levels of automation as foreseen by legislation are discussed. Moreover, the regulatory landscape of automated vehicles is briefly explained considering the legal order of the EU and selected national laws (Germany and the UK). In Sect. 3 we are narrowing down our research on remote driving and remote drivers. The fictional use case is presented and general observations are made on legislation concerning drivers in automated vehicles. With the use case in mind, we explore selected legal issues concerning remote drivers and data protection. The conclusions and recommendations are presented at the end of the chapter.

2 New Forms of Automated Transport

2.1 Automated Vehicles

In relation to a vehicle that performs some or all driving tasks (e.g. steering, acceleration, and braking) the terms such as autonomous vehicle, automated vehicle, self-driving vehicle, or robotic vehicle are used. Due to the existing legislation and the most frequently used terms in scientific articles regarding vehicles that perform some or all driving tasks, we will use two terms in this chapter in particular, automated vehicle and fully automated vehicle. Despite the fact that the above-mentioned terms are used in EU legislation, for completeness, we will also clarify the term autonomous vehicle.

In general, autonomous vehicles can be described as 'computer-controlled vehicles that drive themselves by relying on a number of data resources to access the driving environment and to control the operation of the vehicle' (Collingwood, 2017, pp. 32–45). However, not all vehicles equipped with technologies that can handle some or all of the driving tasks are considered autonomous vehicles.

Autonomous vehicles are able to make decisions independently of human intervention. Furthermore, autonomous vehicles rely on sensor data and artificial intelligence to interpret data, make decisions about vehicle operations, and adapt to changing conditions (Lim & Taeihagh, 2018, p. 3).

The necessity of using artificial intelligence in autonomous vehicles has also been emphasized by the author Ducuing, according to whom, in order to be able to delegate decisions in the field of vehicle operation to an autonomous vehicle, artificial intelligence must be used (Ducuing, 2019, p. 189).

The issue of autonomous vehicles and automated vehicles is not only the subject of research by academics but is also regulated by nonbinding legal acts like resolutions, instructions, opinions of various organizations (UNECE, 2019; Council of Europe, 2020; European Commission, 2020), as well as normative legal acts and normative conventions (Vienna Convention on Road Traffic, 1968).

EU legislation does not define the term autonomous vehicle, but rather the term automated vehicle and fully automated vehicle. The General Safety Regulation[1] (GSR) defines an automated vehicle as a 'motor vehicle designed and constructed to move autonomously for certain periods, without continuous driver supervision but with respect to which driver intervention is still expected or required' (GSR, Article 3 [21]). A fully automated vehicle means a 'motor vehicle that has been designed and constructed to move autonomously without any driver supervision' (GSR, Article 3 [22]). It is obvious from the above definitions that even in the case of a vehicle that moves autonomously without any driver supervision, the person of the driver will be required. However, the question of whether the driver must be in the vehicle or outside the vehicle is different.

Furthermore, fully automated vehicles and automated vehicles must comply with additional specific technical specifications set out in the Commission implementing act. The Commission implementing act will specify technical specifications that relate to (GSR, Article 11 [1]):

1. Systems to replace driver control of the vehicle, including signaling, steering, accelerating, and braking
2. Systems to provide the vehicle with real-time information on the state of the vehicle and the surrounding area
3. Driver availability monitoring systems
4. Event data recorders for automated vehicles
5. Harmonized format for the exchange of data, for instance, for multi-brand vehicle platooning
6. Systems to provide safety information to other road users.

However, those technical specifications relating to driver availability monitoring systems will not apply to fully automated vehicles (GSR, Article 11 [1]).

Automated vehicles and fully automated vehicles must be equipped with specific systems. The requirement that these vehicles move independently will be met mainly through systems replacing the control of the vehicle by the driver, including signaling, steering, acceleration, and braking (GSR, Article 11 [1]).

Due to the legal definitions used in the GSR, we will use the terms automated vehicle and fully automated vehicle for the purposes of this chapter.

[1]REGULATION (EU) 2019/2144 OF THE EUROPEAN PARLIAMENT AND OF THE COUNCIL of 27 November 2019 on type-approval requirements for motor vehicles and their trailers, and systems, components and separate technical units intended for such vehicles, as regards their general safety and the protection of vehicle occupants and vulnerable road users, amending Regulation (EU) 2018/858 of the European Parliament and of the Council and repealing Regulations (EC) No 78/2009, (EC) No 79/2009 and (EC) No 661/2009 of the European Parliament and of the Council and Commission Regulations (EC) No 631/2009, (EU) No 406/2010, (EU) No 672/2010, (EU) No 1003/2010, (EU) No 1005/2010, (EU) No 1008/2010, (EU) No 1009/2010, (EU) No 19/2011, (EU) No 109/2011, (EU) No 458/2011, (EU) No 65/2012, (EU) No 130/2012, (EU) No 347/2012, (EU) No 351/2012, (EU) No 1230/2012 and (EU) 2015/166.

In cases where automated vehicles are equipped with communications technology that enables data transfer with other vehicles, infrastructures, or other networks, the notion should rather be the connected and automated vehicle (CAV) (BSI, 2020, p. 3). Automated vehicles do not necessarily need to be connected, and connected vehicles do not require automation. However, connectivity will be a major enabler for driverless vehicles (European Commission, 2018).

CAVs rely on data created by their in-vehicle technologies, data sent from other vehicles or infrastructure, and traffic and infrastructure data sent from public authorities. Furthermore, they must function properly to observe their conditions and the surrounding environment. Different types of in-technologies are used in CAVs for these purposes. In general, there are three types (Skiho & Shrestha, 2020, p. 20):

1. Sensor technologies. Sensor technologies include radio detection and range (RADAR), light detection and range LiDAR), visible light communication (VLC), infrared systems, etc.
2. Vision technologies. Vision technologies include high-definition (HD) cameras, stereo vision system (SVS), etc.
3. Positioning technologies. Positioning technologies include the Global Positioning System (GPS), radar cruise control, and radar-based obstacle detection (RBOD)

The technologies mentioned above communicate with other internal components of vehicles, such as telematics and actuator, through different networks. These networks include Flexray, Ethernet networks, controller area network (CAN), etc. (Skiho & Shrestha, 2020, pp. 21–23).

2.2 Automation Levels

Automated vehicles are classified according to their level of automation. According to the SAE standard J3016_202104 Taxonomy and Definitions for Terms Related to Driving Automation Systems for On-Road Motor Vehicles (SAE standard), automated cars are divided into six increasing levels of automation. These levels are considered descriptive rather than normative, and technical rather than legal. In general, SAE standard levels primarily identify how the dynamic driving task is divided between a human and a system. At level 0 (no automation), it is performed entirely by a human driver, and at level 5 (full automation), entirely by an automated driving system (International Transport Forum and Corporate Partnership Board, 2015).

Depending on which systems are used for individual levels of automation, the SAE standard distinguishes between the driving automation system and the automated driving system (ADS). Driving automation system is defined as 'the hardware and software that are collectively capable of performing part or all of the DDT on a sustained basis; this term is used generically to describe any system capable of level 1 to 5 driving automation' (SAE standard, 3.6). Unlike this general term for any level

1 to 5 system, an automated driving system is a specific term for a level 3 to 5 system. An automated driving system is defined as 'the hardware and software that are collectively capable of performing the entire DDT on a sustained basis, regardless of whether it is limited to a specific operational design domain (ODD); this term is used specifically to describe a Level 3, 4, or 5 driving automation system' (SAE standard, 3.2).

To understand the concept of automated driving systems, it is important to explain the concept of dynamic driving tasks and the operational design domain. Dynamic driving tasks (DDT) are 'all the operational and tactical functions in real time required to operate a vehicle in on-road traffic, excluding strategic functions such as scheduling the trip and selecting destinations and waypoints, and including, without limitation, the following subtasks (SAE standard, 3.10):

1. Lateral vehicle motion control through steering (operational)
2. Longitudinal vehicle motion control via acceleration and deceleration (operational)
3. Monitoring the driving environment through object and event detection, recognition, classification, and response preparation
4. (operational and tactical)
5. Object and event response execution (operational and tactical)
6. Manoeuver planning (tactical)
7. Enhance conspicuity via lighting, sounding the horn, signaling, gesturing, etc. (tactical)

The domain of operational design means 'operating conditions under which a given driving automation system or a particular feature thereof is specifically designed to function, including, but not limited to, environmental restrictions, geographical and time of day, and/or the requisite presence or absence of certain characteristics of the road or traffic (SAE standard, 3.21).'

The SAE standard lists examples of defined operating conditions, e.g., using a vehicle with an automated driving system only on highways while observing the precisely determined maximum speed, only under good weather conditions and optimal road maintenance conditions (good lane markings, roads not under construction). Other examples are the operation of a vehicle with an automated driving system on a precisely geographically defined military base, during daylight at speeds not to exceed 25 mph. A final example is a commercial truck that uses an automated control system and is designed to pick up parts from a geo-fenced seaport and deliver them along a specific route to a distribution centre 30 mi away. The domain of operational design of the vehicle is limited to daytime operation within the specified seaport and specific roads that form the prescribed route between the seaport and the distribution centre (SAE standard, 3.21).

Level 0 (No Automation)

The human driver performs all the tasks associated with driving. Despite the fact that this level is called 'no automation', we could include the emergency braking system, blind spot warning, and lane departure warning (Frisoni et al., 2016, p. 20). These technologies are considered level 0 because they do not drive the vehicle, but provide warning or immediate action in specific situations.

Level 1 (Driver Assistance)
The driving automation system performs either longitudinal vehicle motion control (acceleration, deceleration) or lateral vehicle motion control (steering). The driver is obliged to monitor the driving environment (e.g. detecting, recognizing and classifying objects and events and preparing to react as necessary) and to perform an appropriate response to such objects and events (SAE standard, 3.19).

Other driving tasks are performed by the driver. An example of a driver assistance system is adaptive cruise control (ACC) or lane keeping assistance system (LKAS) (Frisoni et al., 2016, p. 21).

Level 2 (Partial Automation)
The driving automation system performs longitudinal vehicle motion control (acceleration, deceleration), as well as lateral vehicle motion control (steering). The driver is obliged to monitor the driving environment (e.g., detecting, recognizing, and classifying objects and events and preparing to react as necessary) and to perform an appropriate response to such objects and events (SAE standard, 3.19).

An example of a level 2 system is the highway driving assistant, which is used in KIA, Hyundai, or Genesis vehicles. The driver has his hands on the steering wheel, but when driving on the highway, this system actively controls, accelerates, and brakes the vehicle (Choksey & Wardlaw, 2021). We could also include remote control parking (RCP) at level 2 (UNECE, 2017a).

The systems used in level 1 and level 2 vehicles are usually referred to as Advanced Driver Assistance Systems (ADSA). These systems are activated by the driver and can be deactivated or manually overridden by the driver at any time. The driver still performs dynamic driving tasks and is constantly involved in dynamic steering tasks and therefore must constantly monitor the driving environment and intervene immediately if necessary (UNECE, 2017a).

Level 3 (Conditional Automation)
The vehicle monitors the driving environment through automated driving systems that can perform all driving tasks in certain domains of operational design. A human driver does not need to monitor dynamic driving tasks, but must be able to take control of the vehicle at any time and without prior warning (Frisoni et al., 2016, p. 21). The driver is in the position of the so-called fallback-ready user, who, in the event of a system error related to the performance of dynamic driving tasks or if the automated driving system asks him to do so, must perform the driving tasks in an appropriate manner or achieve a minimal risk condition (DDT fallback).

The minimal risk condition means 'a stable, stopped condition in which a user or an ADS can bring a vehicle after performing the DDT fallback to reduce the risk of a crash when a given trip cannot or should not be continued' (SAE standard, 3.16).

In practice, several standards have been adopted that regulate the issue of fallback-ready user or security operator (BSI, 2021; Centre for connected and autonomous vehicles, 2022).

A practical example of level 3 is an automated driving system that performs dynamic driving tasks during traffic jams on the highway. However, it cannot perform these tasks when it arrives at the crash scene and therefore requests the intervention of the driver (fallback-ready user) to intervene. The driver (fallback-ready user) reacts by taking over the performance of all dynamic driving tasks in order to manoeuver around the crash scene (SAE standard, 3.12).

The vehicle uses different systems or systems that can meet the criteria for artificial intelligence to make decisions based on changing driving situations around the vehicle (Choksey & Wardlaw, 2021).

The automation level 3 system is, for example, an automated lane keeping system (ALKS) that is activated by the driver and maintains the vehicle in its lane at a driving speed of 60 km/h or less by controlling the lateral and longitudinal movements of the vehicle for a longer period of time without the need for further driver intervention (UN Regulation No. 157, 2021). Mercedes-Benz was the first vehicle manufacturer in the world to receive an internationally valid system permit for the automated lane keeping system called DRIVE PILOT in December 2021. This system can be used on 13,191 km of motorway in Germany (Mercedes-Benz, 2021).

Level 4 (High Automation)

Under certain defined operating conditions, the vehicle can perform all driving tasks. A human driver can take control of the vehicle, especially when conditions change predefined use cases (e.g., roadworks, road diversions or when the vehicle driver wishes) (Frisoni et al., 2016, s. 21). A Level 4 automated vehicle requires no human interaction to operate the vehicle, as it is programmed to stop itself in the event of a system failure (Choksey & Wardlaw, 2021).

The automated driving system also performs DDT fallback. In practice, automated driving systems will perform tasks such as pulling onto the road shoulder, turning on hazard lamps, enabling the propulsion system, and summoning roadside assistance (SAE standard, 8.5).

Level 4 systems are currently used mainly in taxi and public transport services. In such cases, the vehicle has clearly defined geographic limits and other conditions within which it can drive. For example, in the case of level 4 taxi vehicles, the vehicles can only drive within the city but no longer outside it. In the case of automated level 4 vehicles that provide public shuttle transport, they can have a precisely defined route that they never leave (UNECE, 2017a).

We could divide automation level 4 into two categories. In the first case, it would be automation level 4 (with a driver), when only the system would perform driving tasks within the operational design domain, but outside of this domain, the driver could take control of the vehicle. As an example, a vehicle whose automation level 4 systems perform all driving tasks only on the highway, however, on roads of the I. and II. class, the driver has to perform the driving tasks. In the case where the driver does not take control of the vehicle, although the system has asked him to do so, these systems must always bring the vehicle to a safe place (UNECE, 2017b).

In the second case, it would be automation level 4 (without a driver), when all driving tasks are performed within operational design domain only by the system

Table 1 Automated vehicle and fully automated vehicle according to SAE standard

GSR	SAE standard
Automated vehicles	Automation level 0, 1, 2, 3, and 4 (with driver)
Fully automated vehicles	Automation level 4 (without driver) and 5

Source: Own arrangement

and the vehicle is operated only within this domain. An example can be a public transport bus that uses automation level 4 systems and drives only within specific streets of the given city. An example of a vehicle of this level is the driverless bus of the EasyMile company from France, which became the first European vehicle authorized to operate on public roads (Bateman, 2021).

Level 5 (Full Automation)

No human driver is required. Automated driving systems handle all aspects of the driving task without the human need to intervene in any scenario at all. The vehicle does not need any pedals or a steering wheel. Automated driving systems make independent decisions. The vehicle can manage situations where unpredictable events occur or the physical environment changes. A suitable example here would consist of a full end-to-end journey (Lim, 2018, pp. 22–34).

In the event that a system error related to the performance of dynamic driving tasks occurs, the vehicle is capable of automatically achieving a minimal risk condition when necessary (SAE standard, 3.25).

A level 5 automated vehicle does not have an operational design domain like level 4 automated vehicles. There is currently no vehicle on the market that achieves level 5 automation.

Level 3 and 4 automated vehicles are sometimes referred to as semi-autonomous vehicles and level 5 automated vehicles as autonomous vehicles (Committee on legal affairs and human rights, 2020).

In the context of terms automated vehicle and fully automated vehicle defined by GSR we could assign these vehicles to levels of automation according to SAE standard as follows (Table 1):

The reason why we could consider a vehicle of automation level 4 (without a driver) and 5 as a fully automated vehicle according to the GSR is that all the defining characters of the concept of a fully automated vehicle are met. Level 4 (without driver) and Level 5 vehicles are designed and constructed to be able to move independently without driver supervision.

Despite the fact that the GSR does not refer to the automation levels according to the SAE standard in the definitions of the terms automated vehicle and fully automated vehicle, in the text of this chapter we will use automation levels according to the SAE standard for practical illustration.

2.3 Regulatory Landscape

2.3.1 Germany

The Act amending the Road Traffic Act and the Compulsory Insurance Act (Autonomous Driving Act) that allows operation of automated vehicles of automation level 4 within precisely defined operating areas on public roads in Germany under specific operating scenarios is effective from July 28, 2021. The Autonomous Driving Act defines a motor vehicle with autonomous driving capabilities as a motor vehicle that can 'perform the driving task independently within the respective defined operating area without a driver intervening' (Road Traffic Act, Article 1e, [2]) and must meet certain technical requirements (Road Traffic Act, Article 1a). A defined operating area means 'public road space determined locally and spatially in which a motor vehicle with autonomous driving function can be operated' if the requirements are met according to the Road Traffic Act (Road Traffic Act, Article 1d). In other words, the defined operating area can be considered the operational design domain according to SAE standard.

The legislation allowing level 4 automated vehicles on public roads in Germany does not contain restrictions on the scope of application. Transportation services represent the area where level 4 autonomous vehicles are most likely to find application. These words confirm the latest trends in automated mobility, in particular driverless bus of company EasyMile from France, which became the first European vehicle authorized to operate on public roads, Waymo One vehicle that can be summoned via an app and will pick up passengers within the Phoenix East Valley area of Arizona or automated buses Toyota e-Palette within shuttle transport used during the 2021 Tokyo Olympics to transport athletes in the Olympic Village.

The Act Amending the Passenger Transport Act that creates the legislative framework for new mobility services such as digital intermediation of rides and car-pooling services was adopted on 5 March 2021. The amendment to the Passenger Transport Act creates new categories of transport: (i) bundled on demand-transportation services and (ii) regular on-demand transportation services. The so-called bundled on demand-transportation services include ride pooling services outside of the scope of public transport (GSK.DE, 2021). Regular on-demand transportation services include demand-driven pooling services of public transport (Ayad et al., 2021).

2.3.2 United Kingdom

The UK was included among the states that adopted legislation on automated vehicles by adopting the Act on Automated and Electric Vehicles from 2018. The legislation in question defines an automated vehicle as a motor vehicle designed or adapted to be capable, in at least some circumstances or situations, of safely driving themselves and may lawfully be used when driving themselves, in at least some circumstances or situations, on roads or other public places in Great Britain. This vehicle must be included in the list prepared and updated by the Secretary of State (Act on Automated and Electric Vehicles, Part 1, Section 1).

Great Britain also sees the potential of services that consist of the transportation of people by automated vehicles. Law Commission and Scottish Law Commission published a joint consultation paper, Automated Vehicles: Consultation Paper 2 on Passenger Services and Public Transport where a new service called Highly Automated Road Passenger Services (HARPS) is defined. Highly automated road passenger services refer to a 'service that uses highly automated vehicles to provide road trips to passengers without a human driver or user in charge. Some services may resemble taxi, private hire, or bus services; others may look and operate differently' (Consultation Paper 2, p. xi). The highly automated vehicle described in the consultation paper in question is level 4 according to the SAE standard.

3 Remote Driver

In this chapter, we will work with the following fictional use-case scenario: *The city of Hightown has implemented new forms of transportation services. Public transport is operated by a bus fleet equipped with automated driving systems. Buses use several pieces of advanced technologies including lidar systems, radars, self-driving sensors, forward facing camera, and camera monitoring interior of the bus with good view on the doors and entering passengers. The fleet is operated by the FleetOperators firm that employs 'remote drivers' ready to intervene during the operation of the automated driving system.*

3.1 Remote Driver Regulation

We consider different concepts and obligations of remote drivers within the theory of levels of autonomy in decision-making. This includes concepts of human-in-the-loop, human-over-the-loop, and human-out-of-the-loop. The concept of human out-of-the-loop reflects the situation where the process is fully automated without or with minimal human intervention. If a person makes positive decisions and the system offers suggestions and recommendations, this should be referred to as human-in-the-loop. In the human-over-the-loop a human oversees and makes interventions when necessary (Chesterman, 2021, p. 53; Merat et al., 2018). These concepts are also fully applicable with regard to remote driving.

The development of technology allows people to control the vehicle remotely, while they can be within the vehicle, within line-of-sight of the vehicle, or beyond line-of-sight of the vehicle. The concept of remote driving and remote driver is regulated by the SAE standard. However, at the EU level or international law level, there is no legally binding legal act that regulates the concept of remote driver.

An example of a system that performs driving tasks on the command of a driver outside the vehicle is a remote parking system. The driver can get out of the vehicle near a parking space and by pressing and holding a special button on the key fob, the vehicle will automatically move into the parking space. The driver is obliged to monitor the driving environment to ensure that no one and nothing enters the vehicle's path during the parking manoeuvre (SAE standard, 3.31.1.2).

Remote driving means 'real-time performance of part or all of the DDT and/or DDT fallback (including real-time braking, steering, acceleration and transmission shifting) by a remote driver' (SAE standard, 3.24).

The SAE standard defines a remote driver as a driver who 'is not seated in a position to manually exercise input devices in vehicle braking, accelerating, steering and transmission gear selection (if any), but is able to operate the vehicle' (SAE standard, 3.31.1.2).

The following situation can be considered as an example in which a remote driver will perform dynamic driving tasks. An automated vehicle of automation level 4 (without a driver) ensures the delivery of goods within the campus. This vehicle had a system failure that performs dynamic driving tasks that forced it to resort to a minimal risk condition by parking on the side of a campus roadway. Subsequently, the remote driver, who is able to control the vehicle using wireless means, returns it to its marshalling yard (SAE standard, 3.31.1.2).

The SAE standard also regulates the issue of driverless operation dispatcher. The concept of driverless operation dispatcher means that a user or users dispatches the vehicle or vehicles equipped with automated driving systems in driverless operation. These persons may also perform other fleet operations functions (SAE standard, 3.31.4).

A remote driver cannot be equated with a driverless operation dispatcher. Although it is not excluded that the driverless operation dispatcher can become a remote driver if he has the means to control the vehicle remotely (SAE standard 3.31.1.2).

3.2 Requirements of Remote Drivers in the Regulation

In 2019, the Draft Resolution on Remote Driving was presented at the 79th session of the Global Forum on Road Traffic Safety (WP.1). The proposal in question also refers to the conclusions of the 75th session, according to which the remote parking system used by the driver outside his vehicle does not endanger road traffic safety, provided that the system complies with the technical regulations of UNECE (2019).

The draft resolution in question was replaced in 2021 at the 83rd session of the Global Forum for Road Traffic Safety (WP.1) by the Informal paper on Remote Driving: Situations when a driver operates a vehicle from the outside of the vehicle. Within the document in question, several requirements were formulated for:

(a) Remote driving systems
(b) Remote driver
(c) Service providers
(d) The developer/manufacturer
(e) Passengers in a remote-controlled vehicle

Regarding remote driver requirements, such driver must have physical and mental capabilities to exercise dynamic control if (UNECE, 2021):

i. He is in control of a non-automated vehicle including those with driver assistance systems.
ii. It is required to resume dynamic control of a conditionally automated vehicle as the safety fallback for the ADS, or
iii. It is expected to resume dynamic control of a highly automated vehicle if the journey continues beyond the parameters of the vehicle's operational design domain.

Furthermore, the remote driver should have the appropriate licenses to use and operate the vehicle in the country where the vehicle is driven. In addition, he should be ready and able to exercise dynamic control and should minimize any other activity that would restrict or impair their ability to resume dynamic control. The remote driver is also able to remotely activate and deactivate the ADS when driving automated vehicles (UNECE, 2021).

The requirements relating to the transportation of passengers by a remotely controlled vehicle are specifically regulated. In such cases, the remote driver must (UNECE, 2021):

(a) Be aware of any passengers inside the vehicle that they are operating in. This includes how many passengers and if any children are on board.
(b) Meet their legal requirements regarding the use of seat belts by passengers.
(c) Ensure that the vehicle is a safe environment for its passengers, including preventing vehicle theft.
(d) Ensure that all waiting passengers have boarded before closing the doors or otherwise verify that the doors have been closed.
(e) Ensure that passengers are safely seated or in an appropriate standing position (where the vehicle allows, for example, on buses), before moving the vehicle.
(f) Support disabled passengers to use the vehicle confidently, comfortably, and safely, including by operating accessibility equipment, providing remote assistance, and communicating audibly and visibly to passengers waiting for and travelling in the vehicle regarding its route and location.
(g) Ensure that the number of passengers/vehicle load on the vehicle does not exceed its limit.
(h) Ensure that cargo, luggage, and passenger possessions are secured to prevent them from falling and posing a safety risk.
(i) Comply with any other relevant domestic requirements set by contracting parties.

Apart from the requirement to hold a valid driver's license, the driver does not have to meet any other specific requirements for professional qualification. However, if remote driving is provided as a service, the employer must ensure that all remote drivers are adequately trained to perform the task under the specific system and conditions used. The informal document in question does not define what kind of training remote drivers have to complete (UNECE, 2021).

Despite the fact that the above-mentioned requirements are regulated by document that is not legally binding, it could serve as good example for state regulators in the field of traffic law.

The remote driver issue is partially regulated at the national level of the member states of the EU in the new German legislation on automated vehicles (Autonomous Driving Act 2021). New legislation allows vehicles with autonomous driving function to no longer require a person to drive the vehicle during operation. However, to ensure compliance with the Geneva Convention on Road Traffic and the Vienna Convention on Road Traffic, the concept of a technical supervisor was created. The technical supervisor is a natural person who 'can deactivate the motor vehicle during its autonomous operation and approve an alternative driving maneuver' (Ebers, 2022, p. 9).

The technical supervisor has the duty to devise an alternative driving maneuver, to deactivate autonomous driving functions, and to communicate with passengers. The technical supervisor is not obliged to constantly monitor the autonomous vehicle. Its duty is to perceive emergency messages from the automated vehicle system and decide whether the vehicle should be deactivated or an alternative driving manoeuver should be initiated (Ebers, 2022, p. 11).

The new law contains the list of tasks of the technical supervisor. Requirements for the technical qualification of a technical supervisor will be specified in a regulation of the Federal Ministry of Transport and Digital Infrastructure.

3.3 Data Protection Regulatory Challenges

3.3.1 Provisions Related to the EU Safety Regulation

On 6 July the law of the EU welcomed into the force another piece of legislation attempting to govern new technologies and fostering the Brussels effect (Bradford, 2020) of the EU law. On this day, the new GSR entered the force. The regulation was adopted with the aim of improving safety and increasing incentives within the EU in the field of automated mobility.

In terms of the scope of the GSR, it applies to a wide range of vehicles, including passenger cars, buses, trucks, and 'to systems, components, and separate technical units designed and constructed for such vehicles' (GSR, Article 1). Such systems represent advanced 'smart' technologies that include advanced driver distraction systems, intelligent speed assistance, emergency stop signals, facilitate installation of alcohol interlocks or driver drowsiness and attention warning systems. Of the essential importance is the obligatory introduction of event data recorders (GSR, Article 6 [1] [g] as one of the advanced systems with potential to aid investigators and operators in detecting causes of potential malfunctions (Vellinga & Ritsema van Eck, 2022, p. 2). The main function of event data recorders is to collect and record 'critical crash-related parameters and information shortly before, during, and immediately after a collision' (GSR, Article 3, [13]). The demonstrative list of potential data that are subject to the obligation is provided in article 6 (4) (a) of the GSR, and the list includes 'vehicle speed, braking, position and tilt of the vehicle on the road, the state and rate of activation of all its safety systems, 112-based eCall system in vehicle, brake activation and relevant input parameters of the active safety and accident avoidance systems on board, with high level of precision and ensured data survivability' (GSR, Article 6 [4] [a]). Event data recorders shall be designed in such a way that they cannot be deactivated (GSR, Article 6 [4] [b]). The recording system itself shall be operated on a closed loop, interoperable with other systems and with several technical and organizational requirements being anonymization and prevention of manipulation and misuse (GSR, Article 6 [4] [c]). Further specifications related to data security shall be provisioned in the EU delegated acts (Draft Commission delegated regulation, 2022, Articles 2 and 3). The recorded data can be made available to national authorities 'for the purpose of accident research and analysis, including for the purposes of system and component type approval' (GSR, Article 6 [4] [d]). Technical specifications of data sharing shall be subject to the EU delegated regulation (Draft Commission delegated regulation, 2022, Article 4). Further limitation is provided by Article 6 (5) with the ban on recording and storing 'the last four digits of the vehicle indicator section of the vehicle identification number or any other information that could allow the individual vehicle itself, its owner or holder, to be identified' (GSR, Article 6 [5]).

In addition, to obligations, several requirements relating to buses are provided in Article 9 of the GSR. Buses must be equipped with a specific lane departure warning system and an advanced emergency braking system (GSR, Article 9 [2]). Additionally, advanced systems that can detect pedestrians and cyclists located near the front or nearside of the vehicle and that provide a warning or avoid collision with such vulnerable road users must also be deployed with such vehicles (GSR, Article 6 [3]). Several requirements are provided in additional provisions regarding specific advanced systems. Only one of them may be possible to disable at the time (GSR, Article 9 [4] [a]) and should automatically start with the start of the vehicle (GSR, Article 9 [4] [b]). Audible warnings may be suppressed under certain circumstances (GSR, Article 9 [4] [c]) and the driver shall be able to override the systems in question (GSR, Article 9 [4] [d]). Vulnerable road users amount to specific considerations as vehicles are designed and constructed 'to improve the direct visibility of vulnerable road users from the driver seat, by reducing to the greatest extent the blind spots in front of and on the side of the driver, while taking into account the specificities of different categories of vehicles' (GSR, Article 9 [5]). Vulnerable road users should be represented by cyclists and pedestrians (GSR, Recital 3). In case the vehicle exceeds the capacity of 22 passengers, the design shall take into account and be accessible for people with reduced mobility, including wheelchair users (GSR, Article 9 [6]).

As evident from the above considerations on GSR, a large number of data, including personal data, shall be processed within vehicles in the scope of GSR (Vellinga & Ritsema van Eck, 2022). It is natural that the GSR itself contains several data protection clauses. The relevance of the data recorded by the driver drowsiness, advanced driver distraction warning systems, and attention warning systems is emphasized (GSR, Article 3). Anonymity requirements are part of the clauses governing event data recorders (GSR, Article 6 [4] [c] and GSR, Article 6 [5]). Advanced systems shall function without any kind of biometric information from drivers or passengers (GSR, Recital 10). These clauses are further supported by the limitations on data sharing with third parties (GSR, Article 6 [4] [d]).

As noted in the literature, the GSR contains several internal clashes among data protection provisions and is described as 'an unnecessarily tangled web' (Vellinga & Ritsema van Eck, 2022). To mention a few inconsistencies, e.g., necessity requirement provisioned in legal definitions of different advanced systems (GSR, Article 3) may be contradictory to specific functions of event data recorders (GSR, Article 5). However, exactly this data may be absolutely essential in determining causes of accidents, thus being in line with the purpose of the event data recorders (Vellinga & Ritsema van Eck, 2022). Furthermore, data anonymisation requirements may render compliance with data-sharing clauses impossible as data protection rules are not applicable to anonymized data (GDPR, Recital 26; Vellinga & Ritsema van Eck, 2022). GSR would benefit from clearer language.

In relation to specific requirements for buses as mentioned in our use case scenario, provisions related to event data recorder mention only owner or holder of the vehicle, e.g., GSR, Article 6 (5) considering ban on processing of identification data. This in practice may mean that the processing of identification data on passengers and pedestrians is not prohibited within purposes foreseen by the GSR (Ivic & Troitiño, 2022). Not to mention that buses shall be equipped with advanced systems and designs that allow persons with disabilities to use the vehicle. Furthermore, it should be considered that public transportation vehicles serve a different function than classic personal automated vehicles as public transport is often associated with public interest and the provision of transport to society. This means that different levels of what is necessary and purposeful may be required considering processing of data.

3.3.2 Provisions Related to the EU Data Protection Regulation

The second important legal element concerning the data protection issues of new forms of automated transport is represented by the EU data protection law consisting of several pieces of legislation. Of the outmost importance for the purposes of this chapter is the General Data Protection Regulation (GDPR)[2] being *lex generalis* for processing personal data in the EU.

From the point of scope, GDPR applies to any processing of personal data via automated, semi-automated means, or manual processing through filing systems (GDPR, Article 2 [1]). The legal notion of processing is broad and includes collection, recording, organization, structuring, storage, adaptation or alteration, retrieval, consultation, use, disclosure by transmission, dissemination or otherwise making available, alignment or combination, restriction, erasure or destruction of persona data (GDPR, Article 4 [1]). The crucial notion for triggering the application of GDPR is 'personal data' as defined in Article 4 (1) GDPR. Personal data represent 'any information relating to an identified or identifiable natural person'. Article 4 (1) GDPR demonstratively enumerates the most common identifiers. The interpretation of identifiability is further provisioned in Recital 26 GDPR and decisions of the Court of Justice of the European Union (CJEU) interpreting the definition of personal data extensively (Court of Justice of the European Union, 2016, 2017). The extensive interpretation of the notion of personal data in GDPR in practice means that 'critical crash-related parameters and information' as foreseen by GSR amounts to personal data (Hamulák, 2018). This is also indirectly confirmed by the guidelines provided by the European Data Protection Board (EDPB) on processing personal data in the context of connected vehicles and mobility-related applications. EDPB acknowledges that 'most data associated with connected vehicles will be considered personal data to the extent that it is possible to link it to one or more identifiable individuals' (European Data Protection Board, 2020, p. 12) illustrating the point on vehicle movement data and vehicle condition data. Specific attention shall be devoted to the geolocation data, biometric data, and data that could reveal offences or traffic violations due to their sensitivity (European Data Protection Board, 2020, p. 12), therefore, requiring deeper considerations (Vellinga & Mulder, 2021).

Furthermore, GDPR differs between intraterritorial scope and extraterritorial scope of application. The intraterritorial scope of GDPR means that the regulation applies 'to the processing of personal data in the context of the activities of an establishment of a controller or a processor in the EU, regardless of whether the processing takes place in the EU or not' (GDPR, Article 3 [1]). On the other hand, GDPR applies also extraterritorially to subjects not established in the EU under two alternative conditions. The first condition relates to the processing of personal data for the offering of goods or services, irrespective of whether a payment of the data subject is required, to such data subjects in the EU. The second alternative represents a situation where the establishment in the third country processes personal data of data subjects in the EU for the monitoring of their behaviour as long as their behaviour takes place within the EU (GDPR, Article 3 [2]). The special regime of territorial application applies by virtue of international public law (GDPR, Article

[2]Regulation (EU) 2016/679 of the European Parliament and of the Council of 27 April 2016 on the protection of natural persons with regard to the processing of personal data and on the free movement of such data, and repealing Directive 95/46/EC (General Data Protection Regulation).

3 [3]). The responsibility and liability of processing personal data primarily lie with the controllers of personal data. Controllers are defined as 'natural or legal person, public authority, agency, or other body which, alone or jointly with others, determines the purposes and means of processing personal data' (GDPR, Article 4 [7]). Specific obligations are also attributed to processors being 'a natural or legal person, public authority, agency, or other body that processes personal data on behalf of the controller' (GDPR, Article 4 [8]). The issue of allocating responsibilities related to vehicles deployed with automated driving systems represents a separate issue discussed elsewhere (Vellinga & Mulder, 2021; Andraško et al., 2021).

Considering applicable data protection legislation including (GSR and GDPR) and the fictional use case above, several challenges related to remote driving shall be considered. First of all, it is necessary to point out that processing of personal data is absolutely essential for remote operation of automated public transport. This is evident from the provisions related to automated driving systems and technologies foreseen by GSR considering the extensive interpretation of the notion of personal data (Purtova, 2018). Additionally, remote driving shall not be considered objectively possible if certain specific activities are not fulfilled. To illustrate this, consider boarding people with vulnerabilities where sensitive personal data may be processed, as the 'driver' needs to be aware of such persons boarding the vehicle. Furthermore, if remote drivers should be responsible for emergency situations in the vehicle such as sudden health issues of passengers (UNECE, 2021). Safety requirements may also require processing of personal data of passengers, especially minors, considering their behaviour in the vehicle (UNECE, 2021).

This leads us to discussion on the processing of specific categories of personal data. Several systems and requirements provided in GSR require the processing of health personal data. GDPR provides the definition of health data in Article 4 (15) which means 'personal data related to the physical or mental health of a natural person, including the provision of health care services, that reveal information about their health status'. Unlike definition of health data in technical norms (see, particularly, ISO 27799), GDPR defines a broader category of data concerning health. In its letter on the definitions of health data, the Article 29 Data Protection Working Party (WP29, the predecessor of the European Data Protection Board) stated that the traditional interpretation of the data in question as data processed by health care providers is now obsolete (Article 29 Data Protection Working Party, 2015). This information has traditionally included data on an individual's diagnosis and treatment, their contacts, or medical and clinical research. However, as part of the dynamic development of technologies, data related to health status also represent data processed within applications that are not operated by healthcare providers. As examples, WP29 lists an individual's intelligence quotient, data on allergies and addictions provided to schools or an employer, visits to support groups (Alcoholics Anonymous or other meetings), or providing information to public authorities that there is a person in the household who is physically handicapped. Health data can also be derived from the purchase or use of specific goods or services, such as purchasing a pregnancy test or sensory monitoring of unusual diseases. At the same time, these data are also processed by the operators of clinical and medical tests or questionnaires for the purpose of determining the state of health (Article 29 Data Protection Working Party, 2015). If we take into account the legislative definition, the views of WP29, and the doctrine (see, e.g., Purtova, 2016), it can be concluded that data on the state of health represent three categories of information. Firstly, traditional information about an individual's state of health, data regarding the individual's state of health at present, but also in the future, including the risk of disease (blood test results, obesity, higher consumption of alcoholic beverages). Secondly, information obtained from sensory devices, data that as such do not reveal information regarding the state of health of an individual, but in combination with others have such potential (e.g. data obtained through mobile applications measuring the number of steps taken during a certain period of time or walking speed). Third, all data are processed to reveal information related to the health status of the individual (dietary or exercise habits). Considering the use case scenario described in the beginning of this section, the second category of data concerning health (information obtained from sensory devices) is of the essence, as the remote driver would have to consider people with disabilities boarding the bus. The question of whether this amount for *processing* of personal data is still an open question. However, it may be argued that if the bus deployed with advanced automated technologies shall be considered a complex data hub with vast collecting and recording functions, then the processing of personal data takes place.

In practice, the processing of special categories of personal data, including health data, is prohibited (GDPR, Article 9 [1]). However, the further provisions contain several exceptions where these personal data may be legally processed. There are three options that can be potentially considered according to Article 9 (2) GDPR according to a specific exception for buses with remote driving. The first option lies in the exception to the processing of personal data for 'reasons of substantial public interest' (GDPR, Article 9 [2] [g]). Such an interest shall be based on the law of the EU or national legal order and subsequent processing must respect proportionality and respect the essence of the right to data protection and provide suitable and specific measures to safeguard the fundamental rights and the interests of the data subject. Controllers of data concerning health processed within bus deployed with automated or advanced systems might argue that safety and inclusiveness of vulnerable subjects as foreseen by GSR are truly substantial public interests. However, the threshold for fulfilling the substantiality of the public interest may prove to be too high. The second option is represented by the exception of manifestly making sensitive data publicly available (GDPR, Article 9 [2] [e]). We agree with the interpretation provided by the EDPB (European Data Protection Board, 2019). that the mere entrance to the area equipped with data collection technologies shall not suffice the requirement of manifestation making data available. As this conclusion has not been tested before the Court of Justice of the EU yet, we are discussing this option. The third option is to require explicit consent of the data subjects (GDPR, Article 9 [2] [a]). However, this may be an inadequate burden for operators of the bus fleet as the threshold for consent is higher than with regular consent (the requirement of consent being 'explicit') and it is not objectively possible to ask for consent for every vulnerable subject who board the bus. The option discussed remains only as a theoretical option. To conclude, processing of data concerning health may prove to be rather challenging for operators of buses or remote drivers due to requirements set up by GDPR.

Further data protection issues are related to event data recorder specifications as prescribed by GDPR. Of the essence is the necessity of processing as personal data (Ramiro Troitiño et al., 2022). The requirement of necessity is reflected in the principle of purpose limitation. According to the basic principle, personal data shall be 'collected for specified, explicit and legitimate purposes and not further processed in a manner that is incompatible with those purposes', with a special exception for further processing for archiving purposes of public interest, scientific, historical, and statistical purposes (GDPR, Article 5 [1] [b]). Correct defining of purposes of processing related to buses deployed with automated driving systems and advanced technologies may be a challenging activity (Vellinga & Ritsema van Eck, 2022). Furthermore, as discussed in section concerning data protection clauses in GSR, the legislation is not always clear and comprehensive, thus requiring controllers to determine purposes and range of processed personal data on them.

Final remarks shall be made on requirements on remote drivers. Apart from the obligations and prerequisites discussed in the previous sections, remote drivers or their employer/contractor may be classified as controllers or processors of personal data. This in practice means that a person responsible for remote driving shall be sufficiently educated (in line with the principle of accountability) and under certain obligations stemming from the EU data protection law. Furthermore, remote drivers must comply with the principle of security and prevent unauthorized misuse or other unlawful processing of personal data.

4 Conclusions

The analysis of the individual levels of automation according to the SAE standard has shown that the deployment of level 4 automated driving systems can significantly affect the existing delivery services. Taxi services, shuttle services and the transport of cargo may be provided by vehicles that do not have a driver in the traditional meaning. However, from the point of view of the EU law, in particular GSR, the person of the driver will always be legally required within automated and fully automated vehicles even though dynamic driving tasks are performed by automated driving systems. In this regard, the concept of remote driver was discussed from the point of view of technical standards, nonbinding legal documents adopted at the UNECE level or national legislation. Despite the fact that the concept of a remote driver is not a new concept from a technical point of view, its legal regulation and especially the requirements for a remote driver are not regulated either at the level of EU law or at the level of international law. Furthermore, there is the need to specifically regulate the requirements relating to the transportation of passengers by a remotely controlled vehicle.

From the point of view of the EU data protection laws, we have discussed the peculiarities of the provisions in GSR and GDPR. GDPR is truly described as 'an unnecessarily tangled web' of requirements for data processing and can cause confusion for manufacturers and operators of automated vehicles. We argued that due to the specific obligations and purposes of buses deployed with advanced automated driving systems and technologies, these vehicles require specific attention in terms of processing necessity. This is mainly due to the fact that the provision of public transport shall be in the public interest. We also discussed options for processing data related to health due to specific tasks of remote drivers. Limited options stemming from restrictive exceptions for processing sensitive personal data according to the GDPR may be impossible for automated vehicle controllers to rely on. Adherence to the principle of purpose limitation may also pose a significant challenge. Finally, the requirements for remote drivers must also include sufficient education and knowledge base in the area of protection of personal data and data security.

References

Andraško, J., Hamuľák, O., Mesarčík, M., Kerikmäe, T., & Kajander, A. (2021). Sustainable data governance for cooperative, connected and automated mobility in the European Union. *Sustainability, 13*, 10610. https://doi.org/10.3390/su131910610

Article 29 Data Protection Working Party. (2015). Letter to the Director of Sustainable and Secure Society Directorate of the European Commission. Published 5 February 2015.

Autonomous Driving Act. (2021). Act amending the Road Traffic Act and the Compulsory Insurance Act – Act on Autonomous Driving, BGBl. 2021 I, 3108 ff.

Ayad, P., Schuster, S., & Koepferich, K. (2021). Germany takes a pioneering role with a new law on autonomous driving. https://www.engage.hoganlovells.com/knowledgeservices/analysis/germany-takes-a-pioneering-role-with-a-new-law-on-autonomous-driving

Bateman, T. (2021). France approves fully autonomous bus for driving on public roads in a European first. https://www.euronews.com/next/2021/12/01/france-approves-fully-autonomous-bus-for-driving-on-public-roads-in-a-european-first

Bradford, A. (2020). *The Brussels effect: How the European Union rules the world*. Oxford University Press.

BSI. (2020). Connected and automated vehicles—Vocabulary BSI Flex 1890 v3.0:2020-10.

BSI. (2021). BSI PAS 1884:2021 Safety operators in automated vehicle testing and trialling—Guide. https://www.bsigroup.com/en-GB/CAV/pas-1884/

Centre for connected and autonomous vehicles. (2022). Guidance—Code of practice: Automated vehicle trialling. https://www.gov.uk/government/publications/trialling-automated-vehicle-technologies-in-public/code-of-practice-automated-vehicle-trialling#safety-driver-and-operator-requirements

Chesterman, S. (2021). *We, the Robots? Regulating artificial intelligence and the limits of law*. Cambridge University Press.

Choksey, J. S., & Wardlaw, Ch. (2021). Levels of autonomous driving, explained. https://www.jdpower.com/cars/shopping-guides/levels-of-autonomous-driving-explained

Collingwood, L. (2017). Privacy implications and liability issues of autonomous vehicles. *Information & Communications Technology Law, 26*(1), 32–45.

Committee on legal affairs and human rights. (2020). Legal aspects of "autonomous" vehicles. https://assembly.coe.int/LifeRay/JUR/Pdf/DocsAndDecs/2020/AS-JUR-2020-20-EN.pdf

Council of Europe. (2020). *Legal aspects of "autonomous" vehicles.*
Court of Justice of the European Union. (2016). Judgment of the Court in Case C-582/14 of 19 October 2016 Patrick Breyer v Bundesrepublik Deutschland.
Court of Justice of the European Union. (2017). Judgment of the Court in Case C-434/16 of 20 December 2017 Peter Nowak v Data Protection Commissioner.
Draft Commission delegated regulation. (2022). COMMISSION DELEGATED REGULATION (EU) .../... of XXX supplementing Regulation (EU) 2019/2144 of the European Parliament and of the Council by laying down detailed rules concerning the specific test procedures and technical requirements for the type-approval of motor vehicles with regard to their event data recorder and for the type-approval of those systems as separate technical units and amending Annex II to that Regulation.
Ducuing, C. (2019). Towards an obligation to secure connected and automated vehicles "by Design"? In A. Vedder et al. (Eds.), *Security and law. Legal and ethical aspects of public security, cyber security and critical infrastructure*. Intersentia.
Ebers, M. (2022). Civil liability for autonomous vehicles in Germany. https://papers.ssrn.com/sol3/papers.cfm?abstract_id=4027594
European Commission. (2018). *Communication from the Commission. On the road to automated mobility: An EU strategy for mobility of the future.*
European Commission. (2020). *Ethics of connected and automated vehicles. Recommendations on road safety, privacy, fairness, explainability and responsibility.*
European Data Protection Board. (2019). Guidelines 3/2019 on processing of personal data through video devices. Version 2.0 Adopted on 29 January 2020.
European Data Protection Board. (2020). Guidelines 1/2020 on processing personal data in the context of connected vehicles and mobility related applications. Version 1.0. Adopted on 28 January 2020.
Frisoni, R., et al. (2016). *Research for TRAN Committee—Self-piloted cars: The future of road transport?* https://www.europarl.europa.eu/thinktank/en/document/IPOL_STU(2016)573434
GSK.DE. (2021). German Bundestag passes reform of the Passenger Transport Act. https://www.gsk.de/wp-content/uploads/2021/03/GSK-Update-Reform-Passenger-Transport-Act.pdf
Hamulák, O. (2018). La carta de los derechos fundamentales de la union europea y los derechos sociales. *Estudios Constitucionales 16*(1), 167–186. https://doi.org/10.4067/S0718-52002018000100167
International Transport Forum and Corporate Partnership Board. (2015). Automated and autonomous driving: Regulation under uncertainty. https://www.itf-oecd.org/sites/default/files/docs/15cpb_autonomousdriving.pdf
Ivic, S., & Troitiño, D. R. (2022). Digital sovereignty and identity in the European union: A challenge for building Europe. *European Studies, 9*(2), 80–109. https://doi.org/10.2478/eustu-2022-0015.
J3016_202104 Taxonomy and definitions for terms related to driving automation systems for on-road motor vehicles.
Lim, H. (2018). *Autonomous vehicles and the law: Technology, algorithms and ethics*. Edward Elgar Publishing.
Lim, H., & Taeihagh, A. (2018). Autonomous vehicles for smart and sustainable cities: An in-depth exploration of privacy and cybersecurity implications. *Energies, 11*(5), 3.
Merat, N., et al. (2018). The "Out-of-the-Loop" concept in automated driving: Proposed definition, measures and implications. *Cognition, Technology & Work, 21*, 87–98 (2019). https://doi.org/10.1007/s10111-018-0525-8
Mercedes-Benz. (2021). https://group.mercedes-benz.com/innovation/product-innovation/autonomous-driving/system-approval-for-conditionally-automated-driving.html
Purtova, N. (2016). Health data for common good: Defining the boundaries and social dilemmas of data commons. In S. Adams, N. Purtova, & R. Leenes (Eds.), *Under observation: The interplay between eHealth and surveillance* (pp. 177–210). Springer International. http://www.springer.com/us/book/9783319483405

Purtova, N. (2018). The law of everything. Broad concept of personal data and future of EU data protection law. *Law, Innovation and Technology, 10*(1). Available at SSRN: https://ssrn.com/abstract=3036355 or https://doi.org/10.2139/ssrn.3036355

Ramiro Troitiño, D., Martín de la Guardia R., & Pérez Sánchez, G.A. (2022). The European Union and its political leaders: Understanding the integration process. *The essence of Europe: Understanding Europe through its designers* (pp 1–4). Springer International Publishing Cham.

Regulation (EU) 2016/679 of the European Parliament and of the Council of 27 April 2016 on the protection of natural persons with regard to the processing of personal data and on the free movement of such data, and repealing Directive 95/46/EC (General Data Protection Regulation). OJ L 119, 4.5.2016, pp. 1–88.

REGULATION (EU) 2019/2144 OF THE EUROPEAN PARLIAMENT AND OF THE COUNCIL of 27 November 2019 on type-approval requirements for motor vehicles and their trailers, and systems, components and separate technical units intended for such vehicles, as regards their general safety and the protection of vehicle occupants and vulnerable road users, amending Regulation (EU) 2018/858 of the European Parliament and of the Council and repealing Regulations (EC) No 78/2009, (EC) No 79/2009 and (EC) No 661/2009 of the European Parliament and of the Council and Commission Regulations (EC) No 631/2009, (EU) No 406/2010, (EU) No 672/2010, (EU) No 1003/2010, (EU) No 1005/2010, (EU) No 1008/2010, (EU) No 1009/2010, (EU) No 19/2011, (EU) No 109/2011, (EU) No 458/2011, (EU) No 65/2012, (EU) No 130/2012, (EU) No 347/2012, (EU) No 351/2012, (EU) No 1230/2012 and (EU) 2015/166, OJ L 325, 16.12.2019, pp. 1–40.

Skiho, K., & Shrestha, R. (2020). *Automotive cyber security introduction, challenges, and standardization*. Springer Nature Singapore.

The European Commission. (2022a). Connected and Automated Transport Studies and reports. https://ec.europa.eu/newsroom/horizon2020/document.cfm?doc_id=46276

The European Commission. (2022b). Key technologies to boost the digitalisation of transport. https://digital-strategy.ec.europa.eu/en/policies/technologies-digitalisation-transport

The National Transportation Safety Board. (2018). *Preliminary Report Highway HWY18MH010.*

UN. (2021). UN Regulation No. 157—Automated Lane Keeping Systems (ALKS). https://unece.org/transport/documents/2021/03/standards/un-regulation-no-157-automated-lane-keeping-systems-alks

UNECE. (2017a, March 14). *Informal document No. 2. Automated driving.*

UNECE. (2017b, September 4). *Reflections about an amendment proposal to the 1968 Convention on Road Traffic.*

UNECE. (2019, July). *Draft resolution on remote driving.*

UNECE. (2021, September 14). Informal paper on remote driving situations when a driver operates a vehicle from the outside of the vehicle. Informal document No. 1.

Vellinga, N. E., & Mulder, T. (2021). Exploring data protection challenges of automated driving. *Computer Law & Security Review, 40*, 105530. https://doi.org/10.1016/j.clsr.2021.105530

Vellinga, N. E., & Ritsema van Eck, G. (2022). The new car safety rules of the GSR in light of the GDPR: An unnecessarily tangled web. *European Journal of Law and Technology, 13*(1). https://ejlt.org/index.php/ejlt/article/view/879

Vienna Convention on Road Traffic. (1968).

Jozef Andraško Associate professor at the Instititue of IT & IP Law, Comenius University in Bratislava, Faculty of Law; Senior researcher at the Palacký University Olomouc, Faculty of law, Czech Republic.

Matúš Mesarčík Assistant professor at the Instititue of IT & IP Law, Comenius University in Bratislava, Faculty of Law; Senior researcher at the Palacký University Olomouc, Faculty of law, Czech Republic.

Part III
Politics

Mapping E-governance in the EU

David Ramiro Troitiño

Abstract New information and communication technologies can make a significant contribution to the achievement of good governance goals. This EU "e-governance" can make public management more efficient and more effective, and bring several benefits. This research outlines the main contributions of e-governance in Europe: improving government processes (e-administration); connecting citizens (e-citizens and e-services); and building external interactions (e-society). Case studies are used to show that e-governance is a current, not just future, reality for developing countries and the European Union. E-governance requires a concrete roadmap for its development and this chapter focuses on the main steps required to implement an effective e-governance within the European Union.

1 Introduction

The European Union is facing a digital transition, reflecting the evolution of the modern culture as a logical step of the institutions adapting to the society. Therefore, the European organization must speed up the digitalization process in order to follow their own citizens, who are already implementing digital tools in their private lives. The European Union, as a supranational organization based on a balance of importance between the Member States (Council), the European citizens (European Parliament) and own process of integration (European Commission), requires a constant effort of adapting to the necessities of the most relevant actors involved in the process of constructing a united Europe.

This contribution integrates the activities of the Jean Monnet Chair "Digital Europe and Future Integration" (DEFI)

D. Ramiro Troitiño (✉)
Jean Monnet Chair on Digital Europe and Future Integration, Tallinn University of Technology, Tallinn, Estonia
e-mail: david.troitino@taltech.ee

D. Ramiro Troitiño et al. (eds.), *Digital Development of the European Union*,
https://doi.org/10.1007/978-3-031-27312-4_19

In addition, E-Governance is a solution for the most relevant problems of a modern State, as the lack of funds, economic resources, or labor force. Any State needs to generate enough money (incomes) to cover the expenditures of the services it provides to the citizens. The society, including private and economic agents, share their wealth to fund the communality of public policies and services. Usually, this transfer of wealth from the private to the public is done via taxes where the members of the society contribute according to their incomes to the global budget. Nevertheless, if the income is not enough, there are traditionally different options as lending money via public debt, reducing the quality and the extension of the services decreasing the cost, or increasing taxes to cover the expenditures. Currently, the digital tools provide with a new option for cutting the cost without affecting the quality of the service, including less taxes. The digital services provided by governments can also be more effective in the use of resources available and workforce. Becoming competitive means fewer resources are required and less workers are involved in the services. Funding issues, in the case of the European Union, go beyond the capacity to afford quality public service because those against the European Union and the integration process usually complain about the national money waste at the European level. This is a common critic with not much ground to support it, but popular among those willing to stop the integration process. Therefore, an effective use of the available economic resources, increasing the quality of the public services provided by the European Union and reducing the expenditures, would fundamentally reduce nationalist criticism of the EU.

The European Union needs to make an effort to identify the fields where a digital approach will increase the satisfaction of the European citizens. The mapping is a crucial task already undergoing under the promotion of the European Commission,[1] whose priorities are:

- eGovernment for public administrations: Cross-border digital public services.
- eGovernment in the Digital Single Market: Information on the EU's electronic exchange of social security information (EESSI), EU rules on social security coordination, public directory of social security institutions.
- Electronic payments and invoicing: The EU's e-Invoicing directive, European initiatives on e-Invoicing, how to check for compatibility with EU rules, exchange of information on e-Invoicing.
- Electronic customs: How the EU plans to replace paper-based customs procedures with electronic ones, EU legislation and initiatives related to electronic customs, strategic plan for electronic customs in the EU.

Obviously, the EU Commission targets are restricted to the areas where the national sovereignty is shared among the Member States of the European Union and the coordination of national policies where the digital convergence could smoothly bring relevant benefits to the European society. Therefore, the priorities in terms of e-governance in the Union are focused on the Digital Single Market and

[1] https://digital-strategy.ec.europa.eu/en/policies/egovernment.

areas associated to it, as payments and customs. Since the beginning of the European integration process, the implementation of a Common Market, with common customs and common trade policy, has been one of the most successful European initiatives. In addition, the economic integration has foster further integration in others fields, as the current society model based on capitalism allows an imbrication of economics with almost all the areas of the society. Nevertheless, the irruption of the digital economy generated a dysfunctionality between the European economy, highly integrated, and the digital economy, still nationally divided. Therefore, the EU priorities on e-governance focus on adapting the digital economy to the European Union common economic frame (Künnapas, 2016).

2 Institutional Support

E-governance requires an active support from the political power and the society. In the case of the European Union, the political power is divided between the European Parliament and the Council of the European Union and the European Council. The first represents the common political interest of the European citizens, and the Councils the national interest of the Member States, balancing European and national interest, and allowing the integration in areas where both levels will benefit alike. The EP counts with its own strategy regarding e-governance[2] financing studies on e-Public, e-participation, and e-voting, topics related to the possibilities of creating a common electoral system in the whole European Union for the European elections. In addition, the EP created a Panel for the Future of Science and Technology (STOA). The specific panel helps the rest of the committees of the Parliament in their decision-making link with new technologies and generates reports of common interest.

The Council annually organizes a ministerial meeting or other high-level conference bringing together experts from around the Union to discuss e-governance aspects from a European perspective. The meeting of Malmö in 2009 ended with a common declaration outlining the relevance of eGovernment for the functioning capacity of the European Union as a common organization for the Europeans. Consequently, the Member States have been actively cooperating with the EP and the European Commission in order to implement an effective e-governance, reflecting the public support to the process (Umbach & Tkalec, 2022).

The path for an effective implementation of e-governance in the European Union requires additional actions to the public support, as IT literacy and know-how. The society needs to be ready to use and enjoy the digital services provided by the European Union. Subsequently, education is a priority to increase the digital capacities of the European citizens. The European Union is implementing a Digital

[2] https://www.europarl.europa.eu/RegData/etudes/IDAN/2015/565890/EPRS_IDA(2015)565890_EN.pdf.

Education Action Plan (2021–2027)[3] to provide the know-how to citizens and business of Europe, aiming to support the adaptation of the education and training systems of Member States to the digital age (Burlacu et al., 2019). This Action Plan was adopted in 2020 and emphasizes cooperation on digital education to address the challenges and opportunities of e-governance and digitalization. Nevertheless, education is a national policy depending on the Member States that is very unlikely to become fully European because of its importance in building the sense of nation and its importance regarding the economic model. Therefore, the Member States want to keep the control over their national educational systems, but at the same time, want to share a reduce share of it regarding e-governance. Coordination allows full national control and common initiatives in the European level. The Digital Education Action Plan is a key enabler to achieve a European Education Area by 2025 contributing to reach the goals of the European Skills Agenda, the European Social Pillar Action Plan and the "2030 Digital Compass: the European way for the Digital Decade." The priorities of the plan are:

- Fostering the development of a high-performing digital education ecosystem.
- Enhancing digital skills and competences for the digital transformation.

The combination of political support and social digital training is a reality in the European Union, facilitating the implementation of a common e-governance agenda (González-Cacheda et al., 2022).

3 European Union Legal Framework on e-Governance

The EU has the capacity to produce its own legislation on the fields managed commonly by the organization. The creation of a stable, and clearly defined, legal frame facilitates the interaction between the different social agents affected by e-governance and provides with security for the development of such a policy.

The European legislation process normally begins by a draft of the European Commission that is sent to the European Parliament and the European Council. Both institutions negotiate modifications and agree on a final legislative initiative. Therefore, the system includes the three main institutions of the European Union and their theoretical support, as Neofunctionalism/Commission, Federalism/EP, and Cooperation/Council. The European legislation approved is over the national legislation in case on conflict, reassuring the principle of EU legal supremacy. The Court system of the EU watches over the sanctity of this code and has the ability to sanction offenders in a unique system in the world that reflects the idea of sharing sovereignty in the European level (Gábriš & Hamuľák, 2021). E-Government involves rethinking the working system of the Union, procedures, and changing manners with the focus of more efficient European services. As a first step, the European Commission is

[3]https://eur-lex.europa.eu/legal-content/EN/TXT/PDF/?uri=CELEX:52018DC0022&from=EN.

taking specific actions for the development of cross-border digital public services in the European Union, including the creation of European interoperable platforms such as a common framework for citizens' electronic identity management, and the fostering of innovation through funding of large-scale pilots. The current free movement of people within the European Union (Schengen area) needs to adapt to the digital transition allowing the citizens to deal easily with public services outside their home country. The initiatives are still cautious in order to avoid rejection from the Member States, but more actions are required in this direction (Botrić & Božić, 2021).

In order to implement an e-governance system in the European Union, the first legal aspect to address is related to the digital identity. It is necessary to define a common digital identity within the Union to guarantee European access to the European public services. Consequently, the European Commission is working on the Electronic Identification, Authentication and Trust Services (eIDAS), because European regulation will provide the foundation for cross-border electronic identification, authentication and website certification within the European Union. Currently, the Member States can notify and recognize, on a voluntary basis, other national electronic identification schemes in their Member States. The recognition of notified electronic identification became mandatory in 2018 and already 60% of Europeans can benefit from the current system. Yet, there is no requirement for Member States to develop a national electronic identification and to make it interoperable with those in other Member States.[4] The essence of the Commission's action is linked with the idea cooperation because the concerns of the Member States are still important and they are afraid of losing control over such a delicate process. The cooperation in the eIDAS allows advancing creating common standards and facilitating further integration when the Member States will be willing to share their national sovereignty in the European level. Therefore, it is a first step in the integration pattern that needs further development to enjoy the benefits of a common digital identity.

In that sense, the European Commission proposed a new regulation on digital identity addressing shortcomings in eIDAS by improving the effectiveness of the framework and extending its benefits to the private sector. It includes the offer, by Member States to citizens and businesses, of digital wallets linking diverse aspects of their correspondent national digital identities. The initiative includes the possibility of consumers being able to access services online without having to use private platforms or unnecessarily sharing personal data. The proposal includes three main pillars:

- Availability to anyone who willing to have it within the Union: Any EU citizen, resident, and business in the EU who would like to make use of the European Digital Identity will be able to do so.

[4] https://digital-strategy.ec.europa.eu/en/policies/electronic-identification.

- Widely use: It will be used to identify users when providing them with access to public and private digital services across the EU.
- Users control: The European Digital Identity wallets allow the holders to choose and keep track of their identity, data, and certificates which they share with third parties.

The European Commission recommended the Member States to establish a common toolbox by September 2022 and to start the necessary preparatory work including the technical architecture, standards and guidelines for best practices. This voluntary preliminary work will allow a smooth entry into force of the Regulation once the European Parliament and the European Council approve it. Therefore, the Commission foresees the approval of their draft and encourage the Member States to start coordinating to avoid delays in the implementation of a common legal frame regarding the basic level of e-governance, the personal digital identification.

The legislative actions need to be complemented with cooperation with the private sector in order to make effective the reforms implemented. There should be a two-way communication between the European institutions and private actors to develop an effective digital society reaching as many as possible. The cooperation with the private sector is required to encourage innovation, smart solutions and provide the best possible service to the Europeans (Eckert & Kovalevska, 2021). Therefore, the European Commission is actively working with private companies to develop research and documents to lead the digitalization path of Europe.[5] The European Parliament is working in a similar way, as STOA ask regularly for external private reports to support the decision-making of the European chamber.

4 Pillars of the European e-Government

The European Union is still defining the foundations of a European e-government model. The organization needs to analyze and define what are the areas that should be integrated in the European level following a digital model. There are obvious choices, as the European digital ID, as identification is the first step to access the rest of the digital services. The European Digital Identity will be available to EU citizens, residents, and businesses wanting to identify themselves or provide confirmation of certain personal information. It will be used for both online and offline public and private services across the EU, allowing the use of a personal digital wallet.[6] It does not mean the implementation of a European identity card, as it is prerogative of the Member States because to hold the EU citizenship is required beforehand to hold the citizenship of any Member State. Therefore, if any EU citizen lose his Member State citizenship, automatically will lose as well the EU citizenship. Nevertheless, the

[5] https://ec.europa.eu/isa2/sites/isa/files/docs/news/10egov_anniv_report.pdf.

[6] https://ec.europa.eu/info/strategy/priorities-2019-2024/europe-fit-digital-age/european-digital-identity_en.

European digital ID will implement the right of every person holder of a national ID card to have a digital identity accepted in all the rest of the Member States, making its validity European. Harmonizing the current national models in the European sphere will allow the access to the European digital services regardless of the location of the user, and is a safe way to control the flow of information shared by the users with services that require sharing of information.

The implementation of the European digital ID is a priority and basic for the digital transition of the European society, but it faces numerous challenges. It is estimated by the European Commission that just 60% of the EU population in 14 Member States are able to use their national e-ID cross-border. In addition, just 14% of key public service providers across all Member States allow cross-border authentication with an e-Identity system, a very small number taking into consideration the creation of a common space with free movement of goods, free movement of capital, freedom to establish and provide services, and free movement of persons.

There are numerous areas where the European digital Id could be used, as any public services such as requesting birth certificates, medical certificates, reporting a change of address, opening a bank account, filing tax returns, applying for a university, in any Member State, storing a medical prescription that can be used anywhere within the European Union, proving the age, renting a car using a digital driving license or checking in to a hotel. In addition, it could be used to apply for a loan, a process that includes bureaucracy signing all the paper documents. By using the European Digital Identity, the applicant just will need to select the necessary documents that are stored locally on his/her digital wallet to reply to any bank's request. These verifiable digital documents will be generated and sent securely for verification to the correspondent bank, continuing with the loan application process.

The European Digital Identity entails different tools to improve the digital services around the European Union, perfecting the system and bringing relevant benefits for citizens and companies. The e-Signature is an electronic format of a person's agreement to the content of a document. The EU plans to integrate it into the digital wallet, helping citizens to sign legal documents and email them, without printing any paper, reducing costs and time through streamlined processes, and helping innovate business procedures. Consequently, the e-Seal will secure the origin and the integrity of a document, avoiding counterfeit and reducing costs and time through streamlined processes and promoting trust in the origin of the document (Kobernjuk & Kasper, 2021).

Another innovation included is the e-Timestamp, an electronic proof that a set of data existed at a specific time. It will proof that citizens have bought any service, enhancing document tracking and achieving greater accountability. The European Digital Identity is also thought for business, generating a Qualified Web Authentication Certificate that will ensure websites are trustworthy and reliable. This digital service linked with business identity will allow to the general users to understand if the websites and apps being used are reliable and safe. This action will foster the confidence of the users and restrict phishing, protecting the reputation of legit online business. Electronic Registered Delivery Service, included in the EU plan, will protect the users against the risk of loss, theft, damage or alterations when sending any item.

Therefore, the European Digital Identity represents a basic step moving forward e-government in Europe with tangible, and immediate, effects on citizens and businesses (Ivic & Lakicevic, 2011). By September 2023, the Member States have to make a Digital Identity Wallet available to every citizen willing to have one.

5 X-Road (Data Exchange), Cybersecurity-by-Design, Data Protection

The implementation of a digital model in the European Union requires a deep collaboration with the Member States digital systems following the model of subsidiarity where Brussels will just take the responsibility of those policies whose effectiveness if higher when commonly managed. The European Union will develop digital policies just in those areas where the benefits are relevant and the Member States are unable to be effective. At the level of e-governance, the European Union needs to create a common European network, being the most effective path for the coordination of the existing national systems to provide European services to the European citizens (Kesa et al., 2020).

Therefore, the data exchange between the Member States is a priority for the right implementation of a common European system. X-Road is a centrally managed distributed Data Exchange Layer (DXL) between different information systems. Member States can exchange information at the European Union level over the digital services. The X-Road requires ensuring confidentiality, integrity, and interoperability between data exchange in all the European Union. The first X-Road system in a Member State was developed and launched by Estonia in 2001. Since its first implementation, X-Road became the spine of the digital development of the Estonian government, the Member State of the Union more advanced in the digital transition. The country already started the digitalization of the State and the society in 1994 with the first draft of the "*Principles of Estonian Information Policy*." A strategic outline for IT development mapping the challenges of the society and the digital solutions. Currently the public services provided by the State meet online most of the citizen necessities, reducing distances, time and increasing effectiveness. The variety of the services offered online required an effort of interconnection between the different public institutions involved. Therefore, the X-Road system allows Estonian's e-services databases, both in the public and private sector, to link up and operate in harmony and effectiveness. The system comprises more than 1000 inter-connected databases covering 1700 services answering more than 50 million enquiries every month.[7]

As the European Union will require a similar system in the European level to implement properly a digital society, the example and the expertise of Estonia are highly valuable as a future model for the EU. Therefore, the European Commission is financing a project to establish a X-Road system between Estonia and Finland. This project creates a joint data exchange platform between Estonia and Finland. As a result, databases in both countries can interface, assist with cross-border services and make e-services accessible to citizens of both countries. It includes so diverse activities as electronically registering a new residence, checking one's personal data from the national databases, declaring taxes electronically, checking the validity of a driver's license and vehicle registration, or automatically registering a newborn child for health insurance.

The binational project started by coordinating the development and interoperability of Estonian and Finish systems. As Finland adopted X-Road as its data exchange system, it is essential that it maintained interoperability with the Estonian version to facilitate the coordination between both systems. The interoperability requires alignment and a roadmap commonly agreed. The previous compartmentalization of data systems, as a registry for health insurance, taxes, or social insurance, increase the challenge of the convergence between both countries because first they have to create a common data space nationally. X-Road enables different public-sector data systems in a single country to understand each other, allowing systems located in both countries to communicate and synchronize information. The system is digital, fast, and secure, being the beginning to connect with other European Union Member States and potentially helping to the creation of a European exchange of information simplified. Therefore, the European Union has substantially funded this project because of its potential expansion to the European level.[8]

A parallel initiative to implement the European e-governance is the security of the system to avoid digital attacks threatening the operability of the system. Therefore, the European Union founded an Agency for Cybersecurity in 2004. European Network and Information Security Agency (ENISA) operates in the field of cybersecurity with the U and the Member States delivering solutions and providing advice. The agency also supports coordinated response to European cybersecurity threats. Additionally, since 2019, ENISA has been drawing up cybersecurity certification schemes and assists the partners involved in the digitalization process to adapt safely to present and future Communitarian legislation.

[7] https://ec.europa.eu/regional_policy/en/projects/europe/x-road-cross-border-co-development-of-national-data-exchange-platform#:~:text=Disclaimer-,X%2DRoad%20%E2%80%93%20cross%2Dborder%20co%2Ddevelopment%20of%20national,up%20and%20operate%20in%20harmony.

[8] https://ec.europa.eu/regional_policy/en/projects/europe/x-road-cross-border-co-development-of-national-data-exchange-platform#:~:text=Disclaimer-,X%2DRoad%20%E2%80%93%20cross%2Dborder%20co%2Ddevelopment%20of%20national,up%20and%20operate%20in%20harmony.

Nevertheless, the main mission of ENISA is to achieve significant common standards on cybersecurity within the European Union Member States as a first step to the implementation of an e-governance system in the European Union.[9]

The last preliminary key issue regarding the foundations of a European Union e-governance system is data protection. The data protection package adopted in May 2016 by the European Union aimed to protect the European Union citizens across the territory of the Union regardless of wherever their data is processed. The most relevant European legislation on this field is the General Data Protection Regulation (GDPR) on the protection of citizens regarding the process of personal data and on the free movement of such data. This EU regulation is an essential step to strengthen individuals' fundamental rights in the digital age and facilitate business by establishing clear rules for companies and public bodies in the Digital Single Market, facilitating the implementation of a European Union e-governance system (Bindu et al., 2019).

The Member States of the Union have founded their own national agencies for protecting personal data in accordance with the EU legislation, avoiding fragmentation that would block deeper integration to generate a common approach regarding data protection in a digital European area (Troitiño et al., 2020b).

In addition, the European Union founded an independent body, the European Data Protection Board (EDPB) to ensure the consistent application of data protection rules throughout the European Union. It includes representatives of the Member States data protection authorities and of the European Data Protection Supervisor. The European Commission participates in the activities and meetings of the Board without voting right to learn from the national positions and present the European interest. The EDPB provides general guidance on key concepts, advises the European institutions on data protection and on drafts for future legislation in the field. In order to increase the compatibility of the national approaches, the EDPB mediates in the disputes between Member States, including the right to impose binding solutions on the contenders.

The European Commission has appointed a Data Protection Officer responsible for monitoring and the application of data protection rules in the European Commission to ensure a coherent internal application of data protection rules.[10]

In terms of legislation, the General Data Protection Regulation of 2018 established data protection rules for all businesses operating in the EU, regardless of their factual location. This legislation generated stronger data protection to allow the citizens to control their personal data, and benefits for the companies with common and clear rules. The European Commission published an evaluation report on the General Data Protection Regulation (GDPR) in 2020.[11] The document includes recommendations for future actions to meet the huge potential of the data protection in the European Union, proposing harmonization of the national rules,

[9] https://www.enisa.europa.eu/about-enisa.

[10] https://ec.europa.eu/info/law/law-topic/data-protection/data-protection-eu_en.

[11] https://eur-lex.europa.eu/legal-content/EN/TXT/?uri=CELEX%3A52020DC0264.

more cooperation with the Member States for the right implementation of the EU legislation, and convergence of data protection rules as a way to ensure safe data flows internationally. In the field of international cooperation, it is remarkable the agreement reached with United States of America (USA) on privacy shield and transfer of passengers' name record data. The European Union is focusing its external actions on data protection on personal data transferred outside the EU determining if a non-EU country has an adequate level of data protection, fostering standard contractual clauses for data transfers between EU and non-EU countries, and designing corporate rules for data transfers within multinational companies (Bradford, 2020).

The EU funding supporting the implementation of the General Data Protection Regulation has been over 6 million euros until 2020, and it is expected to increase in the coming years, showing the relevance of the topic for the European Union.

6 Implementation of e-Services

The European Union needs to redesign the public services offered to avoid the mistake of reproducing the paperwork procedure to the digital world. The vehicle of the service is different, offering numerous additional possibilities that need to be addressed in the design stage in order to enjoy fully the benefits offered. Therefore, the European Union is funding several projects in order to adapt successfully the public services that could provide online in the implementation of an EU e-governance.

The most relevant, and obvious online services that the EU could implement are e-Tax, digital signature, e-Prescription, justice and public safety.

Regarding taxes, the European Union does not have power to generate its own taxes, and just controls the custom taxes, as the borders of the Single Market are European in a practical way. Nevertheless, the Commission has made two legislative proposals looking for a reform on corporate tax rules so that profits are registered and taxed where businesses have significant interaction with users through digital channels. The main aim is avoiding online platforms operating in the whole European Union, obtaining their incomes from customers located in several Member States, to pay their taxes in one specific Member State where their offices are located. Usually, it leads to a concentration of such platforms in specific countries with fewer taxes, avoiding their contribution to the societies where they extract incomes. Therefore, the European Commission is proposing to create a common system within the EU to avoid this kind of usual situations. The measure could eventually be integrated into the scope of the Common Consolidated Corporate Tax Base (CCCTB) for its use in the European level. In the same direction, the European Commission previously has proposed initiative for allocating profits of large multinational groups in a way which better reflects where the value is created.

The second proposal responds to calls from several Member States for an interim tax covering the main digital activities that currently escape tax altogether in the

EU. The interim tax will ensure that the digital economic activities which are currently not effectively taxed would begin to generate immediate revenues for Member States of the Union. Additionally, it will assistance to avoid unilateral measures to tax digital activities in certain Member States, which could lead to a patchwork of national responses which would be damaging for the European Single Market. Therefore, harmonization before the national barriers are constructed, for developing the integration required to reassure the functionality of the European Union (Troitiño et al., 2020a).

The tax will be applied on revenues generated from economic activities where users play a major role in value creation and which are the hardest to capture with current tax rules, such as incomes from selling online advertising space, or digital intermediary activities, which allow users to interact.

The legislative drafts created by the European Commission will be submitted to the Council for adoption and to the European Parliament for consultation in order to have them approved and effectively working in a close future.[12]

As a second pillar, the digital identity has been already analyzed at the beginning of this chapter, as it is a basic step in order to be able to implement an effective e-governance in the European Union level. Users need to be European effectively to use European services.

Regarding electronic cross-border health services, the possibilities are endless in the European level. The eHealth Digital Service Infrastructure (eHDSI) is a groundwork ensuring the continuity of health care for European citizens while visiting any Member State of the organization. The system is based on an exchange of health data to improve the service provided to the EU citizens.

There are many initiatives under study for their implementation because it is a field that will have a huge impact on the EU citizens (Kerikmäe, 2017). Currently there are two electronic cross-border health services implemented in all EU Member States, the ePrescription and eDispensation, and the Patient Summaries. The first action allows EU citizens to obtain their medication, previously prescribed by the health responsible of their own State, in any pharmacy located in another EU Member State. It is possible because of the transfer of information of their electronic prescription. Patient Summaries share at the EU level information on relevant health-related issues such as allergies, current medication, previous illness, surgeries, etc. The digital Patient Summary will provide health specialists with medical characteristics of the patient in their own language, avoiding communication obstacles for a good treatment. By 2025, both services will be gradually implemented in 25 EU countries.

Both of the actions are linked with coordination and sharing information, following the model previously described of X-Road in the European level from a partial perspective based on health issues. It will help to learn about the problems of coordination and share of information, allowing deeper steps in the close future. Next, medical images, lab results and hospital discharge reports, will be shared

[12] https://taxation-customs.ec.europa.eu/fair-taxation-digital-economy_en.

across the European Union Member States providing a complete health record of the patients.

The European Commission is providing a common ICT infrastructure and cross-cutting services to implement properly the system in a common way avoiding national differences that will represent obstacles to the common management of health information. In addition, the European Union is financing by different funds initiatives to increase the connection between the national health systems of the Member States, and has created an expert group for the implementation and functionality of eHealth all over the European Union.[13]

The eHealth represents an important field for the e-governance implementation in the European Union because it can provide better services to the citizens in something as important as their health. It will attract their attention and will increase their commitment towards the European integration because their life conditions will be significantly improved.

Regarding the digitalization of justice systems, its relevance is linked with the democratic model of the European values (Steible, 2021). Therefore, it is a priority for the political aspects of the integration process. Following the subsidiarity structure of the EU, the cross-border situations, can be solved better in the European rather than the national level. Therefore, it is more efficient to coordinate the actions required from the EU level, but this kind of situation are still mostly addressed on paper format and are based on analytical transmission channels. The digitalization of these aspects is the first initiative to rationalize the system at the European Union, increasing the satisfaction of citizens and business alike. The Proposal for a Regulation on a computerized system for communication in cross-border civil and criminal proceedings (e-CODEX system), the e-CODEX Regulation is a relevant tool supporting the digitalization of cross-border judicial cooperation (Kesa & Kerikmäe, 2020).

The EU initiated the digitalization of the justice services with a Multiannual e-Justice Action Plan 2009–2013, outlining the priority fields for coordination. Subsequently an e-Justice Strategy and Action Plan were adopted for 2014–2018 period, giving way to the 2019–2023 e-Justice Strategy and Action Plan. These strategy fosters the coordination of the national judicial systems between themselves and with the European Union court system, producing tangible results as the European e-Justice Portal with valuable information on a comprehensive variety of topics, as well as several of online tools.[14]

The digitalization of justice is a clear example of the benefits of e-governance because it increase substantially the effectiveness of justice supporting an important pillar of the European society. Therefore, the citizens, the central pillar of the society,

[13] https://health.ec.europa.eu/ehealth-digital-health-and-care/electronic-cross-border-health-services_en.

[14] https://ec.europa.eu/info/policies/justice-and-fundamental-rights/digitalisation-justice/general-information_en.

and the business, crucial part of the economy, will feel the improvements, support the process and look forward deeper integration.

7 Conclusions

The European Union has clearly seen the necessity to implement e-governance in the organization. The digitalization is a reality in the current society and the political and economic structures need to adapt to it and extract the benefits associated to the process and decreasing the negative impacts on the society. The Union faces constant threats from a diverse range of actors, including a combination of nationalists, populist, extreme left and right, external forces destabilizing an international actor, and a pleiad of lesser forces. The way of the Union to faces the challenges associated to its own existence is improving the services provided to the citizens and economic agents by implementing effective solutions. Consequently the receivers of these improved services will be satisfied with the EU performance and implications in their existence, supporting the integration process and turning their backs on more radical options. E-governance is an effective approach to achieve this goal.

The adaptation of the EU to the digital transformation of the society is not just a matter of improvement, but of existence. The organization needs to adapt to the social reality to reassure its functionality and interaction with the society. If the digital gap becomes too big, the whole organization will fall because it will not be on the same plane as the society it represents. Therefore, it is also a matter of survival from a functionality perspective.

E-governance represents the present and future of the European Union but faces the same obstacles than the past, national resistance to share sovereignty in the European level. The Member States need to understand the necessity to implement a common system to face the modern challenges of digitalization and increase the welfare of its citizens. The digital world overloads the traditional borders, reducing the effectiveness of the States. The obvious solution is the common management in the European level using a Union with decades of experience and in constant evolution.

References

Bindu, N., Sankar, C. P., & Kumar, K. S. (2019). From conventional governance to e-democracy: Tracing the evolution of e-governance research trends using network analysis tools. *Government Information Quarterly, 36*(3), 385–399.

Botrić, V., & Božić, L. (2021). The digital divide and E-government in European economies. *Economic Research-Ekonomska Istraživanja, 34*(1), 2935–2955.

Bradford, A. (2020). *The Brussels effect: How the European Union rules the world.* Oxford University Press.

Burlacu, S., Alpopi, C., Mitrită, M., & Popescu, M. L. (2019). Sustainable e-governance and human resource development. *European Journal of Sustainable Development, 8*(5), 16.

Eckert, E., & Kovalevska, O. (2021). Sustainability in the European Union: Analyzing the discourse of the European green deal. *Journal of Risk and Financial Management, 14*(2), 80.

Gábriš, T., & Hamuľák, O. (2021). 5G and digital sovereignty of the EU: The Slovak Way. *TalTech Journal of European Studies, 11*(2), 25–47.

González-Cacheda, B., Cancela Outeda, C., & Cordal, C. (2022). Factors for the digitalisation of political parties in Portugal and Spain: A comparative perspective. *Partecipazione e Conflitto, 15*(2), 330–350.

Ivic, S., & Lakicevic, D. D. (2011). European identity: Between modernity and postmodernity. *Innovation: The European Journal of Social Science Research, 24*(4), 395–407.

Kerikmäe, M. S. T. (2017). Challenges in the e-health regulatory policy. *How deep is your law? Brexit. Technologies. Modern Conflicts, 367.*

Kesa, A., & Kerikmäe, T. (2020). Artificial intelligence and the GDPR: Inevitable nemeses? *TalTech Journal of European Studies, 10*(3), 32.

Kesa, A., Kerikmae, T., & Troitiño, D. R. (2020). Inteligencia artificial y protección de datos en Europa¿ Conflicto por defecto? In *Inteligencia artificial: de la discrepancia regional a las reglas universales: integración de percepciones políticas, económicas y legales* (pp. 407–430). Aranzadi Thomson Reuters.

Kobernjuk, A., & Kasper, A. (2021). Normativity in the EU's approach towards disinformation. *TalTech Journal of European Studies, 11*(1), 170–202.

Künnapas, K. (2016). From Bitcoin to smart contracts: Legal revolution or evolution from the perspective of de lege ferenda? In *The future of law and eTechnologies* (pp. 111–131). Springer.

Steible, B. (2021). Hacer respetar los valores de la UE. In *Anuario del Boletín de la Academia de Yuste: Reflexiones sobre Europa e Iberoamérica* (pp. 105–108). Fundación Academia Europea e Iberoamericana de Yuste.

Troitiño, D. R., Kerikmae, T., & Barbosa, P. A. R. (2020a). La inteligencia artificial en el futuro digital de la Unión Europea. In *Inteligencia artificial: de la discrepancia regional a las reglas universales: integración de percepciones políticas, económicas y legales* (pp. 431–444). Aranzadi Thomson Reuters.

Troitiño, D. R., Kerikmae, T., Barbosa, P. A. R., & Shumilo, O. S. (2020b). El Parlamento Europeo y la necesidad de una legislación común en un marco europeo de la inteligencia artificial. In *Inteligencia artificial: de la discrepancia regional a las reglas universales: integración de percepciones políticas, económicas y legales* (pp. 369–384). Aranzadi Thomson Reuters.

Umbach, G., & Tkalec, I. (2022). Evaluating e-governance through e-government: Practices and challenges of assessing the digitalisation of public governmental services. *Evaluation and Program Planning, 93*, 102118.

EU Elections and Internet Voting (i-voting)

David Ramiro Troitiño

Abstract The European Union elections are a reproduction of the national elections, with different electoral systems and national parties dominating the debate. Internet voting provides tools to implement real EU elections in the whole territory of the Union avoiding internal differences. This chapter explores the digital possibilities offered by i-voting to respond to the challenges of the European Parliament, providing a cohesive and European approach to the European elections. Internet voting has the potential to create a cohesive system fostering the union of the citizens of the European Union.

1 Introduction to the European Parliament, Elections and its Challenges

The European Parliament is a central institution in the European Union. Its origins are linked with the first European Community, the Coal and Steel Community (ECSC) that started the integration process deriving to the current European Union. Nevertheless, the original institutions, similar in the global frame to the current institutions of the Union, differed in their content. The executive body was named High Authority, similar to the current European Commission. However, the former institution was substantially more influential. The first Community also included a Council representing the Member States. The parallelism to the current Council of the European Union is obvious, but in its first form was less involved in the process building a common Europe. Other institutions founded within the ECSC were a Court of Justice, and a Common Assembly, the modern European Parliament.

This contribution integrates the activities of the Jean Monnet Chair "Digital Europe and Future Integration" (DEFI).

D. Ramiro Troitiño (✉)
Jean Monnet Chair on Digital Europe and Future Integration, Tallinn University of Technology, Tallinn, Estonia
e-mail: david.troitino@taltech.ee

D. Ramiro Troitiño et al. (eds.), *Digital Development of the European Union*,
https://doi.org/10.1007/978-3-031-27312-4_20

Regarding the Assembly, its associates were selected by the Member States parliaments among its fellows. Usually, the national delegations sent to Luxembourg, first location of the Common Assembly, respected the internal composition of the national parliaments. Consequently, if a national parliament was composed of 50% of socialists and 50% of conservatives, there were sent the same proportion of delegates to the Common Assembly. The same system was implemented in the following European institutional development, the European Economic Community (EEC). Nevertheless, it was a general understanding and some of the Member States did not respect it, as Italy that did not select for the European chamber representatives from the Communist Party, even though they entailed a significant per cent of the parliamentarians in the Italian parliament (Cancela-Outeda, 2020).

Therefore, the Common Assembly had a symbolic power in the institutional frame of the European integration; the citizens of the Member States did not directly elect its representatives. Additionally, it had just advisory powers, but its importance resided in its potential development. The designers of the first European Communities, as Jean Monnet or Robert Schuman, bared in mind the further evolution of the European integration process towards a common political and economic entity. Evidently, the Common Assembly was included in the First Communities to evolve into a common parliament that will provide democratic support to the future common European house. Hence, it was an embryo for the future European Parliament (EP), even when the power given to the new institution was nominal. Further integration would give more power to the institution systematically until the creation of a parliament representing all the Europeans, regardless of their nationality.

The development process was gentle to avoid confrontation with the governments of the Member States, cautious about their national sovereignty and the role of the European Parliament. Even today, the process is not finished and the EP faces many challenges, although its role in the European Union is central and participates actively in its public life.

The first European elections increased significantly the role of the EP as a chamber representing the citizens of the Member States (there were not EU citizens until the Treaty of Maastricht in 1992). The electoral process hold in the Member States, according to their respective electoral legislation, meant the creation of the first European democratic Parliament in 1979. The significance of a chamber responsible of the common sovereignty of the citizens is crucial to understand the evolution of the European Union towards a common structure above the States. In addition, it was the first time that citizens were involved in the process of integrating Europe, besides the enlargement referendums of Denmark, Norway and Ireland in 1972, and the UK in 1975. The Danish citizens supported the participation of their country by 63.3% (turnout 9.1%), the Irish by 83.1% (turnout 70.9%), the Norwegians rejected it with a positive vote of 46.5% (turnout 79%). The French citizens were called for an advisory referendum supporting the enlargement of the European Communities, supporting it with 68.3%of the votes (turnout 60.3%). Finally, the vote of the British citizens reinforced the involvement of their country in the European integration process by a positive vote of 67.2% (turnout 64%).

Following the establishment of the first democratic chamber of a common Europe, the role of the institution became active in pursuing deeper integration. One of its illustrious parliamentarians, Altiero Spinelli, founded an influential crowd name after the restaurant where they hold some meetings, the Crocodile group. The Italian politician was one of the main authors of the Ventotene Manifesto, which greatly influenced the European integration process from a federalist perspective. He also founded the European Federalist Movement, whose members supported the foundation of the United States of Europe in order to solve the nationalist problems of a Europe devastated by the Second World War. The first democratic elected European Parliament entailed numerous idealists pushing for deeper integration in their internal believe they represented the European people. As a relevant example of the atmosphere inside the chamber, there was a plenary session supporting the Spinelli plan for accelerating the integration in Europe towards a federal entity.

It generated a pattern we can trace currently in the European Parliament, as it is an active institution in terms of further integration, increasing its power and its role in the European integration process as it seems itself as the central institution of the future state of Europe.

The popular participation in the first European elections was the highest in the European Union history, as there was a turnout of 61.99% of the electoral body. Nevertheless, the deception of the institution as evident because the expectations of a central role in the Communities affairs never met the reality. The constant conflict between the newly elected, in 1979, European Parliament and the Member States about the primacy in the process integrating the members of the European Communities led to a disenchant and a lack of credibility of the EP as a real alternative power (Ivic, 2010). In addition, the nationality of the members of the European Parliament (MEPs) was still important for the decision-making. As an example, of an early problematic issue dealt within the EP, the reform of the Common Agricultural Policy, all the French MEPs, independently their political affiliation voted against it because it was a policy benefiting France.

After the euphoria of the first elections the turnout in the European elections have been decreasing, with 58.98% in 1984, 58.41% in 1989, 56.67% in 1994, 49.51% in 1999, 45.47% in 2004, 43% in 2009 and 42.54% in the 2014 European elections. Yet, voter turnout in 2019 increased by 8 points compared to the previous European election, to reach 50.66% of the European citizens.

These patterns show a decrease of citizen's interest on the European elections even when the EP is more relevant following the reforms included in the different European treaties approved since 1979.

On 7 June 2018, the Council agreed to reform significantly the EU electoral law and amend old laws from the 1976 Electoral Act in an attempt to increase participation of the Europeans in the elections. The new reform respected the constitutional and electoral traditions of the Member States, accepting the existing differences between citizens within the Union. It clearly avoided confrontation between the Council and the EP in a very sensitive area concerning the democratic customs of each Member State, but rejecting the idea of a common law for a common election in Europe. The main aspects of this reform are linked with forbidding double voting and voting in third countries. The European Parliament gave its consent on 4 July 2018 and the Council adopted the Act on 13 July 2018. Nevertheless, not all Member States ratified the Act prior to the 2019 elections and therefore this election took place in line with the previous rules.

2 Why Is Important Higher Participation?

Despite the rise in the participation in the last European elections, there is a clear negative tendency since 1979 in the turnout. This negative vision deepens if we compare participation between national and European elections, reflecting a notable difference and lack of interest in European affairs among citizens. Therefore, the tendency needs to be reversed in order to reassure the functionality of the European Union itself.

As the integration expands, the political sovereignty transferred to the European Union increases. It affects all the institutions of the organization that witness a relevant increase in their power. The common management of the European society requires common actions and common rules, creating a European sovereignty that needs to be balanced with democratization of the European Union. The holders of the sovereignty are the citizens; they are free, but decided to share their freedom with the rest of the members of the society to create common rules. Therefore, it is required a democracy to respect the will of the majority of the citizens political subjects) that express their opinion casting their votes. It is important to remark that the sovereignty does not belong to the nation, neither to the states, that act as managers of the will of the citizens. As the European Union deepens the integration, more and more common rules are required and implemented, increasing the transfer of loyalty from the state to the European level. Still, this process needs to be confirmed by the real holders of the sovereignty, the citizens. The relevance of the popular participation in the European Union elections resides on this fact; the EP must represent democratically the management of free will of the sum of the particular wills of each citizen (Alemanno, 2020).

The powers of the European Parliament, in the context of the common management o the freedom of the citizens of the European Union are:

1. Constitutional powers and of ratification: Since the Single European Act, the new enlargements and association agreements need the approval of the EP. In addition, the international agreements with important budgetary implications require the approval of the EP.
2. Participation in the legislative process: There are different procedures where the Parliament intervenes in the legislative process of the Union, from merely advisory powers to equal foothold with the Council.
3. Budgetary power: The Parliament stands equally with the Council elaborating the Budget, with the involvement of the European Commission drafting the budget.

The Parliament also monitors its implementation, and provides a discharge on implementation of the budget.

4. Controller on the executive power: The EP has different methods for controlling the European Commission, applying a democratic control over the executive of the Union.
   (a) Investiture of the Commission.
   (b) Motion of censure.
   (c) Parliamentary questions.
   (d) Committees of inquiry.
   (e) Control over Common Foreign and Security Policy.
5. Appeals to the Court of Justice: The European Parliament cannot decide unilaterally if any action is against the Treaties, but can appeal to the European Court.
6. Petitions: The European citizens have the right to make petitions on a subject related with the activity of the Union.
7. Appointing the Ombudsman: The Parliament elects the European ombudsman, who investigates complaints about maladministration related with the EU institutions.

Therefore, a higher participation in the European elections is required due a higher involvement of the EP in the European affairs and its implications in terms of common sovereignty.

3 Federalization as an Answer to the European Problems

The idea of Federalism in Europe can be traced to Kant. The German philosopher wrote in Perpetual Peace about the idea of a federation in order to protect the medium and small states against the superpowers of the time. Nevertheless he was not the only thinker proposing some kind of federal link in Europe in order to prevent wars and devastation emanating from the confrontation between national political entities.

After the II WW the European federalists blamed the sociopolitical structure based on national states for causing the devastation of Europe. They defended that the nation was something cultural based on feelings, and citizenship was something political based on rationality. The union between nation and state was a clear movement to reaffirm the loyalty of the members of the society to the state via the nation. People identified themselves as a part of a nation, it gave them a feeling of belonging, and a high level of integration. Therefore, the state needed the illogical loyalty of the nationals based on feelings to face any sociopolitical challenges. It has proved to be highly effective as a vehicle to attract the loyalty of the members of the society, providing a solid base for the State. Nevertheless, nationalism includes also superiority idea and confrontation with other national groups. In addition, the subjugation of the state to the nation provides the later with uncontrollable access to security forces, economy and other powerful state tools. Consequently, the confrontation between nations exceeds the cultural level achieving a total war with devastating consequences.

The fathers of Europe, the promoters of the integration, the architects of the current European Union, wanted to prevent a new nationalistic confrontation in Europe. Therefore, they proposed to separate nations from states, nationality from citizenship, and started the European Communities.

Obviously, the idea of the fathers of Europe was based on democracy and the creation of a European Parliament where the political views of the European citizens will be represented. Consequently, it is not relevant to be part of a nation to choose your political representative in the EP, but the particular political believes of each citizen exercising the political right of voting. However, the European federalists face the problem of a strong identification of citizens with their own national identity. The French Revolution changed the social paradigm facing similar problems regarding the loyalty of the members of the society to the new regime promoting the French nation. Afterwards, nations have been accepted as something natural in the political life. Therefore, there are enormous difficulties to transfer the loyalty of the citizens from the national to the European level (Hamulak et al., 2021).

The federalist supporters, in terms of European integration, proposed a Bing Bang approach fostering integration from the top to the bottom, creating the European Federal government without the fully identification of the citizens with it. The obvious, for them, benefits will attract later the loyalty of the citizens assuring the viability of the federation itself. The idea is not democratic and is based on fake illusions of future loyalty. This Big Bang federal approach is also linked with the mistrust towards national officials and diplomats because their source of power is the nation. Therefore, their priorities are national. Nevertheless, as a critic, federalism ignores the problem of concentrating power in bigger political entities that could worsen the confrontations they were intended to avoid that still needs to be addressed theoretically to prevent a worse scenario. In addition, federalism in the European Union needs to attract the loyalty of the European citizens towards the European institutions, participation in the European elections is one of the most effective tools to do it.

4 European Union Elections and the European Parliament

The elections to the European Parliament are organized according to the rules of each individual Member State, it creates an evident distortion to create a homogenous group of European citizens with same rights and duties, the base of the European Parliament political meaning. In European elections, most of the Member States function as single constituencies, with the exception of four Member States (Belgium, Ireland, Italy and Poland) that have divided their national territory into a number of regional constituencies. The voting age is 18 in all Member States except Austria and Malta, where it is 16, and Greece, where it is 17. It means that any European citizen, when 16, will have different political rights according to their residence area. Obviously, in order to create a common body of citizens there are required basic common rules. In addition, voting is compulsory in five Member States (Belgium, Bulgaria, Luxembourg, Cyprus and Greece): the obligation to vote applies to both nationals and registered non-national EU citizens as it includes the entire electoral mass.

Nevertheless there are some common provisions for the whole European Union territory as:

1. Right of non-nationals to vote and to stand as candidates: As the EP represents the EU citizens, their vote can be cast in any of the territories of the Member States of the European Union. It establishes a clear pattern of equality between citizens.
2. Electoral system: Besides the already mentioned notable divergences between the electoral systems of the Member States, there are some common standards for the whole EU, as proportional representation and use of either the list system or the single transferable vote system. Member States may also authorize voting based on a preferential list system.
3. Right to stand for election: All the electorate, the EU citizens, can attend the elections in their place of residence even if it does not coincide with their nationality. But even such a Basic right in a common election entails divergences between the different areas of the Union, as for example the minimum age to become a candidate, 18 in most Member States, but 21 in Belgium, Bulgaria, Cyprus, Czechia, Estonia, Ireland, Latvia, Lithuania, Poland and Slovakia, 23 in Romania, and 25 in Italy and Greece.
4. Election dates: There is a common time frame for the elections, from a specific Thursday to the following Sunday, but not a single date for the European elections.

These common rules provide some sense of unity, but lack the cohesive power to generate unity among the European Union citizens, the holders of the sovereignty who transfer it for management to the European level via common elections. The distortion is obvious and needs to be addressed pursuing a higher involvement and participation in the European elections. The Council, by unanimity could implement a common system all over the European Union, but it is unlikely to happen because it represents the interest of the Member States in the EU, not the interest of the European Union citizens.

Regarding the elections, there is an absence of real common political parties even though there are officially nine, the European People's Party, the Party of European Socialists, the Alliance of Liberals and Democrats for Europe Party, the European Green Party, the European Conservatives and Reformists Party, the Party of the European Left, the Identity and Democracy Party, the European Democratic Party, the European Free Alliance and the European Christian Political Movement. These supranational parties work in close cooperation with the corresponding political groups in the European Parliament and they are the result of the alliance of the related national political parties (Schulte-Cloos, 2018). Therefore, their essence is based on cooperation rather than integration, they are the result of the sum of national entities rather than a supranational independent entity. Obviously, this situates the national parties at the centre of power and considerably decrease the capacity of separation between nationality and citizenship.

On 22 November 2012, Parliament adopted a resolution urging the existing European political parties to nominate candidates for the Presidency of the Commission, looking to legitimize the institutions in the eyes of the European citizens. These arrangements were implemented ahead of the 2014 elections, when lead candidates ran for the very first time and Jean-Claude Juncker, was elected as Commission President by Parliament on 22 October 2014. The recurrent idea is foster participation in the European elections, the bait, elect the president of the European executive, albeit tangentially (Graziano & Hartlapp, 2019).

Additionally, the Commission in 2018 put forward the idea of transnational lists for increasing the interest in the European elections and foster the transfer of loyalty from the national to the supranational level. It proposed the creation of a Europe-wide constituency Nevertheless, the EP rejected its earlier implementation and left the issue for further discussions. Therefore, in 2022 while discussing on the reform of electoral law, the EP suggested to bind 28 seats of the chamber with European electorate, maintaining the rest according to the traditional system based on national rules.

Lastly, taking into consideration the idea of European sovereignty based on a common body of European citizens, it would be interesting to create European referendums, calling the EU citizens to decide about EU issues under the control of the European Parliament. It will be another measure to increase the civic participation in the European Parliament elections increasing the legitimacy of the entire European Union. The Member States call for referendums on EU-related issues when should be under the wing of the EP because it is about European Union. These national referendums are not the ratification of the treaties. They should be national because they organize the transfer of policies and powers from the national to the European level, but referendums about enlargements, for example. A single Member State citizens should not have the power to decide by popular vote about something so essential for the entire organization. In addition, enlargements required unanimity in the Council, respecting the national interest of the Member States (Ramiro Troitiño, 2022). Therefore, this duplicity should be eradicated opening a new era of European Union referendums about the most sensitive issues in relation with the development of the organization.

5 Internet Voting

Electronic voting is a voting system using electronic means to either aid or look after of casting and counting ballots. Basically there are two main ways to implement it via voting machines (e-voting) or terminals connected to internet (remote e-voting). A worthy i-voting system must comply with a set of standards established by regulatory bodies, and must also secure the votes to protect the citizens and their decisions.

I-voting is a cheaper system than alternative systems, including other forms of electronic voting, and it allows the emission of the vote from anywhere with internet access. Nevertheless, there is a debate if it fosters participation or not with different studies, still partial and without the entity to generate feasible data. It is also a system with high concerns in relation with security and fraud, but there is constant evolution in the security field with relevant improvements. Logic invites to think that if the investment is higher, the level of security could improve dramatically. Nevertheless, it is a debate to be taken into consideration for the next years as verifiable ballots are necessary because computers can malfunction and the international community is divided about the implementation of i-voting (Kasper, 2020).

In addition, citizens without internet access or lower skills regarding computers, could be excluded. The digital gap could lead to discrimination between citizens based on accessibility or capabilities, but the institutions are investing on it to close this fissure and the amount of internet users in the European Union is high. The successful use of i-voting within the EU in Estonia, has increased the trust on a system whose tendency is positive in terms of public support.

In 2005, Estonia became the first European Member State to offer internet voting to its citizens in a nationwide election. The Estonian State promoted as a way to guarantee the access to all its citizens to the elections regardless their distance from the polling stations. It is important for a country with a very low density of population and 75% of its territory is covered by forest. Parallel, i-Vote was successfully implemented in the Nordic country because a majority of its citizens have access to secure digital authentication and signatures. Since 2005 more than a dozen times i-voting has been used in Estonia, with a growing popularity (around 40% of voters prefer this alternative, historically, no less than 36%, and sometimes nearly 64%, of eligible voters participate in the i-voting system). In addition, the i-voting has expanded to European Union residents in Estonia who could participated in municipal and European elections in Estonia using i-voting with their ID-cards. During the designated pre-voting period, any voter can log onto the system using an ID-card or Mobile ID, and casts their ballot. Then, voter's identity is removed from the ballot before it reaches the National Electoral to protect the secrecy of their vote, ensuring anonymity. The Estonian electoral system is based on the premise that each vote cancels the last. Therefore, each voter has the option to change his/her vote during the pre-voting period.

In terms of security, Estonia uses a secure QR code-based mobile application to ensure that each vote is correctly received by the National Electoral Committee server, providing a secure environment for a healthy electoral system. In addition, the server software is public and external observers are welcome (Neudert et al., 2019).

The functioning system of Estonia, member of the European Union, has attracted the attention for the potential of its innovations. I-voting offers several tools to improve the democratic system in the EP elections but faces the main obstacle in terms of integration since the first European Community, Member States concern regarding national sovereignty.

6 I-voting in the European Parliament Elections

The elections to choose the MEPs encircle a significant territory, a total area of 4,233,255.3 km2 in 27 countries, an estimated total population of about 447 million and even more nations. The electoral process takes place every five years by universal adult suffrage among EU citizens, representing around 400 million people. The citizenship of the European Union is provided to all citizens of the Member States. The 1992 Maastricht Treaty legally created the notion of EU citizen, being additional to state citizenship. It grants EU citizens with rights, freedoms, legal protections available under EU law and the right to participate in the European Parliament elections (among other rights). The notion of EU citizenship is relatively new and still does not have the cohesive power of the Member States citizenship, and is far behind the national personal identification, as clearly BREXIT. Nevertheless, the European Union is working to provide the EU citizens with symbols to increase the empathy with the organization, as a flag, an anthem, free movement of citizens, etc. The target is to reassure the loyalty of the citizens to the institutional organization of the Union within the frame for a new social model. In this context, the EP elections are crucial because they provide democratic support to the whole project and attract the loyalty of the citizens. The Council adopted the last amendments to the 1976 Electoral Act[1] in 2018. It includes provisions on the possibility of different voting methods (advance voting, electronic, internet and postal voting) opening the gate for the implementation of i-voting all over the EU (Troitiño, 2021).

Therefore, it is a priority to enhance its working system to increase its integration capacity and influence over the EU citizens to prevent the collapse of the organization and improve its capacity of integration thanks to the political cohesion of its citizens.

The i-voting can help in several ways to increase the effectiveness of the electoral system in the EP elections. It is important to remark that it does not pretend to substitute the national, regional or local elections, part of the sovereignty of the Member States, but improve the European democratic system for the benefit of all its members, State and citizens likewise.

The first, and probably most relevant effect of the implementation of i-voting, is helping to increase the participation of the citizens in the European electoral process. A healthy democratic society requires an active participation of its citizens in the process electing the managers of the common sovereignty. I-voting, following the example of Estonia, is an effective, and relatively cheap, way to provide the citizens with easy options to exercise their right to vote, as it can be done from almost any devise with internet access, a commodity widespread in the European Union. In addition, the younger Europeans have incorporated internet into their lives as a basic item. Their familiarization with internet solutions makes i-voting the best tool to attract the young population to the European electoral process.

[1] 2018/994 of 13 July 2018.

Nevertheless, there is controversy about the fact if i-voting increases the participation in electoral processes or just facilitates it among those who were going to vote anyway. Switzerland, country not member of the European Union, implemented i-voting in some specific cases in determined cantons. These elections were studied (Germann & Serdült, 2017) and the conclusions revealed that the i-voting had no relevant effect on the electoral participation. A similar research conducted on the Estonian national election of 2009 (Alvarez et al., 2009) reached similar conclusions with reduced influence of i-voting in the turnout.

Contrariwise, other studies reveal a positive influence of i-voting in the citizen participation in elections. A study on online voting in municipal elections in the Canadian province of Ontario (Goodman & Stokes, 2020) revealed an increase of 3.5 percentage points in the turnout. Parallel, a research on the Swiss i-voting (Petitpas et al., 2021) concluded that even if the turnout was not affected by this novelty, it induced occasional voters to participate. Consequently, the discussion about the turnout is alive, requires further research and clarification before the impact of i-voting in elections is properly measured. However, it is a fact that internet age faces a decrease in the civic participation in elections due overload of information, and other relevant reasons. It seems logical that i-voting could reverse this negative tendency attracting youngsters to the political life via their main social tool, internet.

Additionally, there are concerns about the age gap regarding i-voting. The wide implementation of the system could neglect an important part of the European population, used to traditional formats, not familiarized with internet and switch the electoral balance to younger population. Nevertheless, the European population has a relevant educational system that is helping to reduce the age gap. Parallel, the implementation of the i-voting in the European Parliament elections will not mean exclusivity, or the elimination of other alternatives. Therefore, traditional voting will coexist with i-voting until its natural evolution leads to its residuality and the end of this debate.

Regarding higher turnout via i-voting, there is also a concern focus on the population access to internet, as those with higher income will benefit from it, reducing the participation of those without economic means to afford an internet device or good connection. Nevertheless, the European Union already has a significant per cent of its population with access to internet. Over the past 10 years, the amount of households with access to the internet in the European Union has grown to reach nearly 91 percent in 2020. It is a significant 36% growth from 2007, dropping the problem of internet access regarding i-voting. This growth is expected to continue, with 83 percent of Europeans forecast to have mobile internet access by 2025,[2] allowing citizens to use mobile ID (if implemented in the EU level) to participate in the EP elections via i-voting (Kerikmäe & Rull, 2016).

[2]https://www.statista.com/statistics/252753/number-of-internet-users-eu-countries/#:~:text=Number%20of%20internet%20users%20in%20EU%20countries%202020&text=This%20stood%20at%20nearly%2079.1,percent%20of%20the%20entire%20population.

Additionally, the implementation of internet voting in the European Parliament elections could allow the harmonization of the electoral system. Currently there are some common rules, but the respect to the democratic traditions of each Member State, makes the EP lections the sum of national elections rather than a proper European wide election. The introduction of a system based on i-voting will facilitate a common system for all the European Union citizens, regardless their geographical location (not relevant in the internet world). Moreover, the Member States lack democratic traditions on e-democracy as it is a novel aspect in our societies. Therefore, the national exceptions will decrease their relevance in the EU level, fostering the common identity and the cohesion among the European citizens, the main protagonist of the electoral process. The main problem, is the cohabitation of traditional alternatives voting and i-voting, creating a distortion between the rules of the EP elections; solutions are required to avoid relevant distortions to the system. Yet, i-voting provides an effective solution to create a common political frame for the EP elections, increasing the Europeanisms and supporting the democratic life of the European Union beyond the current national limits because eliminating the national barriers to common rules for all the European Union citizens in order to choose their democratic representatives in the European chamber.

The impact of i-voting will foster the European management of the elections. Currently, the elections are carried on by the Member States in terms of logistics, as the EU does not have the capabilities to implement unilaterally the electoral process all over the Unions territories. Therefore, the EU needs its Member States to conduct the EP elections, in a clear logic led by subsidiarity ideas. Nevertheless, if the EU could have an independent EU elections without the involvement of the Member States in the organization, its democratic influence will increase significantly, as the current intermediation will not be required. The European integration process is about a constant evolution and the implementing of an autonomous election is a natural evolution to achieve real European elections. The main obstacle could be the resistance of the Member States to provide such a relevant power to the EU, abolishing their control over the electoral process (Künnapas, 2020). However, the own Member States should be interested in the democratic health of the European Union and the implication of the citizens in the European affairs. The rise of populism in the shape of nationalism or pseudocommunism is a real threat to the EU own existence, as the EU witnessed with the result of BREXIT, consequence of the popular vote. Therefore, in order to protect the EU, the Member States will allow the direct participation of its citizens (also EU citizens) in the European elections, reducing their role as intermediators. I-voting is a tool that could be manage in the European level effectively without major obstacles once the Member States, via the Council, will allow it, but the complete abolition of traditional voting is not still recommended. Cohabitation of both systems could lead to distortions on the European democracy; addressing the problem beforehand is necessary for the implementation of the i-voting in the EU (Schneider, 2019).

Regarding the issue of the lack of real European political parties competing in the EP elections, the i-voting could allow the formation of wider parties in a common online platform within the European Union. A relevant obstacle to create real political parties to concur to the European Parliament with a European programme addressing the European necessities of the European citizens. Therefore, the emphasis on Europeanism requires a shared platform generating the right environment to the rise and development of European political parties beyond their current national constrictions. Internet provides the proper platform and i-voting the required vehicle for an effective implementation. The 2018/994 of 13 July 2018 amendment opened the possibility to give visibility to European Political parties in the ballots; an interesting initiate but still restrained by national borders and the sum of the current national political parties. There are ideas to create independent European political parties beyond the current confederation of national political families (Steible, 2022), but hey need the proper conditions for their growth; i-voting provides a European scope fostering independent democratic interaction between the citizens and the European Union.

Finally, the creation of European referendums consulting the whole body of EU citizens on aspects related to European sovereignty (their common freedom), is currently restrained by national borders because of logistical problems and national concerns on the evolution of the EU towards a federation. Nevertheless, it seems just logical that the sensitive issues depending on the EP should be consulted with all the European citizens. The creation of new policies, or aspects that require the involvement of the Council and the EP need the support of the nationals of the Member States. Therefore, they will be clearly out of any attempt to generate a European referendum, just the European issues exclusively competence of the European Parliament should be, in relevant occasions, consulted with the European citizens via European referendums (Troitiño and Kerikmäe, 2021). The organization of the referendums should be independent from the Member States electoral systems to reassure the independence of the consultation and the obtaining European results and not the sum of national results. It is an important consideration because if the electoral body of a Member State rejects any proposal, there is the danger than populism will use it to attach the European Union as an imposition. If the consultation and results are European, there is no room for these kinds of manoeuvres. The i-voting is the only existing functional way to implement the independent European Union referendums, increasing the loyalty of the citizens to the EU level their identification with the idea of a common Europe and their involvement in EU affairs. In this case, the impossibility to have a dual system maintaining traditional voting, because it will alter the European scope, has serious implications on the feasibility of the matter. Therefore, seems more practical to use i-voting on European advisory referendums, not binding, or with a minimum turnout plus double majority of 50% +1 of all votes (cast or representing majority of the EU citizens), and by a majority of votes cast within a majority of the Member States. The system of double majority already exists within the European Union and could be a way to mitigate the fears of the Member States to the implementation of European referendums in an early stage to evolve to pure European essence in the future. Nevertheless, whichever option is chosen, the vehicle for implementation will be i-voting for its capacity to reach all the European territory and a vast majority of European citizens with a reasonable cost.

7 Conclusions

The European Parliament is facing numerous challenges in the integration of Europe. The institution is under a constant evolution, from an advisory chamber to the main institution of the European Union because of its capacity to represent the democratic will of the EU citizens. This role requires contact with the European people without the mediation of the Member States with a direct link between the manager and the holders of the sovereignty. There are different alternative, as electronic voting (made in machines designed for it) or internet voting, but the last offers more reasonable cost and better level of security. There are examples of functioning i-voting within the EU, as Estonia, proving the system applicable commonly for common issues. Nevertheless, there are still considerable concerns that should be addressed properly to avoid critical situations that could lead to disintegration rather than integration. Therefore, the use of i-voting requires reforms, creative solutions, political bravery and education before its implementation. The path is long, and counts with considerable obstacles, but there is no doubt that i-voting is the future of the European Parliament elections and provides the organization with countless benefits.

Bibliography

Alemanno, A. (2020). Europe's democracy challenge: Citizen participation in and beyond elections. *German Law Journal, 21*(1), 35–40.

Alvarez, R. M., Hall, T. E., & Trechsel, A. H. (2009). Internet voting in comparative perspective: the case of Estonia. *PS. Political Science & Politics, 42*(3), 497–505.

Cancela-Outeda, C. (2020). The post-crisis European Union before the political union: Coordinates and keys of the future institutional architecture. In *The EU in the 21st century* (pp. 117–133). Springer.

Germann, M., & Serdült, U. (2017). Internet voting and turnout: Evidence from Switzerland. *Electoral Studies, 47*, 1–12.

Goodman, N., & Stokes, L. C. (2020). Reducing the cost of voting: an evaluation of internet voting's effect on turnout. *British Journal of Political Science, 50*(3), 1155–1167.

Graziano, P., & Hartlapp, M. (2019). The end of social Europe? Understanding EU social policy change. *Journal of European Public Policy, 26*(10), 1484–1501.

Hamulak, O., Kiss, L. N., & Gábriš, T. (2021). 'This content is not available in your Country'a General Summary on geo-blocking in and outside the European Union. *International and Comparative Law Review, 21*(1), 153–183.

Ivic, S. (2010). The assembly of European regions' Udine declaration: contradictory approaches to European and regional identities. *European Urban and Regional Studies, 17*(4), 443–446.

Kasper, A. (2020). EU cybersecurity governance–stakeholders and normative intentions towards integration. In *The Future of the European Union—Demisting the Debate* (pp. 166–185). The institute for European studies, University of Malta.

Kerikmäe, T., & Rull, A. (2016). Theorising on digital legal (Outer) space. In *The Future of Law and eTechnologies* (pp. 1–9). Springer.

Künnapas, K. (2020). Dysfunctionality from the sovereignty conflict in the ATAD GAAR. *TalTech Journal of European Studies, 10*(1), 97–122.

Neudert, L. M., Howard, P., & Kollanyi, B. (2019). Sourcing and automation of political news and information during three European elections. *Social Media+ Society, 5*(3). https://doi.org/10.1177/2056305119863147

Petitpas, A., Jaquet, J. M., & Sciarini, P. (2021). Does E-Voting matter for turnout, and to whom? *Electoral Studies, 71*, 102245.

Ramiro Troitiño, D. (2022). Margaret Thatcher: British strategy in the European Integration. In *The European Union and its political leaders* (pp. 207–218). Springer.

Schneider, C. J. (2019). *The responsive union: National elections and European governance*. Cambridge University Press.

Schulte-Cloos, J. (2018). Do European Parliament elections foster challenger parties' success on the national level? *European Union Politics, 19*(3), 408–426.

Steible, B. (2022). Emmanuel Macron: The return of France as a driving Force for European Integration? In *The European Union and its political leaders* (pp. 279–294). Springer.

Troitiño, D. R. (2021). La «Década Digital» de la Unión Europea: desarrollos e impactos sobre su ciudadanía y economía. *IDP: revista d'Internet, dret i política, 34*, 1–14.

Troitiño, D. R., & Kerikmäe, T. (2021). Europe facing the digital challenge: obstacles and solutions. *IDP: revista d'Internet, dret i política, 34*, 1–3.

Creating Digital European Citizenship and the Digital European Public Sphere

Sanja Ivic

Abstract The main goal of this chapter is to show the relevance of postmodern theory for the creation of a theoretical framework for digital European citizenship and the digital European public sphere. The European Commission initiated Europe's digital transformation and moved Europe towards creating an "imagined digital political community." The digital transformation requires a rethinking of current conceptions of European citizenship and the European public sphere and the construction of a new theoretical framework for the development of digital European citizenship and the digital European public sphere. European citizenship and identity are dynamic and polyphonic categories. The same can be argued about digital citizenship, which is also a postnational and multilayered form of citizenship. However, both European citizenship and digital citizenship contain different binary oppositions, such as: self/other, urban/rural, European/non-European, and so forth. Postmodernism offers a framework for the development of the digital European public sphere and the concept of digital European citizenship that will overcome binary oppositions and the digital divide. Postmodernism rethinks the basic concepts in the history of philosophy and questions the entire ontological and epistemological regime, which exists as the subtext of the legal system. Postmodernism does not only encompass critical, discursive practices directed towards rethinking existing binary hierarchies and authorities, but also a critical relationship between the very representatives of postmodernism who are differently positioned in these disputes, as postmodernism eludes any coherence and homogeneity.

1 Introduction

Citizenship is a fluid and polyphonic concept that is constantly reinterpreted and modified. Citizenship studies were gaining popularity at the close of the twentieth century. The diverse economic, political, social, and cultural changes brought about by globalization, which produced erosion of national sovereignty, have fueled

---

S. Ivic (✉)
Institute for European Studies, Belgrade, Serbia

D. Ramiro Troitiño et al. (eds.), *Digital Development of the European Union*,
https://doi.org/10.1007/978-3-031-27312-4_21

interest in citizenship studies. Increased migratory patterns resulted in cultural heterogeneity and social milieu differentiation within cultural identities.

The notion of citizenship is still contested (Isin & Wood, 1999). There is no comprehensive and unified definition of citizenship. According to Turner, citizenship has been "a key aspect of Western political thinking since the formation of classical Greek political culture" (Turner, 1993, p. vii). Status and practice are two aspects that are usually described to the concept of citizenship. Citizenship as status refers to a legal aspect of citizenship that encompasses certain duties and rights, whereas citizenship as practice refers to a social and political feature of citizenship that typically entails active engagement in economic, political, cultural, and symbolic life. According to Isin and Wood, "many rights often first arise as practices and then become embodied in law as status. Citizenship is, therefore, neither a purely sociological concept nor purely a legal concept but a relationship between two" (Isin & Wood, 1999, p. 8). The nature of both status and practice is contingent. They emerge in response to specific narratives, interests, and beliefs. Citizenship should not be viewed as an essentialist and universalist idea unaffected by historical, social, or other factors.

Identity is another important aspect of citizenship that should be considered. It represents individual's sense of belonging to a certain political community. There are two dominant points of view regarding the identity aspect of citizenship: 1. the idea of universalist and essentialist nature of individual and social identities; 2. the idea that these identities are historically and socially constructed. The first point of view represents the modern liberal idea of identity, based on binary oppositions such as essential/contingent, citizen/foreigner, nature/culture, national/postnational and so forth. Both elements of those binary oppositions are viewed as static and fixed. The second point of view overcomes binary oppositions and moves towards a postmodern conception of identity as polyphonic, contingent, and flexible category (Ivic, 2016, p. 67).

Identity, according to postmodern philosophers, is created by discourse and is historically and socially constructed. As a result, identity should be viewed as heterogeneous, as it encompasses a variety of voices and views that are continually reinterpreted.

According to Kymlicka (1990), the stability of contemporary democracy depends mostly on citizens, so it cannot rely only on institutions of justice. Consequently, theories of institutional justice should be complemented by developed citizenship theory (Ivic, 2016). Kymlicka (1990) emphasizes that old discussions about justice transform into discussions about citizenship. Nevertheless, it can also be argued that new discussions about citizenship require analysis of the concepts of identity on which citizenship is based.

The modernist conception of citizenship is based on the idea of a stable, fixed identity that gained meaning through numerous binary oppositions, such as we/they, citizen/foreigner, self/Other, national/transnational, and so forth. The modernist conception of identity is based on a fixed, coherent notion of identity, encompassing numerous binary hierarchies, within which the first notion is perceived as dominant because it is thought to originate from reason. The postmodernist conception of citizenship stems from a critique and rethinking of essentialist and universalist notions of identity based on binary oppositions.

As political communities progressed towards postnational and postmodern political systems, new types of citizenship emerged. There are various forms of citizenship, including national, global, postnational, dual, ecological, flexible, postmodern, diasporic and digital. Digital identity and citizenship involve "the urgency for a new approach to understand the economy, culture, and society in which we live 'here and now', characterized by almost instantaneous flow and exchange of information, capital, and cultural communication" (Kasza, 2017, p. 43). Postmodernism offers a framework for developing a new approach to understanding citizenship, identity, and public sphere in the digital age.

2 Developing Digital European Citizenship

European citizenship represents the postnational concept of citizenship, which is not bound by fixed national borders and incorporates numerous identities and affiliations. *The Treaty on European Union* defines EU citizenship as follows: "Every national of a Member State shall be a citizen of the Union. Citizenship of the Union shall be additional to and not replace national citizenship" (European Union, 2012). The European Union establishes the notion of a citizen through membership in various, overlapping political communities at the supranational, national and at the subnational level, which opens the possibility for rethinking the concepts of "citizenship," "identity," and "community" (Kostakopoulou, 1996, p. 344).

The European Union, as a supranational political community, transforms traditional forms of citizenship and identity that are rooted in the nation-state (Caporaso, 1996). The notion of citizenship of the European Union is based on the postnational idea of identity, which is based on the separation of the idea of citizenship from the idea of national identity. Meehan describes the concept of EU citizenship as follows:

> a new kind of citizenship is emerging that is neither national nor cosmopolitan but that is multiple in the sense that identities, rights and obligations associated with citizenship are expressed through an increasingly complex configuration of common Community institutions, states, national and transnational voluntary associations, regions and alliances of regions. (Meehan, 1993, p. 1)

Establishing the concept of citizenship beyond the borders of the nation-state represents a transition from modernity to the postnational conception of citizenship, since citizenship outside the borders of the nation-state (as well as the citizenship that does not bind to the concept of statehood) becomes a political reality. The European Union itself is represented within the framework of contemporary political theory as a postmodern political community, because it is characterized by a weak core, many spatial locations and multilayered politics (Caporaso, 1996, p. 47). The European Union also aims at developing a postmodern conception of sovereignty, which is post-territorial, non-exclusive and non-hierarchical (Karkkainen, 2004, p. 72).

One of the main contemporary goals of the European Commission is the digital transformation of the European Union (European Commission, 2021). The European Union has stated its goal to reinforce its digital sovereignty as a foundation for defending European identity and values in the digital era. "The widespread adoption of the term 'sovereignty'—a concept loaded with legal and political connotations—to describe authority over the digital is a more recent phenomenon" (Roberts et al., 2021, p. 2). In particular, this term has started to be used in the context of the European Union (Roberts et al., 2021, p. 2).

The significance of digital sovereignty for the future of the European Union was emphasized by Viviane Reding, former Vice President of the European Commission, in 2016 (Reding, 2016). Digital sovereignty is presented "as a means of promoting the notion of European leadership and strategic autonomy in the digital field" and defined as "Europe's ability to act independently in the digital world" (Madiega, 2020, p. 1).

According to the European Commission paper on lifelong learning, digital skills are necessary to get European citizenship and "Europe back on track" (European Commission, 2010, p. 3). Digital competencies are described as merging with social participation and the ability to exercise active citizenship (Rahm & Fejes, 2017, p. 22).

The nation was defined by Benedict Anderson as "an imagined political community" (Anderson, 1983, p. 49). The European Union can also be seen as a "imagined political community" from a poststructuralist and postmodernist perspective, one that is continually reconstructed and articulating its goals and purpose (Aydin-Düzgit, 2012, p. 8). The European Commission initiated Europe's digital transformation and moved Europe towards creating an *imagined digital political community*.

> In the State of the Union Address in September 2020, President [of the European Commission] von der Leyen announced that Europe should secure digital sovereignty with a common vision of the EU in 2030, based on clear goals and principles. The President put special emphasis on a European Cloud, leadership in ethical artificial intelligence (Troitiño et al., 2020), a secure digital identity for all, and vastly improved data, supercomputer and connectivity infrastructures. (European Commission, 2021)

However, in order to create an *imagined digital community*, the concepts of digital citizenship and human rights should be developed. The European Union is trying to advance democracy through the development of digital European citizenship and identity (European Commission, 2021). However, "the EU needs a clear-headed set of policy priorities with clear and achievable targets. But it also needs a Bill of Digital Rights capable of establishing the political and ethical basis on which jurisprudence might be later built" (Gak, 2021).

The European Commission proposed an inter-institutional European Declaration on Digital Rights and Principles in 2022 (Benifei et al., 2022). The proposed principles are meant to serve as a guide for the course of Europe's digital transformation, highlighting Europe's role as a forerunner in terms of digital policies. According to Benifei et al. (2022) the European Union needs an updated social contract for the digital age. European citizens need a digital economy based on rules and principles that are necessary for a free, inclusive and just digital society. Political declarations setting guidelines are important. Yet they fall short of actual digital rights to fight and stand for (Benifei et al., 2022). A Charter outlining these rights would guide the EU's legislative work as the digital transformation progresses (Benifei et al., 2022).

The Charter of Fundamental Rights of the European Union, which was adopted in 2000, establishes the fundamental rights of EU citizens (European Union, 2000). Nonetheless, it is a document from a time when the digital economy of today was still a long way off. As a result, present efforts to protect European citizens in the digital era are dispersed across a plethora of initiatives, resolutions, and legislative acts (Benifei et al., 2022).

According to Benifei et al.,

> It is time to introduce a European Charter for Digital Rights. The Charter should go beyond merely stating general principles. Rather, it should update EU rights to the digital era. There are promising examples of what this could look like. In early 2021, the Lisbon declaration on Digital Rights was signed under the Portuguese Presidency. Later that year in July, the Spanish Government adopted the Digital Rights Charter that highlighted 28 concrete areas for digital rights. (Benifei et al., 2022)

Benifei et al. (2022) argue the proposed digital Charter should be developed not only for Europeans but also by Europeans. The greatest method to ensure the formation of a truly European charter is to hear citizens' voices and thoughts on the subject (Benifei et al., 2022).

One of the main goals of the European Union is to strengthen its digital sovereignty as a basis for protecting European identity and values in the digital age (Roberts et al., 2021). The idea of digital sovereignty is tied to the idea of digital citizenship.

The relation between the concepts of EU citizenship and digital citizenship is not sufficiently explored yet as contemporary research on digital citizenship mostly focuses on citizenship education (see: Bennett et al., 2009; Coleman, 2008; Westheimer & Kahne, 2004; Choi, 2016) and digital rights and responsibilities (Pangrazio & Sefton-Green, 2021; Meréchal, 2015; Karppinen & Puukko, 2020; Custers, 2022; Daskal, 2018). As both concepts of digital and European citizenship represent postnational, multilayered and complex categories, the concept of European citizenship is "applicable in a networked and digitalized society" (Choi, 2016, p. 566).

The term "digital citizenship" is not new; it has been used in both European and American contexts to emphasize the relevance of the Internet in citizens' ability to engage in society (Choi, 2016). However, "without practical implementable framework," it remains a symbolic phrase (Choi, 2016). Despite the necessity of encouraging socially responsible citizenship in the Internet age, there is a scarcity of study on how to define digital citizenship or digital citizens (Choi, 2016, p. 565).

In the expanding Internet age, it is critical to understand how digital citizenship can be defined and how it is changing, as well as the deeper penetration of digital technology into everyday activities for better citizenship education (Choi, 2016, p. 589). Because it is not a single dimension or a sudden abrupt change in what citizenship implies, digital citizenship should be analyzed in connection with existing concepts of citizenship (Choi, 2016, p. 589).

According to Gleason and von Gillern (2018, pp. 201–202), the following approaches to studying digital citizenship are the most common: (1) the normative approach, which emphasizes rights and responsibilities in the context of ethical and safe Internet use, (2) the participatory approach, which emphasizes inequality in access, (3) linking online and offline participation to empower one another and (4) developing "new forms of political participation through online interaction and communication" (Bozilovic & Pavlovic, 2021, p. 210).

According to Rahm and Fejes (2017) and Ohler (2010), the concept of digital citizenship includes: digital access, digital literacy, digital security, digital communication, digital law, digital rights and responsibilities, digital health, digital etiquette, and digital commerce.

According to Abbas, "for citizens to become digital, digital literacy, ethics in digital behaviour and safety in the use of the Internet, must be introduced, and private information in exchange for public information must be recognised" (Abbas, 2020, p. 1494). Citizens must have digital competence to advance to the level of digital citizenship, which is described as a set of skills that enable the use of digital devices, communication applications, and networks to access information and improve management (Abbas, 2020, p. 1494). These skills enable the creation and exchange of digital information, as well as communication and collaboration, and problem-solving, in order to accomplish productive and creative growth in life, work, and social activities (Abbas, 2020, p. 1494). According to Nunes (2017) and O'Hara (2013), digital citizenship entails understanding social, political, and cultural issues related to the use of ICT (information and communication technology) as well as applying appropriate behaviors to this understanding and the guiding principles: ethics, security, responsibility and legitimacy in the use of the Internet, social networks, and technologies available. Digital citizens have the right to have access to and allocation of information and communication technology in order to improve their digital skills, access information on the Internet in a transparent, secure and private manner, and engage using technical means.

The digital society requires a digital citizen, one who is both digitized in records and archives and possesses the digital abilities required to adapt to societal changes (Rahm & Fejes, 2017, p. 22). The intersection of digitalization and citizenship enactments generates "digital inclusion" as normative, desirable, and non-political, as well as "digital exclusion" as marginalization, which adult education must address (Rahm & Fejes, 2017, p. 22).

Isin and Ruppert argue:

> We insist that because digital acts that bring digital citizens into being are transversal, making rights claims traverses multiple political borders and legal orders that involve 'universal' human rights law, international law, transnational arrangements, and multiple state and non-state actors. The rights of the political subject emerging across such borders and orders, and their aggregation and integration, are distinctly and irreducibly transversal and cannot be contained within existing orders and borders. We also insist on naming these political subjects as 'citizens' and, in so doing, connect this political subjectivity to the

genealogies of citizenship that embody both openings and closings. (Isin & Ruppert, 2020, p. xiii)

The established studies on digital technologies, the Internet, and citizenship tend to examine digital citizenship as a unidimensional concept. For example, from an educational perspective, some argued that when students learn online etiquette and communication skills, they become good digital citizens. (Choi, 2016, p. 585) However, both concepts of EU citizenship and digital citizenship are multilayered and complex and cannot be perceived as unidimensional concepts.

Postnational and postmodern political communities, such as the European Union, imply notions of citizenship and identity as dynamic, polyphonic and contextual categories. Representatives of postmodernism argue that "identity processes are fundamentally ambiguous and always in a state of flux and reconstruction" (Collinson, 2006, p. 182). As a result, identity is fragmented and fluid concept. The assumption behind fluid identity is that the subject is created by discourse. The notion of identity cannot be considered as static or stable, and it is always in the process of reinterpretation.

The concept of digital citizenship includes various binary oppositions such as: wealthy/poor, young/old, urban/rural and so on. "The idea of the 'digital divide' refers to the growing gap between the underprivileged members of society (Kerikmäe et al., 2019), especially the poor, rural, elderly, and handicapped portion of the population who do not have access to computers or the internet; and the wealthy, middle-class, and young (. . .) living in urban and suburban areas who have access" (Stanford University, n.d.).

EU citizenship can also be perceived as a polyvalent, controversial concept, which reflects both inclusive and discriminatory elements, as it does not overcome binary oppositions such as citizen/foreigner, European/non-European, global/local, universal/particular, and so forth.

The postmodern political thought offers tools for deconstructing these binary oppositions. According to Scott:

> We need a refusal of fixed permanent quality of binary oppositions, a genuine historization and deconstruction of the terms (. . .) We must find ways (however imperfect) continually to subject our categories of criticism, our analyzes to self-criticism. If we employ Jacques Derrida's definition of deconstruction, this criticism means analysing in context the way any binary opposition operates, reversing and displacing its hierarchical construction, rather than accepting it as real or self-evident or in the nature of things. (Scott, 1988, pp. 40–41)

Postmodern theory is relevant to rethinking the concepts of European citizenship and digital citizenship and for developing the concept of digital European citizenship. The era of postmodernism is marked by an increase fragmentation and differentiation in the field of culture, the end of the grand narratives about social and political legitimacy, by advocating the ideas of diversity and heterogeneity, by globalization of culture through information and communication technologies, by rejecting the idea of a coherent identity and weakening the industrial society that is replaced by "post-industrialism" (Isin & Wood, 1999, p. 7).

Proponents of poststructuralism and postmodernism reject the idea firmly defined, static boundaries, viewing them as metaphors and mental constructs. These theorists reject not only the modernist understanding of identity as conscious, stable, and rational category, but also do not accept the modernist understanding of difference, emphasizing that, representatives of modernism define *difference* as a homogeneous and universalist category. Jacques Derrida (1974) emphasizes that difference, like identity, should be understood as a fluid and heterogeneous category (Hamulák, 2018). Fragmentary determination of identity and difference transcends the monolithic understanding of ethnic, national, civic, and other identities. Identities are created from different narratives, experiences and perspectives and are continuously revised in touch with multiple, often opposing voices.

Postmodernism offers the possibility of rethinking the basic concepts represented in the history of philosophy and questions the whole ontological and epistemological context, which is in the subtext of legal discourse (Benhabib, 1994). Postmodernism includes not only critical, discursive practices aimed at rethinking and overcoming existing binary hierarchies and power relations, but also a discussion between different representatives of postmodernism, who advocate different views, because postmodernism strives to overcome coherence and homogeneity (Butler, 2002). The postmodernist approach offers an expanded, alternative idea of the political (Ivic, 2016). Instead of seeing power as a set of prohibitions, Foucault suggests that it should be seen as the formation of a wide range of identities, meanings, identities, beings, knowledge, and acts (Rabinow, 1984). "The meaning is placed in the discursive practices that are produced, disputed, and transformed in socio-historical actions, rather than in a *sui generis* scheme of timeless categories" (Ivic, 2016, p. 58).

Representatives of postmodernism argue that fixed and homogeneous identities lead to violence and repression (Derrida, 1992), and they dispute the validity of any identity that excludes otherness (Butler, 1992). Postmodernism can represent not only the context in which both EU citizenship and digital citizenship can be studied, but also a source of criticism of essentialist claims of both EU citizenship and digital citizenship, as an anti-foundationalist view that includes multiple and shifting identities.

3 Creating the Digital European Public Sphere

The postmodern perspective also offers a framework for developing a new digital European public sphere. From a digital standpoint, the concept of European citizenship emphasizes the importance of the EU forging a direct link with EU citizens.

In 2005, the European Commission acknowledged the existence of a democratic deficit in the Union (European Commission, 2005). The European Commission has organized a variety of initiatives and programs to encourage transnational deliberation among European citizens since 2005 (Ivic, 2016).

The democratic deficit in the European Union is frequently claimed to be caused by a lack of European political parties, representational accountability, and a fully functioning public sphere (Eriksen, 1999) Because of the dilemma of democratic legitimacy, citizens are expected to obey legislation that they have not authorized. As a result, the EU has become a "Europe of elites and experts," rather than "Europe of citizens."

The previous European Commission's attempts of creating a direct link of the EU with EU citizens were not successful. In these attempts, *European values* and *European public sphere* were not recognized as contingent, shifting and heterogeneous concepts.

In 2005, then Vice-President of the European Commission Margot Wallström announced the *Plan D for Democracy, Dialogue, and Debate* (European Commission, 2005). This initiative was built on the idea of including EU Member States and citizens in a discussion about Europe's future. However, the European Commission's *Plan D for Democracy, Dialogue and Debate* (The Plan D) was not successful, because it employed static and homogeneous concepts of European public sphere and European values (European Commission, 2005). The Plan D defines local, regional, and national public spheres as homogeneous categories (European Commission, 2005). However, "actors within the public sphere" should not be "restricted to territorial division" (Bruell, 2007). According to Bruell, "this means that the public sphere is not a materialized arena restricted to the national, regional and local level"(Bruell, 2007). Public spheres are multilayered and heterogeneous categories that transcend borders.

The European Commission established a very narrow definition of the public sphere, which is viewed as "an information providing instrument" (Bruell, 2007). Plan D does not allow for the realization of diverse and contradictory projects, which is a fundamental quality of a democratic public sphere. As a result, the public realm can be "misused as propagandistic organ" (Bruell, 2007). The failure of the European Commission's Plan D, on the other hand, can serve as a positive example for other European efforts that should reject homogenizing assumptions and purely instrumental conceptions.

One of the most recent European Commission's initiatives to bring the European Union closer to its citizens is reflected in launching the project "New Narrative for Europe" in 2013 (Barroso, 2012). However, this project was not effective because it did not sufficiently recognized the polyphonic nature of the European public sphere and did not take into account the issue of solidarity. The growth of populism and extreme nationalism, according to José Manuel Barroso, then-President of the European Commission, is a threat to Europe's future (Barroso, 2012). Barroso also stated that the European Union is experiencing a financial, economic, social, and political crisis, which is causing public distrust, and that a new narrative for Europe is required (Barroso, 2013). Barroso also emphasized the importance of dispelling the myth that the European Union can respond to solutions at the national level (Barroso, 2013).

The relevance of European science and culture in the broadest sense was underlined in the "New Narrative for Europe" project. Former European

Commission President José Manuel Barroso has called on artists and intellectuals to help create a new narrative for Europe. The goal of this effort was to demonstrate that the European Union's nature is not solely economic (Barroso, 2012). The project also aimed to encourage young people to engage in critical debate about European values. However, the project "New Narrative for Europe" did not sufficiently develop and emphasize the heterogeneous nature of the European public sphere and the role of developing transnational solidarity. That is why it was not successful and did not fulfill its goal to bring the European Union closer to its citizens.

The crises faced by the European Union after the launch of the "New Narrative for Europe" project—Europe's migration crisis, Brexit and the pandemic crisis, have shown the need to create a new narrative for Europe based on a polyphonic idea of European public sphere and transnational solidarity, which would contribute to greater responsibility and efficiency of the European Union and bring it closer to its citizens.

The digital transformation of the European Union (European Commission, 2021) can lead to building a transnational public sphere and bring the EU closer to its citizens. According to Gak (2021), the closest thing the European Union has to a true multilingual, multilayered European public sphere is social media platforms and digital information tools: the metaphorical European agora. Gak argues that:

> the rule of European law has lost sovereign power to the rules defined by tech companies' terms of service. To a large degree, this means that EU citizens lead their lives as second-class digital citizens of these digital states, while the EU becomes a sort of digital province subject to rules established in remote jurisdictions over which it has little or no influence and with which it has no possibility of competing. (Gak, 2021)

Algorithms used to govern EU citizens' information and social interactions should be imbued with European values (Gak, 2021). The existing algorithms that structure huge digital platforms prioritize a market-oriented model (Gak, 2021). The challenges posed by climate change likewise call for a fundamentally different approach to consumption. According to Gak (2021):

> If the European Green Deal is to be so tightly tied to digitalisation as the European Commission suggests, the algorithmic infrastructure of the major platforms cannot be geared towards increasing the scale of environmentally degrading activity. An algorithmic regime is the most fundamental way to inculcate EU values in the digital public sphere. (Gak, 2021)

The most profound potential in the building of a public sphere has been demonstrated by social media platforms (Gak, 2021). Gak argues that the creation and promotion of an EU public sphere would be a watershed moment in the evolution of a future Europe based on transnational collaboration and participation. EU projects, on the other hand, have no chance of succeeding until the tech giants' monopolistic methods are stopped (Gak, 2021).

Building a postmodern public sphere is relevant for constructing and promoting digital European public sphere. The postmodern condition implies ambiguity and heterogeneity, and it necessitates a reassessment of modernist practice. Representatives of postmodernism reconstruct the concepts of citizenship, identity, values, and the public sphere, saying that they cannot be based on the Enlightenment's political legacy of denying particularity and diversity. The public sphere is not regarded to be a single entity, but as heterogeneous. Thus, it cannot be based on a single model or set of values.

> Consequently, it is even rather insolent to use the term "public" in relation to politics, since the sphere of common deliberation has now been enlarged to include a wide range of groups and individuals that were previously simply ignored. Fragmentation and eclecticism have turned the political category of state-centered citizenship into a chaotic hotchpotch of values and accounts of civic life. This may not have made the notion of citizenship trivial, but it is certainly making it more problematic. (Van Ham, 2001, p. 163)

According to Bridges, "a postmodern civic culture must represent and affirm citizenship as an ideal that is contingent, particularistic, and culturally constructed" (Bridges 1994, p. 28). The postmodern condition implies ambiguity and heterogeneity, and it necessitates a reassessment of modernist practice. The postmodern condition requires rethinking the notion of citizenship and the concept of the public sphere in spatially complex and diverse ways. The concept of digital sovereignty should acknowledge that numerous agents might wield digital sovereignty at the same time, implying that sovereignty is something that can be shared. Digital sovereignty and citizenship should not be defined by fixed borders and values, instead digital sovereignty should be defined "as something that can be held across political communities and spatial networks" that are not restricted to the nation-state (Roberts et al., 2021, p. 7). However, the concept of digital sovereignty is not completely detached from the notion of sovereign (Roberts et al., 2021, p. 7; Agnew, 2005; Couture & Toupin, 2019). For example, "companies like Facebook and national governments are redefining, through their interactions and equilibria, the sense in which an agent can hold and exercise sovereignty" (Roberts et al., 2021, p. 7).

4 Conclusion

"New technologies help to dismantle the same vision of the world, which have supported in the past—'leading to the exclusion/alienation of the self, that existentially alone seems to be irretrievably lost for itself'- and as such search for a new sense of the relations/links (with the world) around thus shared and reduced identity" (Kasza, 2017, p. 46). Constructing digital European citizenship involves postmodernist discourse. Postmodernism applied to European studies refers to emphasizing the importance of the ideas of polyphonic and fluid identities, as solving the problem of internal exclusion in the European Union. The postmodern idea of identity applied to the idea of digital European citizenship implies decentralization, as well as the recognition and reconstruction of essentialist elements (leading to the digital divide and homogenization of the ideas of European identity and values, the European public sphere and so forth). "The development of new ICT technologies radically transformed the spatiality of social interaction by introducing the phenomena of simultaneity or any chosen time frame in social practices, regardless of the location of the actors engaged in the process" (Kasza, 2017, p. 43). This paradigm shift requires developing a new postmodern public sphere and postmodern understanding of citizenship.

References

Abbas, A. A. (2020). Digital citizenship behaviours. *International Journal of Innovation, Creativity and Change, 13*(4), 1492–1506.

Agnew, J. (2005). Sovereignty regimes: Territoriality and state authority in contemporary world politics. *Annals of the Association of American Geographers, 95*(2), 437–461. https://doi.org/10.1111/j.1467-8306.2005.00468.x

Anderson, B. (1983). *Imagined communities: Reflections on the origin and spread of nationalism*. Verso.

Aydin-Düzgit, S. (2012). *Constructions of European identity: Debates and discourses on Turkey and the EU*. Palgrave Macmillan.

Barroso, J. (2012). *State of the Union 2012 address*. European Parliament. https://ec.europa.eu/commission/presscorner/detail/en/SPEECH_12_596

Barroso, J. (2013). *A new narrative for Europe*. European Commission. https://ec.europa.eu/commission/presscorner/detail/en/SPEECH_13_357

Benhabib, S. (1994). Democracy and difference: Reflections on the metapolitics of Lyotard and Derrida. *Journal of Political Philosophy, 2*(1), 1–23. https://doi.org/10.1111/j.1467-9760.1994.tb00013.x

Benifei, B., Schaldemose, C., Garcia del Blanco, I., Cutajar, J., Munoz Gálvez, L., Leitão-Marques, M. M., Kumpula-Natri, M., Tang, P., & Wölken, T. (2022, February 4). Towards an EU charter for digital rights: Open letter to the presidents of the European commission, the European council and the European parliament. *Euractiv*. https://www.euractiv.com/section/digital/opinion/towards-an-eu-charter-for-digital-rights-open-letter-to-the-presidents-of-the-european-commission-the-european-council-and-the-european-parliament/

Bennett, L., Wells, C., & Rank, A. (2009). Young citizens and civic learning: Two paradigms of citizenship in the digital age. *Citizenship Studies, 13*(2), 105–120. https://doi.org/10.1080/13621020902731116

Bozilovic, J., & Pavlovic, N. (2021). Digital citizenship—A conceptual and practical framework. *Facta Universitatis—Series: Philosophy, Sociology, Psychology and History, 20*(3), 203–215. https://doi.org/10.22190/FUPSPH2103203B

Bridges, T. (1994). *The culture of citizenship. Inventing postmodern civic culture*. SUNY.

Bruell, C. (2007). "Debating Europe"—Identity and citizenship in a European democracy. NECE workshop: The impacts of national identities for European integration as a focus of citizenship education. https://www.bpb.de/medien/128025/PZMCL9.pdf

Butler, C. (2002). *Postmodernism: A very short introduction*. Oxford University Press.

Butler, J. (1992). Contingent foundations: Feminism and the question of postmodernism. In J. Butler & J. W. Scott (Eds.), *Feminist theorize the political*. Routledge.

Caporaso, J. A. (1996). The European Union and the forms of state: Westphalian, regulatory or postmodern? *Journal of Common Market Studies, 34*(1), 29–52. https://doi.org/10.1111/j.1468-5965.1996.tb00559.x

Choi, M. (2016). A concept analysis of digital citizenship for democratic citizenship education in the internet age. *Theory & Research in Social Education, 44*(4), 565–607.

Coleman, S. (2008). Doing it for themselves: Management versus autonomy in youth e-citizenship. In W. L. Bennett (Ed.), *Civic life online: Learning how digital media can engage youth* (pp. 189–206). MIT Press.

Collinson, D. (2006). Rethinking followership: A post-structuralist analysis of follower identities. *The Leadership Quarterly, 17*(2), 179–189. https://doi.org/10.1016/j.leaqua.2005.12.005
Couture, S., & Toupin, S. (2019). What does the notion of "sovereignty" mean when referring to the digital? *New Media & Society, 21*(10), 2305–2322. https://doi.org/10.1177/1461444819865984
Custers, B. (2022). New digital rights: Imagining additional fundamental rights for the digital era. *Computer Law & Security Review, 44*. https://doi.org/10.1016/j.clsr.2021.105636
Daskal, E. (2018). Let's be careful out there . . . : How digital rights advocates educate citizens in the digital age. *Information, Communication and Society, 21*(2), 241–256. https://doi.org/10.1080/1369118x.2016.1271903
Derrida, J. (1974). *Of grammatology*. John Hopkins.
Derrida, J. (1992). *The other heading: Reflections on today's Europe* (P.-A. Brault and M. B. Naas, Trans.). University of Indiana Press.
Eriksen, E. O. (1999). The question of deliberative supranationalism in the EU. *ARENA working papers* WP 99/4. www.arena.uio.no/publications/wp99_4.htm
European Commission. (2005, October 13). *Plan D for democracy, dialogue and debate*. https://eur-lex.europa.eu/legal-content/EN/TXT/HTML/?uri=LEGISSUM:a30000.
European Commission. (2010). *A digital agenda for Europe*. http://eurlex.europa.eu/legalcontent/SV/TXT/?uri=URISERV%3Asi0016
European Commission. (2021, March 9). Digital compass: The European way for the digital decade. https://ec.europa.eu/info/sites/default/files/communication-digital-compass-2030_en.pdf
European Union. (2000). *Charter of fundamental rights of the European Union*. https://www.europarl.europa.eu/charter/pdf/text_en.pdf
European Union. (2012). *Consolidated version of the treaty on European Union*. https://eur-lex.europa.eu/resource.html?uri=cellar:2bf140bf-a3f8-4ab2-b506-fd71826e6da6.0023.02/DOC_1&format=PDF.
Gak, M. (2021, December 9). A ten-point manifesto for a digital European citizenship. *Open Democracy*. https://www.opendemocracy.net/en/technology-and-democracy/a-ten-point-manifesto-for-a-digital-european-citizenship/
Gleason, B., & von Gillern, S. (2018). Digital citizenship with social media. Participatory practices of teaching and learning in secondary education. *Educational Technology & Society, 21*(1), 200–212.
Hamulák, O. (2018). La carta de los derechos fundamentales de la union europea y los derechos sociales. *Estudios constitucionales, 16*(1), 167–186.
Isin, E., & Ruppert, E. (2020). *Being digital citizens*. Rowman & Littlefield.
Isin, E. F., & Wood, P. K. (1999). *Citizenship and identity*. Sage.
Ivic, S. (2016). *European identity and citizenship: Between modernity and postmodernity*. Palgrave Macmillan.
Karkkainen, B. C. (2004). Post-sovereign environmental governance. *Global Environmental Politics, 4*(1), 72–96. https://doi.org/10.1162/152638004773730220
Karppinen, K., & Puukko, O. (2020). Four discourses of digital rights: Promises and problems of rights-based politics. *Journal of Information Policy, 10*, 304–328. https://doi.org/10.5325/jinfopoli.10.2020.0304
Kasza, J. (2017). Post modern identity: "In between" real and virtual. *World Scientific News, 78*, 41–57.
Kerikmäe, T., Troitiño, D. R., & Shumilo, O. (2019). An idol or an ideal? A case study of Estonian e-Governance: Public perceptions, myths and misbeliefs. *Acta Baltica Historiae et Philosophiae scientiarum, 7*(1), 71–80.
Kostakopoulou, T. (1996). Towards a theory of constructive citizenship. *Journal of Political Philosophy, 4*(4), 337–358. https://doi.org/10.1111/j.1467-9760.1996.tb00056.x
Kymlicka, W. (1990). *Contemporary political philosophy*. Clarendon Press.

Madiega, T. A. (2020). Digital sovereignty for Europe. *European Parliamentary Research Service.* https://www.europarl.europa.eu/RegData/etudes/BRIE/2020/651992/EPRS_BRI(2020)651992_EN.pdf

Meehan, E. (1993). *Citizenship and the European community.* Sage.

Meréchal, N. (2015). Ranking digital rights: Human rights, the Internet and the fifth estate. *International Journal of Communication, 9,* 3440–3449. https://doi.org/10.15307/fcj.mesh.009.2015

Nunes, M. (2017). Selfies, self-witnessing and the "out-of-place" digital citizen. In A. Kuntsman (Ed.), *Selfie citizenship* (pp. 109–117). Palgrave Macmillan.

O'Hara, K. (2013). Welcome to (and from) the digital citizen. *IEEE Internet Computing, 17*(1), 92–95. https://doi.org/10.1109/mic.2013.17

Ohler, J. B. (2010). *Digital community, digital citizen.* Corwin Press.

Pangrazio, L., & Sefton-Green, J. (2021). Digital rights, digital citizenship and digital literacy: What's the difference? *Journal of New Approaches in Educational Research, 10*(1), 15–27. https://doi.org/10.7821/naer.2021.1.616

Rabinow, P. (Ed.). (1984). *The Foucault reader.* Pantheon Books.

Rahm, L., & Fejes, A. (2017). Popular education and the digital citizen: A genealogical analysis. *European Journal for Research on the Education and Learning of Adults, 8*(1), 21–36. https://doi.org/10.3384/rela.2000-7426.rela9113

Reding, V. (2016). Digital sovereignty: Europe at a crossroads. EIB Institute. https://institute.eib.org/wp-content/uploads/2016/01/Digital-Sovereignty-Europe-at-a-Crossroads.pdf

Roberts, H., Cowls, J., Morley, J., Taddeo, M., & Floridi, L. (2021). Safeguarding European values with digital sovereignty: An analysis of statements and policies. *Internet Policy Review, 10*(3). https://doi.org/10.14763/2021.3.1575

Scott, J. W. (1988). *Gender and the politics of history.* Columbia University Press.

Stanford University. (n.d.). Digital divide. https://cs.stanford.edu/people/eroberts/cs181/projects/digital-divide/start.html

Troitiño, D. R., Kerikmae, T., Barbosa, P. A. R., & Shumilo, O. S. (2020). El libro blanco sobre inteligencia artificial: análisis y comentarios sobre mercado, valores y cooperación europea. In *Inteligencia artificial: de la discrepancia regional a las reglas universales: integración de percepciones políticas, económicas y legales* (pp. 303–318). Aranzadi Thomson Reuters.

Turner, B. S. (1993). Preface. In B. S. Turner (Ed.), *Citizenship and social theory.* Sage.

Van Ham, P. (2001). *European integration and the postmodern condition: Governance, democracy, identity.* Routledge.

Westheimer, J., & Kahne, J. (2004). What kind of citizen?: The politics of educating for democracy. *American Educational Research Journal, 41*(2), 237–269. https://doi.org/10.3102/00028312041002237

The European Commission, the Council, and the European Parliament: Differentiated Theoretical Frame for the Digital Revolution

David Ramiro Troitiño

Abstract The European institutions are determined to foster the "Digital Decade" of Europe, based on common digital sovereignty and common standards according to European values and cultural principles, rather than following those of international powers. The priorities of the European institutions focus on data, technology, and infrastructure. Therefore, the combination of effectiveness and social focus is a priority in the process of digitalization. The European Union is making a considerable effort to implement digital solutions, but the process needs the leadership of the EU institutions in order to bring benefits for citizens and companies, respecting the European way of life. Each of the institutions selected for this research, European Commission, European Council and Council of the EU, and European Parliament, answer to different theoretical traditions and stimuli that are analyzed in order to clarify their role in the digitalization process of the European Union.

1 Introduction

The chapter is developed from three conceptual axes, European Commission/Neofunctionalism, European Parliament/Federalism, and Council/Intergovernmentalism. The purpose of the research is to analyze the digitalization process of the European Union from a holistic and current perspective. This theoretical/practical frame defines the definition of the research question used:

Do the institutions follow the theoretical guidelines, on which their functionality is based, in relation with digital integration?

This contribution integrates the activities of the Jean Monnet Chair "Digital Europe and Future Integration" (DEFI)

D. Ramiro Troitiño (✉)
Jean Monnet Chair on Digital Europe and Future Integration, Tallinn University of Technology, Tallinn, Estonia
e-mail: david.troitino@taltech.ee

D. Ramiro Troitiño et al. (eds.), *Digital Development of the European Union*,
https://doi.org/10.1007/978-3-031-27312-4_22

Therefore, it is necessary to analyze whether the Commission follows a neo-functionalist model based on technical integration. Alternatively, if the Council promotes cooperation on a digital level. Is the European Parliament working toward the creation of a digital political identity for European citizens? Does the EP advance at the political level within a common digital framework?

Alternatively, if, on the contrary, the main European institutions, overwhelmed by the digital challenge, have opted for other options to accelerate the process of European digitization from a common approach.

Sources have been conditioned by the novelty of the subject matter, which, on the one hand, means a reduced bibliographical selection, and which often treats the research topic tangentially. The partial deficiency in bibliographic material is compensated by the abundance of working documents published by the institutions studied. This literature link with digital documents supposes a considerable pillar to face the investigation with guarantees. The selection of the selected working documents has been based on several premises, such as the institution that publishes it, the theme, the importance of the digital future of Europe and its dissemination.

2 The European Commission

The European Commission considers that digital technology is changing the lives of humanity globally. Therefore, it considers that a common digital strategy is required within the European Union to mitigate the negative effects of digitization and enhance its positive effects. The European citizens (Ivic, 2012) and companies that enjoy the Single Market are under the jurisdiction of the European Union, and therefore affected by the work of the Commission, the institution in charge of ensuring the management of common actions. Therefore, the European Commission clearly opts for a digitization model based on practical solutions to the needs of citizens and businesses, which, at a theoretical level, supposes a clear commitment to the neo-functionalist model (Nugent & Rhinard, 2019).

Within the global digitization process, the European Commission has identified three specific objectives as priorities at a practical level:

1. Data management
2. Development of European technology
3. Creation of the necessary infrastructures

They are ambitious objectives related to the creation of the technological framework and management of the data generated (Rieder, 2018), to protect the privacy of citizens and facilitate greater competitiveness for European companies, including protection against unscrupulous States able to use the weaknesses of the digital system for their own benefit, or for their national companies. Therefore, the European Commission considers a priority to guarantee the digital security, of both companies and citizens, against the illegal use of technological tools.

The justification given by the Commission itself to assume these powers is related to the consolidation of European digital sovereignty, a vague concept in its theoretical development, although with relevant practical implications. The Commission links digital sovereignty with European values, which are enshrined in Article 2 of the Treaty on European Union entailing:

- Respect for human dignity
- Freedom
- Democracy
- Equality
- The rule of law
- Respect for fundamental rights, including the rights of persons belonging to minorities

Respect for human rights is one of the fundamental obligations of the EU, and therefore all the actors involved in the process of European construction, including the Commission itself and the Member States, must assume these principles.

The Commission supports the creation of the European digital identity, available to EU citizens, residents, and companies that wish to identify themselves or confirm certain personal information (Steible, 2018). It can be used to access services, both public and private, online or offline, throughout the EU, generating a common system for all Europeans. The practical benefits are numerous, such as:

- Access to public services (requesting a birth or medical certificate or communicating a change of address)
- Opening a bank account
- Filing a tax return
- Applying for a university place at any Member State
- Storage of a medical prescription that can be used anywhere in Europe
- Proof of age and identity
- Rental of a car using a digital driving license
- Registration at a hotel.

At the level of integration, the advantages are really notorious since the European digital identity means a big step toward the common identity of European citizens. The digital world facilitates cohesion between citizens with different national affiliations. A common identity is created because the foundations are laid to attract the loyalty of citizens toward the European project, creating the necessary social framework for group identification at a European level. Thus, social compatibility among citizens is encouraged. As the Commission guides the integration process in a common direction, avoiding deviations in the process for national reasons, pursuing a common digital identity seems correct. In addition, it is a sector that is underdeveloped at the national level and that transcends borders, so the common benefit is evident.

Within a more technical framework, the European Commission seeks to strengthen the Union's position in clearly defined areas, such as Artificial Intelligence, seeking to offer citizens the necessary confidence to adopt these technologies, while simultaneously supporting European economic agents to develop tools for use (Outeda, 2020). For which the Commission has drawn up a concise roadmap that includes:

- Creation of a legal framework on AI
- Promote excellence in AI

Artificial Intelligence, as an autonomous technological element, generates numerous unknowns from a political theoretical point of view, with conflicting aspects around the sovereignty and individual freedoms of citizens. The Commission, as guardian of the treaties, has been given the role of protector of European companies and citizens by applying the principle of subsidiarity (the most effective level to solve problems should be in charge of managing the same). Therefore, AI completely overflows traditional borders and state controls, with the European Union managing a fundamental field for common social and economic development. The capacity to face the challenges from a European perspective is greater than from fragmented positions at the national level due to state limitations to face the immensity of the digital world.

Artificial Intelligence (AI) is mainly based on the ability to generate useful data for use in generating patterns, giving data protection crucial relevance. Consequently, the Commission has presented a European data strategy, which seeks to make the EU a leader in the sector, since the creation of a single data market will allow data to flow freely throughout the EU globally for the benefit of companies, researchers, and public administrations.

The digital law service is another of the pillars of the Commission's strategy. The Law consists of a series of regulations on online intermediation services, which are used by millions of Europeans every day. The obligations of the different agents correspond to their respective role, size, and impact in the digital ecosystem. This action will open new opportunities to provide digital services across internal borders, guaranteeing the protection of users. It will allow the creation of a single market for digital services, adapting it to the current Single Market. Economic integration has been one of the great successes of the European Union, the freedom of economic exchanges, together with the freedom of movement of workers and services, has led to the creation of the largest trading bloc in the world. Successful economic integration has provided tangible benefits to Europeans, increasing citizens' sense of belonging and loyalty to Europe. Likewise, it has reduced tensions between European nations by obstructing mercantilist options and discrimination between Europeans, with very specific measures for equal treatment between all the agents involved.

This action is complemented by the Digital Markets Law, which seeks to control the dominant position of certain platforms so that there is more equitable competition, meaning an equivalent to the active antitrust laws in the Single Market. Controlling the big platforms, generally from the United States, also seeks to facilitate competition from small and medium-sized European technology companies, which currently lack the necessary muscle to face the big technology corporations (Caena & Redecker, 2019). Therefore, a common problem for Europeans, which cannot be resolved individually by the Member States due to their size and lack of cross-border efficiency, but which a united Europe can tackle with a certain guarantee of success.

The expansion of the economic model to the digital world also has important consequences for European integration and strengthens the role of the Commission. The Neo-functionalist model is based on the integration of fields with minimal political exposure, which do not attract too much media attention. Contrary they have the potential to promote further integration. This option argues that economic integration will promote political integration in order to control by democratic means the economic agents. Therefore, the digital economy presents unlimited options for the future, which at a certain point will require political control. Consequently, this political regulator must also be European, facilitating the creation of a deeper political framework within the European Union.

In reference to the integration of the digital economy at a European level, the European Commission will assume a relevant role in the organization's international relations. Like the Single Market, the digital market will involve the elimination of barriers between Member States and the creation of common external barriers, which will be managed by the Commission, the guarantor of the commonwealth. It gives the organization a very active role in an economic area that is constantly growing, and whose expectations are very high in terms of generating economic resources. In short, it will represent an important external power for the European Commission, which it will manage independently under the supervision of the European Council, one more step toward a united Europe.

The last of the Commission's priorities in the digitization process refers to the process itself. By Europeanizing the digital economy, Artificial Intelligence, administrative services or digital identity, common frameworks are created that need common protection. In this sense, the Commission, as a managing body, seeks to promote European cybersecurity that protects the European digital fabric. A European system protected by national institutions incapable of crossing the borders of the States could not have the same efficiency as a level of European protection managed in a common way. The accumulation of national resources would also rationalize spending and increase the effectiveness of initiatives. Therefore, it makes perfect sense that, by creating a European digital space, it should be protected in the same dimension, the European one.

3 The European Parliament

The role of Parliament in the institutional framework of the European Union is clearly focused on common European sovereignty and is visualized through the legislative process. At the beginning of the integration of Europe, the so-called Common Assembly of the Coal and Steel Community had no more meaning than that of a further evolution planned. Since then, each European Treaty has increased the powers and relevance of Parliament, until reaching the current levels of power, similar to those of the European Council in the legislative and budgetary process. Perhaps the most relevant milestone has been the democratic election of members of parliament in 1979, thus providing the institution with basic political support for its very essence. The representation of European citizens, who in turn are bearers of the citizenship of a Member State, has been in conflict with those States closest to intergovernmental theses, which consider that European sovereignty does not exist and that national parliaments are the last recipients of the individual freedom of the members of society. Thesis that defends that democracy needs a demo, and currently there is no cohesive European demo (Ramiro, 2022a). Despite this, the vast majority of member states have supported the evolution of the EP and its growing power, as reflected in the Treaties negotiated and approved by the capitals of Europe.

Despite the evident evolution of the powers of the EP, there are still numerous obstacles in its operation to reach an optimal dimension, which the digitization process could remit, mainly by improving European democracy and bringing citizens closer to parliament. Although, technological evolution cannot be imposed autonomously, since the prerogative of granting itself new powers does not correspond to the European Parliament, but to the governments of the Member States. It seems sensible to think that in the coming years the Council will face the digital revolution with an adaptation of the European institutions to them, for which a new Treaty will be necessary. However, the difficulty of approving new treaties, due to the unanimous requirement, means that the technological aspects will probably be added to the Treaties when it is essential. It will be included in a text focused also on other issues, such as an adaptation to the possible expansion of the organization to the Balkans, or a Treaty to profoundly reform the Union, perhaps with policies such as defense, fostered by the conflict in Ukraine. Therefore, the uses of the possibilities of new technologies for online democracy linked to the European elections will have to wait for an atmosphere favorable to change in a more general sense.

Currently, the EP focuses on the legislative process related to digital issues, and the preparation of reports to ask the Commission for concrete actions on specific content. Obviously, this means that on many occasions the EP and the Commission share objectives when it comes to digitalization. Although the negotiation with the European Council defines the final contents of the initiatives of a legislative nature.

Digital transformation is one of the current priorities of the European Parliament, which is why it is working in this direction to promote the adaptation of European companies and citizens, sharing the priority with the European Commission, to the new technological ecosystem. Therefore, the adaptation of companies to change, and the learning of citizens through digital education are priorities. The pedagogical factor is essential to provide qualified labor to European companies, facilitate digital consumption by citizens and digitize public services, while respecting fundamental rights and values. In addition, the European Parliament links the technological revolution with the European ecological transition and climate neutrality in 2050 due to the possibilities it offers to face the environmental challenge in a more effective way. The ambition of the European Parliament is great, since it combines different objectives with digitalization, opening the doors to a European citizenry educated in the face of the technological challenge, preparing the ground for the implementation of online democracy in the not-too-distant future (Tsakalidis et al., 2020).

Once, citizens have acquired the necessary skills, society will be ready for a system of digital democracy that connects citizens with the European Parliament directly and allows the implementation of a single European electoral system, promoting the creation of European political parties and increasing citizen participation in European elections. Despite the importance of these actions, the EP has not yet fully raised them and focuses on more practical and plausible actions in accordance with current conditions.

In May 2021, the European Parliament approved a report on the configuration of Europe's digital future in which it asked the Commission to respond to the challenges presented by the digital transition. The adopted resolution was a clear response to the Commission's strategy to shape Europe's digital future. MEPs called on the Commission to address the challenges posed by the digital transition, such as digital skills or connectivity. The resolution was adopted by 571 votes in favor, 17 against, and 105 abstentions. Therefore, a significant part of the Chamber was in favor of shaping the digitization of the European economy and society, emphasizing the need for strong institutional support for digital innovation. Consequently, a joint legal frame must be provided, a clear path for European technological evolution, facilitating exchanges, promoting a technology that citizens support, accessible and respecting European freedoms and values. The EP has focused its support for the digital economy on European SMEs, which need adequate institutional support to adapt and benefit from new technologies, be it through trial facilities, better access to data, easier regulatory requirements or financing. The federalist approach is clear, protection of citizens and small and medium-sized enterprises with strong European roots, which through their digital Europeanization become autonomous agents in the process of European construction.

In reference to Artificial Intelligence, the European Parliament maintains that AI solutions could lower existing barriers and reduce the fragmentation of the Internal Market, support the European digital economy and its competitiveness, also contributing to security, education, health care, transport, and the environment (Cohen et al., 2020). Parallel, MEPs advocate for a clear legal framework for AI as a prerequisite for its proper use that respects European values. Parliament is not only concerned about technological aspects but also how new developments could affect European citizens and their way of life (Hamulak, 2015). Therefore, the EP concerns about European citizens, the source of its power and legitimacy; a relevant example to link the parliament's position to federalist options in the digitization process.

Technological developments based on algorithms depend on their programmers, on the political and social environment where they are developed. Therefore, implementing Chinese or North American rules in AI would mean accepting the influence of their social and political models within Europe. Faced with this situation, the EP advocates taking the lead in this field, in order to protect Europe from foreign influences, from other systems that could transform the European model. EP supports the protection of the European citizen against other corporate models, and internationally expand the level European protection.

This theoretical frame has been implemented with important legislative initiatives, such as the civil liability regime for artificial intelligence, or the regulation on ethical aspects of artificial intelligence, robotics, and related technologies. At the same time, the European Parliament is working on new legislation on digital services, which will especially affect digital platforms and their control and influence over the content they host. It has approved a European Cybersecurity Center and to prevent the spread of terrorism and digital threats as a priority area of action (Kajander et al., 2020).

EP's own initiative on matters relating to:

- Intellectual property rights for the development of artificial intelligence technologies.
- General guidelines for the use of AI in the civil and military sector.
- The influence of artificial intelligence in education, culture, and the audiovisual sector.
- Recommendations to remove barriers to the functioning of the digital Single Market.
- Improve the use of artificial intelligence for European consumers.

Reflect the priorities of the European institution. These recommendations drawn up by the European Parliament are fundamental in the internal functioning of the Union since it is the only directly democratically elected body in the EU. However, unlike the parliaments of the Member States, it has virtually no formal right of legislative initiative, which rests within Commission, and, to a limited but increasing extent, includes the European Council and the Council of the EU. Parliament's "own initiative reports" are therefore a widely underestimated initiative, and a little-recognized tool for informally shaping the EU's political agenda by creating a cooperative environment with the rest of the institutions involved in the legislative process. In this sense, the effort being made by the EP to influence the EU digitization process based on the numerous self-initiative recommendations mentioned here is relevant.

Other proposals that the European Parliament is currently working on its own initiative referring to artificial intelligence in criminal law and its use by police and judicial authorities in criminal matters. In addition to a report on artificial intelligence in the digital age, derived from the work of the Special Committee on Artificial Intelligence in the Digital Age of the EP.

As far as the recommendations are concerned, the EP has also given its opinion on tax matters, always very delicate in the inter-institutional relations with the Council, since they influence the self-financing of the Member States and their capacities to finance their social systems, an essential part of their national differentiation. In any case, the EP's recommendation seeks to adapt the fiscal framework to the technological reality, updating the tax burden to an economy that is increasingly linked to the digital world. The European Parliament proposes the introduction of a minimum effective corporate tax rate and new taxation rights (Künnapas, 2016). Consequently, taxes will be paid where value is created and not where taxes are lowest. The big technology companies are taking advantage of the legal loopholes generated by the prevailing dissociation between the Single Market and the digital economy. Therefore, an adaptation of the system is urgently needed to avoid the pernicious effects generated by the lack of legal adaptation to technological advances in the economic world (Kerikmäe & Särav, 2017).

In reference to the financing of the digitization process in Europe, the EP is also involved from a budgetary point of view, since its powers in approving the European Union budget are substantial. The budgetary power of the European Parliament has historically been the institution's way of influencing the process of European integration. Therefore, the amendments to the budget introduced by the EP influence the direction of European policies, and the EP's priorities in terms of digitization are reflected through this financial instrument. On a practical level, the recovery plan for the economy after the crisis generated by COVID includes digital transformation among its contents; 20% of the 672,500 million euros allocated. In addition, the Horizon Europe investment programs, focused on research and innovation, and the Connecting Europe Facility, dedicated to infrastructure, allocate significant funds to digital advances. Parliament has approved the implementation of the Digital Europe Program, which is the main financial instrument for direct and exclusive use in the digitization process. Its objective is to invest in digital infrastructures so that strategic technologies are aligned with the priorities expressed by the different European institutions. The European Parliament calculates that the program contributes 7588 million euros in five areas of action:

- High-performance computing: 2227 million
- Artificial Intelligence: 2062 million
- Cybersecurity: 1650 million
- Advanced digital capabilities: 577 million
- Implementation of digital technologies in the economy and society: 1072 million

The budgetary effort gives a clear reference to the EP's priorities in terms of digitalization, and supports the conclusions reached through the legislative initiatives and the recommendations issued by the parliament.

4 The European Council and the Council of the European Union

Both institutions represent the interests of the Member States within the European Union and base their operations mostly by consensus and cooperation, seeking common positions before imposing majority solutions, although not unanimous. Therefore, its intergovernmental character is part of the essence of its operation (Ramiro, 2022c). Obviously, it is often not possible to reach common positions among so many members, but dialogue is an important part of the Council's work.

Among the powers of the Council, highlights its ability to direct the process of European construction through the definition of political priorities stands out. Power that they have frequently used in relation to the digitization process of the European Union. The Council of the EU at the end of October 2021, included in its conclusions a request for an agreement between the European Council (dependent on the Council of the EU at a practical level) and the EP, in reference to legislation related to the digitization of the European Union. The fields indicated in the Council's conclusions include the regulations on roaming, the Digital Services Law and the Digital Markets Law. In addition, advances are requested in legislative files related to sectoral data spaces, artificial intelligence, a European framework for a digital identity and a European ecosystem of microchips. In agreement with the other two institutions involved in the legislative process, the Council also supports advances in cybersecurity, in the hope that common management will increase the effectiveness of actions aimed at protecting the European digital space. These conclusions clearly mark the digital agenda for the coming years, due to their influence on other institutions of the European Union and on the governments of the Member States (Verina & Titko, 2019).

In reference to the Council, due to its theoretical link with Intergovernmentalism, the references after the March 2021 meeting to European digital sovereignty are surprising. The institution shares the Commission's view on the matter and is open to the joint development of the concept through the Digital Compass 2030. Conversely, the reference is quite general and does not have practical consequences, therefore, it is very likely that sovereignty European digital network will be a source of conflict in the near future between the European institutions due to the high political component that is associated with it.

At the fiscal level, the Council has clearly positioned itself within a framework based on international cooperation, with a possible global tax that provides a global solution to the problem of paying taxes by large technological companies that derive their income from their online activities, escaping from the control of public administrations. It is an agreement between governments worldwide to solve a global problem through implementation at the state level without supranational controls. The Council even threatens unilateral actions at the European level to put pressure on the rest of those involved at the international level, although a common tax system does not seem feasible due to what it means at the level of integration in Europe. A confrontation with the Commission and the EP is expected regarding the proposal for the creation of a European digital tax for 2023, because the Member States participating in the G20 have reached an agreement at the end of October 2021 to impose a global tax minimum of 15% to companies to discourage them from using accounting resources in order to evade taxes in tax havens. Although it is not yet clear if the United States Congress will endorse the agreement or if its implementation will be effective.

The political indications of the Council of the EU, which brings together the presidents or prime ministers of the participating countries in the EU, have great weight in the European Council, a more practical institution made up of ministers, and is in charge of the legislative and budgetary process in collaboration with the European Parliament. Likewise, since there are no European political parties, MEPs belong to national political parties, often in the governments represented in the

Council, so their influence over the EP is relevant. However, the independence of parliamentarians is significant, with examples in which they have not followed the guidelines of the leaders of their political parties.

The Council's intergovernmental option with respect to the European digitization process is clear, although the negotiations with the Commission and Parliament divert this model toward a compromise solution.

5 Conclusions

In the digital world, there have not been yet major conflicts between the European institutions, although it is possible that they will occur in the coming years in reference to two specific fields, European digital sovereignty and digital taxation. The transnational nature of the digital world makes the EU the most appropriate framework for its management and control, despite the traditional attitudes of the Member States regarding common digital policies with relevant social implications.

On the rest of the fields included in the digitization process, especially the Single Digital Market and Cybersecurity, there is a union of common objectives among the European institutions facilitating effective and relatively rapid common actions.

In any case, the European institutional framework is designed to incorporate different visions and actors into the process of European integration. Their priorities do not have to coincide, but the negotiation process implies the inclusion of different approaches in a common framework. Decision-making in the EU has often been criticized for its slow pace and compromise between positions leading to often-unambitious results. Nevertheless, it is also the strength of the Union, as it is a free Union seeking the greatest possible benefit for the greatest number of participants involved.

The axiomatic goodness of digital integration, which is obviously more effective at a European level than the national administration, overwhelmed by the impossibility of control beyond traditional borders, means that the speed of digital integration is not affected by relevant dogmatic controversies between the actors involved in decision-making.

It is even possible that digital aspects facilitate integration in other fields, such as the creation of a common European cybersecurity, which can positively influence the development of a common defense in the European Union, or the creation of a European police force that fights against European crimes.

The research reveals a clear respect for the three institutions included in the study with their initial theoretical parameters. The Commission's position on the digitization process is clearly neo-functionalist, despite influencing the concept of digital sovereignty (Ramiro, 2022b). The European Parliament seeks to complement the actions of the Commission with a federalist approach that affects European citizens and companies, temporarily setting aside the more federalist aspects related to the creation of a single electoral system and a true European digital democracy. Finally, the Council (in its two versions) is promoting global agreements based on cooperation, following an intergovernmental approach. The actions regarding digitization reveal that the traditional position of European institutions follows the established patterns despite being a new topic.

The functionality of the European Union itself faces a joint challenge leading to common actions to solve the problems posed by digitization, predicts a predominance of neo-functionalist theses in the medium term, which ultimately lead to long-term federalist options to give democratic legitimacy to digital Europe. Therefore, the process, despite current traditional models at an institutional level, has a marked European meaning dictated by digitization itself, which will largely depend on the flexibility of the Member States to adapt to an irreversible process. The rise of populism and nationalist positions in Europe is a fact that may affect the digitization process, probably slowing it down.

Bibliography

Caena, F., & Redecker, C. (2019). Aligning teacher competence frameworks to 21st century challenges: The case for the European Digital Competence Framework for Educators (Digcompedu). *European Journal of Education, 54*(3), 356–369.

Cohen, I. G., Evgeniou, T., Gerke, S., & Minssen, T. (2020). The European artificial intelligence strategy: Implications and challenges for digital health. *The Lancet Digital Health, 2*(7), e376–e379.

Hamulak, O. (2015). Is charter of the fundamental rights of the EU taking social rights seriously? *European Studies: The Review of European Law, Economics and Politics, 2*, 14–28.

Ivic, S. (2012). EU citizenship as a mental construct: Reconstruction of postnational model of citizenship. *European Review, 20*(3), 419–437.

Kajander, A., Kasper, A., & Tsybulenko, E. (2020, May). Making the cyber mercenary–autonomous weapons systems and common article 1 of the Geneva conventions. In *2020 12th International Conference on Cyber Conflict (CyCon)* (Vol. 1300, pp. 79–95). IEEE.

Kerikmäe, T., & Särav, S. (2017). Paradigms for automatization of logic and legal reasoning. *Law and Logic: Contemporary Issues.* Duncker & Humblot, 205, 222.

Künnapas, K. (2016). Maksukohustuse täitmise preventiivne tagamine enne maksukohustuse tuvastamist: Ettevaatuspõhimõte maksumenetluses. *Tartu Ülikool kirjastus.*

Nugent, N., & Rhinard, M. (2019). The 'political' roles of the European Commission. *Journal of European Integration, 41*(2), 203–220.

Outeda, C. C. (2020). La inteligencia artificial: Una lectura desde la política. In *Inteligencia artificial: De la discrepancia regional a las reglas universales: Integración de percepciones políticas, económicas y legales* (pp. 319–344). Aranzadi Thomson Reuters.

Ramiro Troitiño, D. (2022a). George of Poděbrady: The role of external stimulus to the European integration. In *The European Union and its Political Leaders* (pp. 5–16). Springer.

Ramiro Troitiño, D. (2022b). Jean Monnet: Neofunctionalism at work in the European integration. In *The European Union and its Political Leaders* (pp. 121–140). Springer.

Ramiro Troitiño, D. (2022c). François Mitterrand: French leadership in the European Union. In *The European Union and its Political Leaders* (pp. 233–242). Springer.

Rieder, G. (2018). Tracing big data imaginaries through public policy: The case of the European Commission. In *The Politics of Big Data* (pp. 89–109). Routledge.

Steible, B. (2018). EU support to domestic prosecution of violations of International Humanitarian Law. *UNIO–EU Law Journal, 4*(1), 51–66.

Tsakalidis, A., Gkoumas, K., & Pekár, F. (2020). Digital transformation supporting transport decarbonisation: Technological developments in EU-funded research and innovation. *Sustainability, 12*(9), 3762.

Verina, N., & Titko, J. (2019, May). Digital transformation: Conceptual framework. In *Proc. of the Int. Scientific Conference "Contemporary Issues in Business, Management and Economics Engineering '2019", Vilnius, Lithuania* (pp. 9–10).

Artificial Intelligence: A Reading from European Politics

Celso Cancela Outeda and Bruno González Cacheda

Abstract At least since 2018, Artificial Intelligence (AI) has been in a state of real effervescence. At the end of 2022, it became popular thanks to an advanced AI called Chat GPT. The proliferation of countless AI tools, in commercial versions, applicable to various professional, public and private spheres, has fuelled the debate on its evolution, impact, use or regulation. It is an exponential, disruptive, cross-cutting technology that raises ethical, legal and political questions and challenges. This chapter attempts to provide a political reflection on the EU's regulatory approach to AI. First, it outlines the general context in which the EU should address the regulatory treatment of AI. Second, it looks at the threats and risks generated by AI which, to a large extent, are conditioned by the adoption of a techno-optimistic or techno-pessimistic reading. Thirdly, it examines models of governance and public intervention in the area of AI. Finally, it reviews the response offered by European institutions to the challenges posed and the regulatory model adopted in the field of AI.

1 Introduction

Generally speaking, Artificial Intelligence (AI) refers to both a field of research and a technology that enables machines to replicate human reasoning, the ability to learn and solve complex tasks (Russell & Norvig, 2016). It is based on algorithms that already make decisions about different areas of our lives. This development of AI makes it necessary and urgent to reflect on its potential and transformative capacities, as well as its consequences and impacts in different spheres (political, economic, social, legal, ethical, etc.). Sometimes, analysts and experts present optimistic scenarios that almost border on the prodigious. However, on other occasions, from

This contribution integrates the activities of the Jean Monnet Chair "European Institutions and Citizens: Adapting the EU to the digitalization challenges" (EICAEUD)

C. C. Outeda (✉) · B. G. Cacheda
Universidade de Vigo, Vigo, Spain
e-mail: ccancela@uvigo.gal; bruno.gonzalez.cacheda@uvigo.gal

D. Ramiro Troitiño et al. (eds.), *Digital Development of the European Union*,
https://doi.org/10.1007/978-3-031-27312-4_23

a pessimistic point of view, concerns and fears are expressed regarding its implications in the economic, ethical, legal, social or political spheres. In this sense, according to Abdala et al. (2019), on the positive side, it can facilitate citizen participation, improve service delivery and efficiency through human-AI collaboration in the workplace and policy innovation, increase productivity geometrically and add 15.7 trillion dollars to the global economy by 2030; on the negative side, the development of "killer robots" (lethal autonomous weapons), the Cambridge Analytica case or Amazon's facial recognition errors show its impacts on the functioning of democratic systems, or fundamental rights. In short, like other technologies, AI also openly shows its ambivalent character.

If we consider AI from a political perspective, numerous and diverse applications appear in fields such as international relations, political theory, public policy and public management, among others. For example, it affects relations between the US, China, Russia and the EU, AI governance models, the evolution of job profiles in public administrations, democratic participation, the development of public policies or the provision of public services (Ramió, 2019). The list could be extended to include the reliability of AI systems (introduction of bias or discrimination based on gender or race in the coding of algorithms), the guarantee of civil rights and political values (privacy, freedom and equality), the worsening of social inequalities, market concentration (monopolies or oligopolies) or the widening of the digital divide. Therefore, attention should not be limited to the condition of mere users or consumers (attention to and assessment of the effectiveness and efficiency of the practical uses and applications of AI), but should, at least in Western societies, adopt a political perspective that considers its impact on rights and freedoms (civil and political), that assesses the opportunities and risks for society and that is committed to a governance model (self-regulation or public regulation).

It is precisely the question of governance that is proving to be of great importance. Currently, the debate on AI has been more oriented towards its ethical aspects and self-regulation through standards, protocols or ethical codes. Supporting this position, given that AI is in its infancy, some experts have argued that it should not be regulated in order to foster its potential. But then the need for regulation appeared. At the 2020 Davos Forum, representatives of companies such as Microsoft and Google argued in favour of regulation.[1] It was pointed out that the absence of regulation resulted in a lack of legal certainty and technological trust, which slows down the development of AI applications (Sousa, 2020).

We are now, therefore, at a stage where AI must be approached politically and political responses must be given, which requires reflection on the regulation and organisation of AI based on political principles and public values. Here is where the difficulties begin. As Abdala et al. (2019) point out, "the discussion about the best path towards regulation is divided between those who support making vertical, sectoral and specific rules (e.g. AI Sector Deal in the UK), and those who prefer cross-cutting regulations, depending on the type of technology [...]. The first group considers that specific policy domains such as health or education have their own

[1] In 2017, Elon Musk came out publicly in favour of regulating the use of AI.

trajectories, regulatory frameworks and risks, so that a national AI body would find it difficult to meet the requirements of sectoral expertise needed to achieve a nuanced regulatory model. The second group thinks it is necessary to develop shared standards and to ensure interoperability, for example, of privacy systems".

Within the EU, AI raises the question of competence for its regulation: should it be at state or European level, should it be an area of exclusive competence of the EU or shared with the member states? In practice, the present development of AI shows a scope, content and nature that go beyond the political-institutional framework of member states (Ramiro et al., 2020).[2] In fact, the EU institutions -exclusive stakeholders of Big Tech- have taken the initiative and have been developing a regulatory framework since the middle of the previous decade; thus, in 2018, the High-Level Expert Group on AI was activated and it elaborated the European Strategy for AI.[3] We will return to this idea later. From the perspective of European integration, new technologies such as AI will favour the expansion of EU competences. As Ramiro et al. (2020) write, this "is a fertile field because of its potential, which to a certain extent is not yet regulated by the member states, making it a fertile field for common initiatives, for the European treatment of issues that transcend the traditional borders of the states".

This chapter provides a political view on the EU's regulatory approach to AI. First, we will outline the general context in which the EU should address the regulatory treatment of AI. Second, we will look at the threats and risks generated by AI, which, to a large extent, are conditioned by the adoption of a techno-optimistic or techno-pessimistic reading (Kerikmäe et al., 2019). Thirdly, we will study models of governance and public intervention in the area of AI. And finally, we will look at the response offered by European institutions to the challenges posed and the regulatory model adopted in the field of AI.

2 What Is at Stake

The governance and regulation of AI in the EU requires consideration of some of its various implications (political, economic, legal, technological, etc.). AI has developed quickly and become extraordinarily important in less than a decade. It is affecting our societies at an accelerating pace. It is often accompanied inseparably by big data and machine learning. We are generating an ever-increasing amount of data that manifests attitudes, preferences, views and personal opinions. Public and

[2]Ramiro writes that "European countries individually lack the capacity and resources to effectively regulate the actions of actors beyond their control by spilling over the geographical reach of state power. Therefore, the need for a broader regulatory environment with effective power to enforce the rule of law is a necessity for Europe" (Ramiro et al., 2020).

[3]Communication from the Commission to the European Parliament, the European Council, The Council, The European Economic and Social Committee and the Committee of the Regions, Artificial Intelligence for Europe, [COM(2018) 237 final] (last access 10–05–2022).

private actors collect and use it in multiple ways thanks to AI. On the one hand, private companies appropriate this data to increase their profits and business opportunities; on the other hand, public actors can use it for safety and security applications and the public sector to improve public services. It is clear that AI offers novel opportunities in many fields of public interest such as education, training, health, safety and security or public services or democratic participation. However, both its private and public uses raise risks and threats. Because it is at the forefront of datification processes, this technology impacts on both fundamental individual values (privacy, freedom or equality) and collective values (equity, justice, security, safety, inclusion, accountability and democratic control) (Cancela-Outeda, 2020).

More specifically, aspects such as data ownership or inalienability, regulation of data flows, neutrality and explainability of algorithms and machine learning, sensitive applications such as facial recognition, etc. require regulatory treatment. Consequently, policy strategies must be adopted to ensure the integration of this technology and certain (legal and ethical) values and principles according to the White Paper on Artificial Intelligence: a European approach to excellence and trust (-European Commission, 2020). On the one hand, it is necessary to consider its possible negative impacts on fundamental rights and democracy (protection of citizens from potential abuses originating in the public and private spheres) and, on the other hand, to investigate the exploitation of possible opportunities in the field of political participation or the provision of public services. It is already possible to consider some incipient experiences based on AI. In New Zealand, there is a virtual politician named Sam created by artificial intelligence.[4] In the Tokyo district of Tama, one of the human mayoral candidates proposed that an AI-powered robot would make district decisions.[5] These possible uses of AI in political practice raise some questions: could AI take over the political activity of a local community? Could it participate in elections with its own electoral programme? Will it replace politicians or will it live alongside them to assist them in their decision-making? How will AI affect representative democracy, i.e. which criteria should be considered, that of the majority or that of AI? Of course, there is no shortage of arguments in favour of the use of AI in the political arena: the contribution of objectivity (elimination of subjectivity, primacy of the general interest, etc.), the avoidance of errors in decision-making or corrupt practices, and the capacity to offer efficient solutions to a large number of global problems (famine, poverty, education, health, etc.).[6]

In relation to this, Ramió argues that "there is almost unanimity in the specialised literature that artificial intelligence can generate enormous social, economic and

---

[4] Sam, http://www.politiciansam.nz/about.html (last access 10–05–2022).

[5] Another example is the candidate robot Michihito Matsuda. https://elpais.com/internacional/2018/04/18/mundo_global/1524045163_744119.html (last access 01–05–2022).

[6] According to European Tech Insights 2019, a survey conducted by the Center for the Governance of Change in several European countries indicated that 25% of respondents were willing to be governed by an AI. Available https://docs.ie.edu/cgc/European-Tech-Insights-2019.pdf (last access 10–05–2022).

sustainability benefits in the short term and also at a labour level, but in the longer term (fewer working hours and minimum universal income) [. . .]. Solving these major global challenges can be a very effective way to achieve greater development and legitimacy of democratic systems". Presumably, "this impact [of AI] on politics can be a deeper, more collaborative and deliberative democracy that uses technology through direct participation systems that can capture collective intelligence. With much more transparent political power, more accountable and more subject to expert citizen scrutiny, etc. But, beware, it could also have a totally opposite drift and call into question the values and essence of the democratic system" (Ramió, 2018).

Access to mass data on people's daily behaviour is so extensive that doubts or criticisms can no longer be confined to the concept of privacy and its effects. From this emerge other types of challenges that affect the very foundations of the modern-liberal order. They are challenges that impact the political integrity of societies and the future of democracy (Mäskine & Kasanen, 2020). Several authors have already referred to these issues. In particular, Sadin considers that the massive collection and processing of data implied by AI, coupled with the algorithmic systems that are responsible for managing the course of our individual and collective existences, imposes a digital soft-totalitarianism (Sadin, 2018). Similarly, Harari argues that "in its current form, democracy will not survive the fusion of biotechnology and info-technology. Either it will successfully reinvent itself in a radically new way, or humans will end up living in 'digital dictatorships'" (Harari, 2018).

Algorithms that support AI can identify fears, desires and needs, and this information can be used against citizens. Thus, the abusive use of surveillance and manipulation devices could render representative democracy unworkable and create an "informational dictatorship" (Giraldi, 2019). In this sense, it will be necessary to pay attention to the political (and social) effects that may rise from the personalisation of content through AI, linked to the echo chamber effect in shaping public opinion, to the radicalisation that may drift towards violent extremism, to the preservation of informational and media pluralism connected to freedom of expression, or to digital tools of mass surveillance. In democratic systems, AI can contribute to the improvement of certain elements such as participation, deliberation, assistance in the design of public policies, control and accountability, although the generation of social inequality that can cause political polarisation, populism or authoritarianism cannot be excluded.

While bringing and enhancing socio-economic benefits, AI creates new risks for society as a whole. It is a quickly changing, cutting-edge technology under the monopolistic control of a group of countries that has already generated fierce competition between China, the US and the EU.[7] European authorities are then under the influence of other pressures to regulate the field of AI. Today, there is

[7] "There are currently two countries that can exercise effective power in the field of new technologies in general, and artificial intelligence in particular, which are the United States of America and the People's Republic of China, whose massive size allows them some effective control over the development and implementation of applications related to artificial intelligence" (Ramiro et al., 2020).

already a real competition between state actors that want to lead in this technological field. The words of Russian President Vladimir Putin who, in 2017, declared that "whoever leads the race for artificial intelligence will rule the world" are illustrative. The statement supports the sense that a new "arms race" involving research centres and universities, private companies and states is underway. In reality, a global competition, often labelled the "AI Race",[8] is taking place between three global players: China, the US and the EU (Antonov & Kerikmäe, 2020). This trio of actors is joined by other countries that have taken particular initiatives in this area (Girasa, 2020; Robles, 2020). All are already aware of the importance and value of AI as an instrument of global geo-economic and geopolitical competition and of its domestic and international transformative potential (Arteaga, 2019).

Nowadays, the EU lags behind in the technology race which could further increase its dependence in this area. In this context, the EU is aiming for a balanced regulatory approach that addresses several objectives: global technological leadership, harnessing its economic and social benefits in various sectors and activities (climate change, environment, health, public sector, finance, mobility, security, agriculture and so on) or developing and operating in line with values, principles and fundamental rights. This does not seem an easy task. Indeed, in the *Report on artificial intelligence in the digital age,*[9] the European Parliament has called for "light-touch regulation" that, on the one hand, counteracts the foreseeable negative effects of this technology and, on the other hand, allows the necessary capacity for innovation and the development of European technology companies. The importance of AI development and the need for regulation is recognised, but without impeding the expansion of the economic sector on a European scale.

Apart from this area, different approaches to AI are detectable: economic performance for the US, control of people for China, and the human-centric and democratic approach for the EU. In the regulatory sphere, there are also notable contrasts between the US, which is more inclined towards a laissez-faire and minimal regulation (or self-regulation) approach, China, which is inclined towards state protagonism through expansive regulation, and the EU, which "has emphasised the need for a human and ethical digital economy" (Newman, 2020; Ramiro et al., 2020). Beyond the point of view set out by Newman (2020), the approach and governance model articulated by the EU so far will be analysed in the following sections.

Connected to this issue, in Western systems another aspect linked to the use of AI, mainly by private companies, is the preservation of competition in the digital market. Governments are trying to counterbalance the power of Big Tech, i.e. the power of large technological monopolies or oligopolies. This concentration of digital power,

---

[8]Harari, Y. "Who Will Win the Race for AI?", *Foreign Policy*, 2019. Available at https://foreignpolicy.com/gt-essay/who-will-win-the-race-for-ai-united-states-china-data/ (last access 10–05–2022).

[9]Report on artificial intelligence in a digital age (2020/2266(INI)). Available at https://www.europarl.europa.eu/doceo/document/A-9-2022-0088_EN.html (10–07–2022).

in addition to eventually compromising the functioning of democratic systems, can be a factor in strangling innovation and creativity (Ortega, 2021).

Although with some delay compared to the US or China and following a mainly reactive approach (measures on guaranteeing privacy or cybersecurity and avoiding excessive concentration of the digital market), the EU has been shaping its own agenda. In particular, the European Commission, after assuming the strategic importance of AI (and big data) at both the economic and strategic level, has since 2014 adopted several initiatives for its promotion and regulation (EU data and AI policy). In addition to the protection of civil rights, it seeks to promote European digital and technological sovereignty in a way that translates into increased global competitiveness of the EU economy. Generally speaking, digital sovereignty implies, on the one hand, the ability to manage and control the use of European data and, on the other hand, to have its own resources (human capital, data, technology and infrastructure) and the possibility to set its own rules. In particular, in order to face digital competition from the US or China, the EU is already promoting European innovation and the preservation of control over the management of the European digital space, as well as the attribution of responsibilities to the actors involved.

In this sense, the Digital Services Act (DSA), which is in the process of being formally approved, aims to ensure that competition rules are also respected in digital markets, especially by Big Tech (Amazon, Apple, Google, Microsoft, Alibaba, Huawei), while establishing criteria for the accountability of online platforms and the protection of internet users and their basic rights. The aim is to reduce the enormous dominance of these technology companies over the digital market, while forcing them to be more accountable and transparent. It is expected to reduce the publication of illegal content, establishing the responsibility of the platforms with regard to the publications made in their virtual spaces, increasing the transparency of their algorithms and giving users greater control over the data and information they share.

Alongside the Digital Services Act, political agreement has also been reached on the Digital Markets Act (DMA), which will provide for a new regulatory regime applicable to certain platforms that provide digital services. Basically, it aims to avoid actions by dominant platforms in the internal market that affect competition. In the coming years, several regulations are expected to be approved, such as the Data Governance Act, included in the European Data Strategy, and the Data Act, the AI Act, the Chips Act, and the Cyber Resilience Act. This regulatory development aims to enable the EU to play its own role in the digital economy (reducing external dependence) and to create a framework of legal certainty.

On the other hand, both private companies and public bodies can use AI tools provided that big data are available at European level in a sort of common data space. This requires common standards, interoperable data collection and storage, processing and aggregation infrastructures in specific domains. In other words, a single data market needs to be established in which data circulates freely throughout the EU and is accessible to businesses, universities, research centres and public administrations. Data, the indispensable underpinning of AI, is a key element in the digital transformation being driven by the EU and its member states. To this end, access to the growing volume of data being produced must be ensured and the technical capacity for its use must be available. In short, they constitute essential material for innovation and economic growth (digital or data economy). In the coming years, AI, accompanied by big data, will drive innovation and bring benefits to individuals and the European economy as a whole. It will generate economic growth and help tackle global problems such as climate change, local ones such as urban mobility or personal ones such as health. In relation to this, the White Paper on Artificial Intelligence (European Commission, 2020) highlighted the need to ensure access to data in line with the FAIR (Findable, Accessible, Interoperable, and Reusable) principles. Ultimately, artificial intelligence is part of the European digital agenda. This is extremely complicated because it has complex links with multiple policy areas both internally (education, research, industry, competition, fundamental rights and so on) of the EU and externally (Newman, 2020). At present, European authorities are taking a proactive approach to this issue, as will be discussed in the next section on governance systems.

3 Governance, Techno-optimism and Interrelated Challenges

As noted above, the growing importance and expansion of AI is taking place in the midst of an eminently academic debate that oscillates between techno-optimistic and techno-pessimistic positions. Today, the hegemonic position corresponds to the level of maturity of AI-related applications. Thus, discussions on emerging technologies, characterised by a significant degree of uncertainty about the impacts and outcomes of future developments, tend to be dominated in their early stages by techno-optimistic arguments. As Treré and Barranquero (2013) underline, the emergence of new technological applications is historically associated with the predominance of techno-optimistic visions based on the possibilities and solutions of new technologies. However, each cycle of techno-mitification is followed by another of techno-demitification once the use of technologies becomes trivialised and enters the realm of the commonplace and the ordinary (Treré & Barranquero, 2013). In relation to this issue, Ulnicane et al. (2021) point to the exaggeration that can be observed in policy documents and reports regarding the possibilities and potential for social, economic and political transformation of AI.

In any case, despite the current hegemony of techno-optimistic positions, the dangers and threats surrounding the expansion of AI-related applications are gaining ground. Following the logic set out by Treré and Barranquero (2013), this fact would be associated with the current application of AI in the fields of the internet and the business world (Lee, 2020). To respond to these dangers, governance systems face a variety of interrelated challenges (Ulnicane et al., 2021). First, the uncertainty involved in debating and making decisions on issues that, to a large extent, relate to future developments and applications. Secondly, the complexity arising from the large number of stakeholders involved in formulating alternatives and proposals that satisfy most of the actors involved (Newman, 2020). Finally, and as a central issue, the risks to society that poor governance of AI may engender (Radu, 2020).

Beyond specific and sectoral elements, we have already mentioned some cross-cutting AI-related problems and threats. Thus, as Ulnicane et al. (2021) note, one of the concerns that occupies most space in political and scientific documents related to AI has to do with the possibility of a growing concentration of power at the social, political and economic levels. From our point of view, this structural issue is currently exemplified in various aspects of AI in its internet and business applications. Thus, the structure of the data market that serves various key functions linked to AI (Srnicek, 2018) or the concentration of information and resources related to the key variables that shape AI by a small group of actors highlights the high degree of concentration of power that AI currently generates. As Ulnicane et al. (2021) point out, the development of the key elements of AI are in the hands of five large technology companies.[10] In this regard, and as a consequence of the volume of resources and information derived from the oligopolistic position of the *big five* (Borras & Edler, 2014), Guihot et al. (2017) point to the limited regulatory role of public institutions. In the same way, and beyond the risks of institutional capture, Taeihagh (2021) emphasise the asymmetry of information between public institutions and large corporations. This would make it difficult to generate new and effective regulations. In this way, the regulatory role of the state is being de facto substituted through the configuration of AI implemented by large private corporations. As an example of the growing power and influence of Big Tech on public institutions, the Corporate Europe Observatory (2021) states in a recent report:

> Ten years ago, the picture of EU lobbying was different with sectors like finance or pharma dominating it. But this has changed over the last decade with Big Tech overtaking them in terms of spending, reach, and influence. But it is not just Big Tech's lobby firepower that is a problem: its business models threaten to undermine democratic decision-making in our societies. The huge concentration of economic and lobby power is poison to our democracy (p. 42).

Beyond the concentration of power and the hoarding of data, a fundamental input for the development of AI, the literature and analytical literature point to a number of specific issues. According to Jobin et al. (2019), there is global convergence on the ethical principles (transparency, fairness, equity, non-maleficence, accountability and privacy) that should guide the development of AI and mark the threats and challenges arising from AI. However, according to these authors, the interpretation of these principles in terms of the implementation of concrete rules generates important divergences between the different public and private actors involved. In contrast, principles such as sustainability, dignity and solidarity appear very infrequently in the documents analysed (Jobin et al., 2019).

[10]Under the acronym GAFAM, the five largest global technology companies are listed: Google, Apple, Facebook, Amazon and Microsoft.

Related to the issue of the data and the algorithms that articulate the AI, Yeung (2016) points to the rapid erosion of people's right to privacy as a consequence of the continuous monitoring of human activity for the collection of digitally generated data. In this view, the current model of privacy self-management structured around the paradigm of notice and consent provides insufficient privacy protection (Yeung, 2016). She also points to the dangers of certain state regulations aimed at improving efficiency in the provision of public services at the cost of sacrificing basic rights such as privacy (Yeung, 2016).[11]

In addition to the issue of surveillance and privacy, the management of information collected by the algorithms that underpin AI systems raises a number of other social and political issues. The first relates to the consequences of the absence of algorithmic transparency. This deficit can lead to the spread of practices that represent corporate economic interests by sacrificing principles linked to the general interest or less well-resourced collectives (Ulnicane et al., 2021). Yeung (2016) adds a double concern from a democratic point of view. First, she mentions the capacity of algorithmic systems to generate discrimination and exclusion. In her work she illustrates this claim through experiences linked to racist practices. Secondly, algorithmic opacity limits society's ability to control the practices of the economic groups that dominate the AI systems that influence their daily behaviour. In this way, the discriminatory practices described become difficult to verify (Yeung, 2016). Finally, lack of transparency, opacity and algorithmic complexity limit the availability of quality information for effective decision-making by governments (Taeihagh, 2021).

Secondly, as a result of the complexity and lack of human control over the consequences and outcomes of AI algorithmic configurations, several authors (Taeihagh, 2021; Yeung, 2016) point to the difficulty in determining responsibility and accountability in the event of damage or social problems arising from the activity and data-driven decision-making of AI systems (Janssen et al., 2020). Finally, yet scarcely touched on, it is necessary to mention the environmental problems associated with the implementation of AI systems as a result of the high energy consumption derived from the massive computational resources on which they are based (Jobin et al., 2019).

Below we will analyse, from a theoretical point of view and on the basis of the most recent literature, the main models of public intervention implemented by different public bodies and institutions to regulate the different aspects that shape AI.

[11] The author cites as an example the decision of the UK authorities to make the medical data of 1.6 million patients from three National Health Service hospitals available to Google-linked AI arm DeepMind (Yeung, 2016).

4 AI and Public Intervention Models

Understanding and managing the above risks is crucial to optimise the realisation of the benefits associated with the technological modernisation implied by AI models (Taeihagh, 2021). According to Taeihagh (2021), the study of governance and public intervention models in the area of AI remains an underdeveloped field of research (Taeihagh, 2021). Among the materials published to date, we should mention the work of Radu (2020). Following the qualitative analysis of various national strategies launched between 2017 and 2019, Radu (2020) detects some common trends and patterns in AI governance models. First, he highlights the strong alignment between industry and government interests at the national level. Second, he points to the prioritisation of the development of ethical guidelines over the design and implementation of mandatory standards. This is reflected in a vague definition of the roles of public and private actors in AI governance. Under the label of hybrid governance, Radu (2020) refers to a regulatory model characterised by non-hierarchical interaction between public and private institutions, embodied in the development of ethical standards and the creation of supervisory institutions and agencies. In this regard, the author compares the current framework of AI governance and regulation with the development and expansion of the internet: "The general reluctance to regulate AI at an early stage is reminiscent of the approach chosen for regulating the Internet only when security and legal problems became widespread" (Radu, 2020).

In the same way, Taeihagh (2021) indicates that the hegemonic model of AI governance is based on soft regulatory or self-regulatory mechanisms, as opposed to hard regulatory frameworks. Thus, published documents are characterised by the inclusion of non-binding ethical standards, norms and guidelines. In addition to presenting difficulties in integrating the views of different stakeholders, this model adds the risk of industry interests taking over and capturing AI regulation and standards (Guihot et al., 2017). According to Jobin et al. (2019), the risk of regulatory capture takes shape through lobbying and the inclusion of representatives close to industry in expert committees set up by public institutions. In this regard Radu (2020) notes:

> the majority of the nations included in this analysis envisioned the creation of special AI Councils or Data Committees to monitor AI adoption and implementation processes. Oversight bodies or AI councils driving the AI policy mandates were generally dominated by representatives of academia and private sector. Overall, NGOs and rights groups were not equally represented (e.g. UK, Canada). AI strategies rarely included end-users of these technologies as a specific group for policy dialogue (p. 189).

Moreover, the approach based on self-regulation or soft regulation allows the industry to take advantage of its dominant position vis-à-vis other actors in the accumulation and management of data and through algorithmic design. This regulatory and governance scheme could result in an increase in private industry profits at the cost of sacrificing certain democratic principles or the interests of broad social groups (Jobin et al., 2019). In this respect, the findings presented in the report

Table 1 Typology of governance modes

	Driven by state actors	Driven by societal actors
Hierarchical, dominated	Command and control	Oligopoly
Heterarchical, non-dominated	State as primus inter pares	Self-regulation

Source. Borrás and Edler (2014, p. 31)

produced by the Corporate Europe Observatory (2021) in relation to the Big Tech discourse and the regulatory models discussed in Brussels are eloquent. In particular, the narrative is built around a fundamental idea: regulation can negatively affect the development of innovation and thus be negative for economic actors and customers:

> The corporate sector in general has been fundamentally challenging regulation for years, and Big Tech has played an important role here. This attitude is reflected in the simple cliché that "regulation stifles innovation". Behind it lies a problematic understanding of politics: state action favouring public interest is devalued. Instead, corporate interests are presented as central for the well-being of society. Companies' one-sided lobbying is thus legitimised, and every regulatory project is put under particular pressure to justify itself (p. 30).

Following the governance model proposed by Borrás and Edler (2014) articulated on the nature of the actors and the type of coordination,[12] the AI regulation scheme proposed by Radu (2020) would not fit within the categories of Table 1 insofar as the hybrid governance model is articulated on the absence of command and the alignment of interests between public and private institutions. However, the hegemony of soft, ethical and voluntary regulations (Taeihagh, 2021; Jobin et al., 2019) could be interpreted as a consolidation of the dominance of economic actors such as the corporations that make up GAFAM. In this respect, and following the analytical scheme proposed by Borrás and Edler (2014), the hegemonic model of governance and regulation at the international level of AI in its most prominent cross-cutting aspects, such as data management and algorithms, would be driven by a limited group of large companies. In line with this assertion Ulnicane et al. (2021) argue that:

> the current governance characterized by oligopoly of a small number of large companies is indicated as one of the reasons for problems such as lack of consideration of societal needs and concerns. To address these problems, governance frame in AI policy documents assigns more active and collaborative roles to the state and society (p. 1).

Before analysing governance and the European model of regulation of the most important cross-cutting aspects of AI, in the following section we will try to complete the information regarding the models of regulation and governance implemented in different policy areas in recent times.

[12] According to Borrás and Edler (2014) the nature of the actors will be public or private. In addition, the coordination models between actors will be hierarchical or non-hierarchical.

5 European Governance and Regulation of AI Within the Framework of Hegemonic Intervention Models

Having reviewed the main threats and models of AI regulation and governance, we will analyse, from a cross-cutting point of view, the response of European institutions to the challenges mentioned and the regulatory model adopted along the lines set out above. Before discussing the substantive content of the documents under study, we consider it pertinent to approach the process that has given rise to the AI regulatory proposal in order to complete the analysis of the AI governance model at the European level.

The European Parliament gave the first warning signal of the need to address the new reality associated with the spread of AI-related applications. Its resolution of 16 February 2017 with recommendations to the Commission on civil law rules on robotics already explicitly referred to AI. The Commission committed to adopting a European approach to AI. In 2018, the High-Level Expert Group on AI was activated and it presented the *European Strategy for AI.*[13] In early 2020, the White Paper on Artificial Intelligence and the European Data Strategy was published.[14] The European Commission is also working on updating the Coordinated Plan on AI[15] to ensure the security and fundamental rights of individuals and businesses, while strengthening investment and innovation in all EU countries (political, technical, legal, ethical requirements). In addition, to support the development of AI, the Commission plans to allocate resources from the Digital Europe and Horizon Europe programmes. Recently, in early May 2022, the European Parliament, through the Special Committee on Artificial Intelligence in the Digital Age, adopted the *Report on Artificial Intelligence in the Digital Age.*[16]

In 2020, the European Council urged that "the opacity, complexity, bias, a certain degree of unpredictability and partially autonomous behaviour of certain AI systems, to ensure their compatibility with fundamental rights and to facilitate the enforcement of legal rules" be addressed (European Commission, 2021). At the same time, the European Parliament, in addition to adopting several resolutions related to AI, in 2020 called on the European Commission to propose legislative measures in order to seize the opportunities of AI while taking into account the ethical challenges it poses.

In response to the demands raised, in 2021 the European Commission presented the proposal for a Regulation on harmonised rules on artificial intelligence (Artificial

---

[13] Artificial Intelligence for Europe (COM(2018) 237 final). Available at https://ec.europa.eu/transparency/documents-register/detail?ref=COM(2018)237&lang=en (last access 8-05-2022).

[14] White Paper on Artificial Intelligence: a European approach to excellence and trust, COM(2020) 65 final https://ec.europa.eu/info/sites/info/files/commission-white-paper-artificial-intelligence-feb2020_es.pdf (last access 12-05-2022).

[15] Coordinated Plan on Artificial Intelligence COM(2018) 795 final https://eur-lex.europa.eu/legal-content/EN/TXT/HTML/?uri=CELEX:52018DC0795&from=ES (last access 01-05-2022).

[16] See Report on artificial intelligence in a digital age.

Intelligence Act),[17] which responded to a European approach. The document is based on a general objective and four specific objectives. The general objective is worded as follows: "the general objective of the proposal, which is to ensure the proper functioning of the single market by creating the conditions for the development and use of trustworthy AI in the Union" (European Commission, 2021). In addition, three of the four specific purposes appear to be geared towards the structuring of a European AI-related market. Although one of them states the need to create a secure, value- and rights-based European market, we can observe an economic and market orientation. Only the fourth of the specific objectives is directly related to the object analysed in this paper: "enhance governance and effective enforcement of existing law on fundamental rights and safety requirements applicable to AI systems" (European Commission, 2021). In fact, this proposal was the result of a consultation process with the different actors and stakeholders involved in AI regulation. Thus, in February 2020, for four months, and following the publication of the White Paper on AI (2020),[18] an online public consultation was launched.[19] As a result, 1216 contributions were received, mainly from individual citizens (372), companies (222), academic institutions (152), business associations (131), NGOs (128), public administrations (73) and trade unions (22). By states, contributions from Germany (251), Belgium (162), France (117), Spain (105), the United Kingdom (76), the United States (60), the Netherlands (60) and Italy (60) stand out (White Paper on AI, 2022).[20]

In order to draw up the final proposal based on the objectives set out above, the European Commission evaluated up to four regulatory formulas based on different degrees of public intervention. The options considered were as follows:

- Option 1: EU legislative instrument setting up a voluntary labelling scheme.
- Option 2: a sectoral, "ad-hoc" approach.
- Option 3: Horizontal EU legislative instrument following a proportionate risk-based approach.
- Option 3+: Horizontal EU legislative instrument following a proportionate risk-based approach + codes of conduct for non-high-risk AI systems.

[17] Proposal for a Regulation of the European Parliament and of the Council laying down harmonised rules on Artificial Intelligence (Artificial Intelligence Act) and amending certain Union legislative acts, COM(2021) 206 final.

[18] The basis for the White Paper on AI and the European AI Strategy was elaborated and discussed by a group of 52 experts advising the European Commission. In 2019, the European Commission endorsed the conclusions of this expert group for ethical and reliable AI (European Commission, 2021).

[19] In addition to the consultation period, we should mention the constitution of the European AI Alliance, which brings together more than 4000 actors and stakeholders. Discussions and debates take place within the Alliance, culminating in an annual AI Assembly.

[20] Apart from formal governance mechanisms, we should not lose sight of the enormous lobbying efforts made in Brussels in recent years by the Big Tech sector. According to Corporate Europe Observatory (2021), this sector is a leader in this area and has overtaken the pharmaceutical and financial sectors in the last decade.

- Option 4: Horizontal EU legislative instrument establishing mandatory requirements for all AI systems, irrespective of the risk they pose (European Commission, 2021).

After examining the various options, Option 3+ was adopted, based firstly on a structured approach to define different levels of risk for AI applications and then intervening on those AI systems that fall into the high-risk category. Those excluded from this label will be governed by voluntary codes of conduct. Likewise, the requirements governing high-risk applications will refer to the following items: "data, documentation and traceability, provision of information and transparency, human oversight and robustness and accuracy and would be mandatory for high-risk AI systems" (European Commission, 2021). This risk-based approach to AI regulation was developed following a minimum articulation and development, with the aim of not obstructing economic development and innovation associated with AI, as stated in the regulatory proposal drafted by the European Commission in various sections:

> To achieve those objectives, this proposal presents a balanced and proportionate horizontal regulatory approach to AI that is limited to the minimum necessary requirements to address the risks and problems linked to AI, without unduly constraining or hindering technological development or otherwise disproportionately increasing the cost of placing AI solutions on the market (p. 3).

In this way, the discourse and justification of the regulatory model chosen by the European Commission is in line with the approach defended by the Big Tech sector, as stated in the report prepared by Corporate Europe Observatory (2021), in which it points out the negative effects that could be derived from "excessive" regulation. This could be due to the influence and power of Big Tech over public institutions, or it could simply be a coincidence of ideology and approach. Returning to the Option 3+ model, we must look at the levels of risk established for decision-making related to AI systems.

As can be seen in Fig. 1, four levels of risk are established. At the highest level are the so-called unacceptable risks, which include the prohibition of AI systems based on social scoring models promoted by governments or toys and devices that include voice assistance systems and that could encourage dangerous behaviour. Next, under the heading of high risk, are a series of technologies and systems applied to areas considered critical such as transport, education, health, employment, access to public and private services or the administration of justice and democratic processes. In this case, AI systems related to these high-risk areas are subject to a number of pre-commercialisation obligations.[21] Finally, included in the last two categories of minimal and limited risk, and subject to self-regulation based on codes of conduct,

[21] Developers of high-risk systems should provide complete and detailed information for assessment by the authorities. They must also provide clear and adequate information to the user. In addition, the results of such applications will be recorded in order to ensure traceability. It should be noted that biometric identification systems will be considered high risk and will be subject to strict requirements (European Commission, 2022).

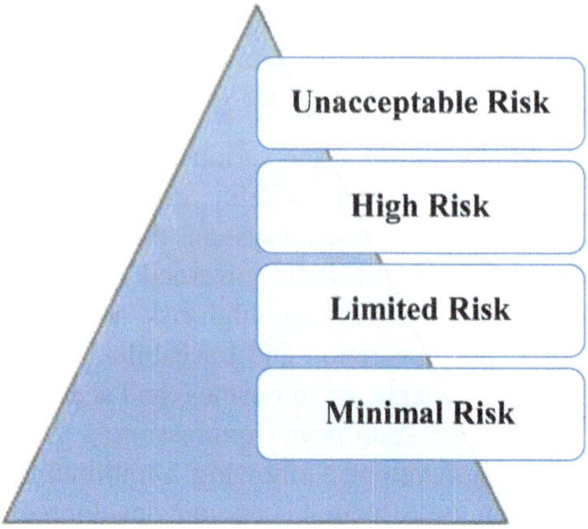

Fig. 1 Levels of risk for the regulation of AI systems. Source. European Commission (2022): Digital Strategy

are most of the AI systems currently operating in Europe (European Commission, 2022).

After observing the main elements that make up the process and the legislative proposal drafted by the European Commission, we consider that the model of interaction between actors and regulation analysed would fit the definition of hybrid governance enunciated by Radu (2020). Firstly, we must consider the process that led to the proposal presented by the European Commission, characterised by non-hierarchical interaction between public and private institutions, centred on the European AI Alliance (European AI Alliance, 2022) and the consultation period opened after the presentation of the White Paper on AI. The influence of private actors could be related to the discursive coincidence observed between the rationale and approach adopted by the European Commission (2021) and the Big Tech sector, which is highly active in Brussels through lobbying activities (Corporate Europe Observatory, 2021). Secondly, the approach exhibited in the rationale behind the European Commission (2021) proposal determines an orientation in its objectives that is embodied in a vision that emphasises the economic dimension of AI over social and democratic issues. The Option 3+ regulatory model adopted combines different intervention models according to the levels of risk of the different AI applications, and although it predisposes a higher level of intervention in applications of unacceptable and high risk, the fact is that at a quantitative level most of the AI systems already in place in the EU states would be framed within the lower risk levels. In this way, responsibility would fall mostly on economic actors, through codes of conduct, and on individuals and citizens through informed consent. This approach would coincide with Radu's (2020) assertion that the governance and regulation of AI would follow a model similar to that developed in the general area of the internet. This model of soft and ethical regulation would be currently globally hegemonic in the regulation of AI (Taeihagh, 2021; Jobin et al., 2019).

Given the model of governance and regulation promoted by the European Commission, we must ask ourselves whether it will be effective in establishing an economic sector in which the dominant position of GAFAM companies is not problematic for the fundamental rights of citizens and the functioning of democratic systems. According to authors such as Taeihagh (2021) and Jobin et al. (2019), soft regulatory models such as Option 3+ could consolidate the dominant position of large corporations that currently dominate the data market and the shaping of the algorithms through which multiple AI systems are developed. Beyond the declaration of the risks associated with some of the AI applications and systems, it is essential to design a governance model that addresses the dominant position of certain actors that could be decisively conditioning AI governance models. Thus, at a democratic level and from a cross-cutting perspective to the different AI systems, a different point of view is needed for the regulation of the following aspects:

1. Data processing that guarantees citizens' privacy. Informed consent might not be effective insofar as the asymmetry of information between the parties would hinder the proper functioning of this system.
2. Regulation and control of algorithms so that they do not introduce biases and discriminations that go against the principles set out in the Charter of Fundamental Rights of the European Union.

6 Conclusion

In this chapter we have analysed the process and proposal developed by the European Commission (2021) to regulate and harmonise the application and use of AI at the European level. In doing so, we have taken as our starting point the broad context and the cross-cutting threats to society and democracy that arise from the application of AI systems. In particular, the concentration of power in large corporations and Big Tech, algorithmic opacity and discrimination of social groups or the degradation of the right to privacy have been identified as potentially pernicious. Having identified the cross-cutting threats and risks, we have used various conceptual and theoretical tools recently used in research on the governance and regulation of AI at international and European level to analyse the governance process and the proposal drawn up by the European Commission (2021). The assessment and analysis of the process of AI regulation and the European Commission's proposal allows us to frame the adopted model within the theoretical framework described as hybrid governance. This conclusion is based on the following characteristics:

1. Non-hierarchical relationship and interaction between public and private actors and coincidence in approach and rationale.
2. Combination of hard and soft regulatory measures, based on the prohibition of certain systems considered dangerous and the provision of codes of conduct and informed consent.

However, we must bear in mind that AI governance and regulation is currently a rapidly developing field subject to constant change and innovation. Thus, we cannot rule out a shift in the coming years towards other governance formulas that imply a greater verticality and protagonism of public institutions.

References

Abdala, M., Lacroix, S., & Soubie, S. (2019). *La política de la inteligencia artificial: sus usos en el sector público y sus implicancias regulatorias, CIPPEC*. Available at https://www.cippec.org/wp-content/uploads/2019/10/185-DT-Abdala-Lacroix-y-Soubie-La-pol%C3%ADtica-de-la-Inteligencia-Artifici....pdf (last access 10-10-2022).

Antonov, A., & Kerikmäe, T. (2020). Trustworthy AI as a future driver for competitiveness and social change in the EU. In D. R. Trotiño et al. (Eds.), *The EU in the 21st century* (pp. 135–154). Springer.

Arteaga, F. (2019). Contexto estratégico de la inteligencia artificial. *La inteligencia artificial aplicada a la defensa*. Real Instituto Elcano, 153–172.

Borrás, S., & Edler, J. (2014). The governance of change in socio-technical and innovation systems: Three pillars for a conceptual framework. In S. Borrás & J. Edler (Eds.), *The governance of socio-technical systems*. Edward Elgar Publishing.

Cancela-Outeda, C. (2020). La inteligencia artificial: Una lectura desde la política. In *Inteligencia artificial: de la discrepancia regional a las reglas universales: Integración de percepciones políticas, económicas y legales*/coord. por Tanel Kerikmae, María Claudia Solarte Vasquez, Matti Rudanko, David Ramiro Troitiño, Thompson Reuters, 319–344.

Corporate Europe Observatory. (2021). *The lobby network: Big Techs web of influence in the EU. Corporate Europe Observatory and Lobby Control*. https://corporateeurope.org/en/2021/08/lobby-network-big-techs-web-influence-eu (last access 10-06-2022).

European AI Alliance. (2022). *European AI Alliance*. https://futurium.ec.europa.eu/en/european-ai-alliance.

European Commission. (2020). *White paper on artificial intelligence: A European approach to excellence and trust*. Available at https://ec.europa.eu/info/sites/default/files/commission-white-paper-artificial-intelligence-feb2020_en.pdf (last access 10-06-2022).

European Commission. (2021). *Proposal for a regulation of the European Parliament and of the council laying down harmonised rules on artificial intelligence (artificial intelligence act) and amending certain union legislative acts*. Available at https://eur-lex.europa.eu/legal-content/EN/ALL/?uri=CELEX:52021PC0206 (last access 6-06-2022).

European Commission. (2022). *Digital strategy European commission*. Available at https://digital-strategy.ec.europa.eu/en/policies/regulatory-framework-ai (last access 3-06-2022).

Giraldi, E. (2019). *Digitalización, política e inteligencia artificial ¿Qué futuro podemos esperar?*. NUSO n° 283. Available at https://nuso.org/articulo/digitalizacion-politica-e-inteligencia-artificial/ (last access 8-05-2022).

Girasa, R. (2020). *Artificial intelligence as a disruptive technology. Economic transformation and government regulation*. Palgrave Macmillan.

Guihot, M., Matthew, A. F., & Suzor, N. P. (2017). Nudging robots: Innovative solutions to regulate artificial intelligence. *Vanderbilt Journal of Entertainment & Technology Law, 20*.

Harari, Y. N. (2018). *21 Lecciones para el siglo XXI*. Debate.

Janssen, M., Brous, P., Estevez, E., Barbosa, L. S., & Janowski, T. (2020). Data governance: Organizing data for trustworthy Artificial Intelligence. *Government Information Quartely, 37* (3), 1–8. https://doi.org/10.1016/j.giq.2020.101493

Jobin, A., Ienca, M., & Vayena, E. (2019). The global landscape of AI ethics guidelines. *Nature Machine Intelligence, 1*(9), 389–399.

Kerikmäe, T., Troitiño, D. R., & Shumilo, O. (2019). An idol or an ideal? A case study of Estonian e-Governance: Public perceptions, myths and misbeliefs. *Acta Baltica Historiae et Philosophiae scientiarum, 7*(1), 71–80.

Lee, K. (2020). Superpotencias de la inteligencia artificial: China, Silicon Valley y el nuevo orden mundial, Deusto. [English version AI Superpowers, Houghton Mifflin Harcourt].
Mäskine, J., & Kasanen, E. (2020). Las grandes compañías tecnológicas y los desafíos a la democracia liberal. In T. Kerikmae, M. C. S. Vasquez, M. Rudanko, & D. R. Troitiño (Eds.), *Inteligencia artificial: de la discrepancia regional a las reglas universales: Integración de percepciones políticas, económicas y legales* (pp. 99–127). Thompson Reuters.
Newman, A. (2020). Digital policy-making in the European Union. Building the new economy of an information society. In H. Wallace et al. (Eds.), *Policy-making in the European Union* (pp. 275–296). Oxford University Press.
Ortega, A (2021). *China, EEUU y la UE: Tres formas de confrontar las Big Tech.* Available at https://www.realinstitutoelcano.org/china-eeuu-y-la-ue-tres-formas-de-confrontar-las-big-tech/ (last access 5-07-2022).
Radu, R. (2020). Steering the governance of artificial intelligence: National strategies in perspective. *Policy and Society, 40*(2), 178–193.
Ramió, C. (2018). *¿Puede la inteligencia artificial secuestrar a la democracia? (II),* Blog https://www.administracionpublica.com/puede-la-inteligencia-artificial-secuestrar-a-la-democracia-ii/ (last access 5-07-2022).
Ramió, C. (2019). *Inteligencia artificial y administración pública. Robots y humanos compartiendo el servicio público,* Madrid, Catarata.
Ramiro, D. et al. (2020). El Parlamento Europeo y la necesidad de una legislación común en un marco europeo de la inteligencia artificial. In *Inteligencia artificial: De la discrepancia regional a las reglas universales: Integración de percepciones políticas, económicas y legales*/coord. por Tanel Kerikmae, María Claudia Solarte Vasquez, Matti Rudanko, David Ramiro Troitiño, Thompson Reuters, 369–384.
Robles, M. (2020). La gobernanza de la inteligencia artificial: contexto y parámetros generales. *Revista Electrónica de Estudios Internacionales.* https://doi.org/10.17103/reei.39.07
Russell, S., & Norvig, P. (Eds.). (2016). *Artificial intelligence: A modern approach* (3rd ed.). Essex.
Sadin, E. (2018). *La silicolonización del mundo. La irresistible expansión del liberalismo digital.* Caja Negra.
Sousa, S. (2020). El impacto de la confianza en los desafíos de la IA. In *Inteligencia artificial: De la discrepancia regional a las reglas universales: Integración de percepciones políticas, económicas y legales*/coord. por Tanel Kerikmae, María Claudia Solarte Vasquez, Matti Rudanko, David Ramiro Troitiño, Thompson Reuters, 129–148.
Srnicek, N. (2018). *Capitalismo de plataformas.* Caja Negra.
Taeihagh, A. (2021). Governance of artificial intelligence. *Policy and Society, 40*(2), 137–157.
Treré, E., & Barranquero, A. (2013). De mitos y sublimes digitales: Movimientos sociales y tecnologías de la comunicación desde una perspectiva histórica. Redes.com, *8*, 27–47.
Ulnicane, I., Knight, W., Leach, T., Stahl, B. C., & Wanjiku, W. G. (2021). Framing governance for a contested emerging technology: Insights from AI policy. *Policy and Society, 40*(2), 158–177.
Yeung, K. (2016). Algorithmic regulation and intelligent enforcement. In M. Lodge (Ed.), *Regulation scholarship in crisis?,* CARR Discussion Paper n°. 84, 50–62.

SPRINGER NATURE

GPSR Compliance

The European Union's (EU) General Product Safety Regulation (GPSR) is a set of rules that requires consumer products to be safe and our obligations to ensure this.

If you have any concerns about our products, you can contact us on ProductSafety@springernature.com

In case Publisher is established outside the EU, the EU authorized representative is:

Springer Nature Customer Service Center GmbH
Europaplatz 3
69115 Heidelberg, Germany

The manufacturer's authorised representative in the EU is Springer Nature Customer Service Centre GmbH, Europaplatz 3, 69115 Heidelberg, Germany. If you have any concerns regarding our products, please contact ProductSafety@springernature.com

Printed and bound by CPI Group (UK) Ltd, Croydon, CR0 4YY
30/09/2026
02225561-0012